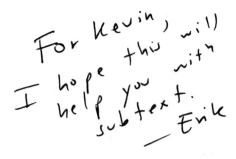

For Kevin, I hope this will help you with subtext. — Erik

Daydreaming in
Humans and Machines

A Computer Model of the Stream of Thought

Erik T. Mueller

ABLEX PUBLISHING CORPORATION
NORWOOD, NEW JERSEY

Library of Congress Cataloging-in-Publication Data

Mueller, Erik T.
    Daydreaming in humans and machines.

    Bibliography: p.
    Includes index.
    1. Fantasy—Computer simulation.   2. Fantasy.
3. Artificial intelligence.   I. Title.
BF408.M84   1988   154.3   89-6483
ISBN 0-89391-562-9

Ablex Publishing Corporation
355 Chestnut St.
Norwood, NJ 07648

Contents

List of Figures

List of Tables

For Robert and Diana

ACKNOWLEDGEMENTS

This book is a revised version of a dissertation submitted to UCLA in March 1987 for the degree of doctor of philosophy in computer science.

First, I would like to thank my advisor Michael Dyer. This work would not have existed without him: He created the Artificial Intelligence Laboratory at UCLA that enabled me and my fellow students to pursue our research. He encouraged me to pursue daydreaming—a topic few others would have agreed to. In the early stages of my research, he helped me clarify the significance of daydreaming as a subject of investigation. Throughout the almost four years of work on this topic, he provided ideas, guidance, and constant support. I also thank him for emphasizing the importance of presenting our work to the outside world—both at conferences and program demonstration sessions in the lab.

Second, I want to thank the other members of my committee for their valuable comments and suggestions: Margot Flowers, Judea Pearl, Kenneth Mark Colby, and Robert Bjork. Kenneth Mark Colby was especially generous with his time. I thank him for many useful meetings and discussions.

Third, I would like to thank the other members of the lab: Sergio Alvarado, Stephanie August, Charles Dolan, Richard Feifer, Michael Gasser, Seth Goldman, Jack Hodges, Michael Pazzani, Alex Quilici, John Reeves, Ron Sumida, Scott Turner, and Uri Zernik. Uri, Scott, and John read earlier drafts of this book and provided helpful comments. I especially thank Uri for the many interesting and productive discussions we had on our respective research topics.

Fourth, I want to thank Janene McNeil from the UCLA psychology department who supplied me with a collection of daydream diaries she collected from her subjects in a pilot study. I also thank those friends who told me about or sent me their daydreams.

Fifth, for their efforts in keeping the Apollo workstations in the lab up and running, I would like to thank Eve Schooler and Seth Goldman. I also thank Bill Davis, Doris McClure, and other members of the Center for Experimental Computer Science at UCLA for providing additional computing support. For lab administrative assistance, I thank Anna Gibbons and Anne Finestone.

Sixth, I want to thank the following friends during my graduate school years on the West Coast: Mark Joseph, Ann Gardner, Karen Lever, Patty Liu, Andy Lust, Johanna Moore, Raya Palm, Ellen Perlman, Tony Thijssen, and Carrie Young.

Seventh, I would like to thank my grandmother, Ilus Lobl, my great-aunt Ellie Bermowitz, and my great-uncle Jack Bermowitz.

Finally, I want to thank my wonderful parents, Robert Mueller and Diana Mueller, and my fantastic sister, Rachel Mueller-Lust.

This research was supported by a fellowship from Atlantic Richfield, a grant from the Hughes Artificial Intelligence Center, an IBM Faculty Development award, and a grant from the W. M. Keck Foundation.

Preparation of the revision for publication as a book was made possible by the Analytical Proprietary Trading unit of Morgan Stanley & Company in New York. I would like to thank Barbara Bernstein and Carol Davidson at Ablex for their efforts, and anonymous reviewers for helpful suggestions.

Chapter 1

Introduction

The field of artificial intelligence is concerned with the construction of computer systems which exhibit intelligent behavior in order to augment and complement our own human intelligence, and to understand more about how the human mind works. Consider the human phenomenon of *daydreaming*: the spontaneous activity—carried out in a stream of thought—of recalling past experiences, imagining alternative courses that a past experience might have taken, and imagining possible future experiences. Why is daydreaming an interesting topic of study for artificial intelligence? On the face of it, daydreaming may seem like a human shortcoming, a useless distraction from a task being performed which would be undesirable in computer systems. But how can we ignore an aspect of human thought which appears so frequently in everyday life? For example:

- While driving home from work with the car radio on, you suddenly realize you have completely missed the last few minutes of the news report. Instead of listening, you have been mentally replaying an event of the day and altering it to how you would have liked it to come out. You think about what you might have said or done, but didn't.

- An important job interview is coming up. You have trouble concentrating on your current task. Instead, you find yourself imagining the future event again and again. You go over what you and others will say or do in that situation, and you imagine things go—or don't go—the way you would like them to go.

- You are reading a book and are startled by the noisy cuckoo clock. You realize that you have been lost in thought for minutes, following a chain of experiences and associations triggered by the sentence or word you read just before drifting off.

1

- You are sitting in a classroom and are bored with the material. You miss the question which the instructor has just asked you because, in your fantasy, you have been off driving in a convertible on the French Riviera, wearing a scarf, your hair waving in the wind.

- When things don't go as planned, you make yourself feel better by imagining how if they *had* gone as planned, you might have been even worse off.

Most likely, some or all of the above experiences are familiar to the reader. Since daydreaming is such a pervasive aspect of human experience (Pope, 1978; J. L. Singer, 1975; Klinger, 1971), the following questions naturally arise: Why do we daydream? How is daydreaming useful to us? What are the advantages (and disadvantages) of daydreaming? If daydreaming serves a useful function in humans, then might daydreaming not also be useful in a computer system? Do we want our computers to daydream? Can a computer even be fully intelligent *without* being able to daydream?

1.1   The DAYDREAMER Program

This book presents a theory of human daydreaming implemented as a computer program called DAYDREAMER. This program models the daydreaming (and to a lesser extent, the overt behavior) of a human in the domain of interpersonal relations and common everyday occurrences.

Ideally, DAYDREAMER would be a robot capable of moving around, performing actions, and communicating with humans and other robots in the real world; this would enable the program to interact with and learn from a fertile and often unpredictable environment just as human daydreamers do. However, in order to construct such a robot, difficult problems in robotics, vision, and speech understanding would have to be solved first (not to mention obtaining and constructing the necessary hardware). Since these problems are outside the scope of our work, DAYDREAMER is instead embedded within a *simulated* world which models the real world of humans.

As input, DAYDREAMER takes descriptions of events in this world. As output, DAYDREAMER describes both actions which it performs in this world as well as its internal "stream of thought"—sequences of events in imaginary past or future worlds—or daydreams. (A more precise definition of daydreaming will be provided later on.)

Several sources of verbal reports of daydreaming—henceforth called *protocols*—have been tapped in the construction of our theory and program: (a) previously published protocols (Pope, 1978; Klinger, 1971; G. H. Green, 1923; Varendonck, 1921), (b) unpublished retrospective reports (see Ericsson & Simon, 1984) collected from subjects by McNeil (1981), and (c) protocols collected by the author from people via the methods of immediate retrospective report, retrospective report of memorable or classic daydreams, and typing on a computer terminal concurrently with daydreaming.

DAYDREAMER was constructed in order to produce daydreams similar but not identical to the obtained protocols. The daydreams produced by the program result from its particular knowledge structures and processes, which were designed to correspond to a hypothetical female daydreamer in her twenties.

DAYDREAMER operates in two modes: *daydreaming mode* and *performance mode*. In daydreaming mode, the program continuously daydreams until it runs out of things to daydream about or is interrupted. In performance mode, the program accepts input states or actions and produces output actions.

When DAYDREAMER is first started, the daydreamer has a job and is not romantically attached. Below is a portion of the I/O actually produced by DAYDREAMER, showing:

1. input world states or actions (indicated by "Input" and displayed in a slanted type style)

2. output actions which modify the state of the world (indicated by "External Action" and displayed in a bold type style)

3. the internal "stream of thought" of the program (displayed in a sans serif type style)

Both input and output are in the form of English sentences.

LOVERS1

I want to be going out with someone. I feel really interested in going out with someone. I want to be entertained. I feel interested in being entertained. I have to go see a movie.

External action: **I go to the Nuart.**

Input: *Harrison Ford is at the Nuart.*

What do you know! I have to have a conversation with him.

External action: **I tell Harrison Ford I would like to know the time.**

He does not think much of me because I am not well dressed. I fail at him thinking highly of me. I feel really embarrassed.

External action: **I introduce myself to him.**

Input: *He introduces himself to me.*

He has to be interested in me. He cannot be going out with anyone. Maybe he thinks I am cute.

External action: **I tell him I like his movies.**

He thinks highly of me. He is interested in me. Maybe he is not going out with anyone. Maybe he wants to be going out with me.

External action: **I tell him I would like to have dinner with him at a restaurant.**

Input: *He declines.*

I fail at going out with him. I feel really angry at him.

External action: **I watch a movie at the Nuart.**

I want to be going out with someone. I feel really interested in going out with someone.

External action: **I go home.**

I succeed at being entertained. I feel amused.

In the above experience, called LOVERS1 for future reference, DAYDREAMER first activates two *goals*: one to form a romantic relationship, and another to satisfy a need for entertainment. She acts first on her entertainment goal by performing an external action: going to see a movie at the Nuart Theater. There she accidentally meets the actor Harrison Ford whom she has always liked; his presence at the theater is provided as external input. Acting on behalf of her goal to form a relationship, she asks him if he knows the time. Now she feels embarrassed because she is not wearing her best clothes. In any case, she introduces herself to him and he introduces himself to her. She then tells him she enjoys his films and asks him if he would like to have dinner sometime. Unfortunately, he turns her down. Her goal to form a relationship thus fails and she is angry at him. She then watches the movie, resulting in the success of her entertainment goal. After this experience, DAYDREAMER amuses herself with the following revenge daydream:

REVENGE1

I study to be an actor. I am a movie star even more famous than he is. I feel pleased. He is interested in me. He breaks up with his girlfriend. He wants to be going out with me. He calls me up. I turn him down. I get even with him. I feel pleased.

DAYDREAMER then tries to make herself feel better by generating the following rationalization daydream:

RATIONALIZATION1

What if I were going out with him? He would need work. I remember the time he had a job with Paramount Pictures in Cairo. He would go to Cairo. Our relationship would be in trouble. I would go to Cairo. I would lose my job at May Company. I feel relieved.

DAYDREAMER is relieved about being turned down by Harrison Ford because if she had ended up in a relationship with him, she would have lost her job. Another rationalization is then generated:

RATIONALIZATION2

I remember the time my being turned down by Irving led to a success by being turned down by him leading to succeeding at going out with Chris. I go to Mom's. He goes to Mom's. I am going out with him. My being turned down by Harrison Ford leads to me succeeding at going out with Chris.

In the above daydream, DAYDREAMER imagines that as a result of being turned down by Harrison Ford, she goes to Mom's Bar and meets someone else. Further rationalizations are produced:

RATIONALIZATION3

Anyway, I was well dressed because I was wearing a necklace. I feel less embarrassed. Anyway, I do not think much of Harrison Ford. I feel less embarrassed. I rationalize being turned down by him.

In the above daydream, DAYDREAMER reduces her embarrassment about her clothes by (a) thinking about the nice necklace she was wearing, and (b) deciding she does not like Harrison Ford anyway.

ROVING1

I remember the time Steve told me he thought I was wonderful at Gulliver's. I feel pleased. I remember the time I had a job in the Marina. I remember the time Steve and I bought sunglasses in Venice Beach.

In this daydream, DAYDREAMER diverts attention from negative emotions associated with being turned down by Harrison Ford to a pleasant past experience. She is then reminded of two other related experiences. Further daydreams resulting from LOVERS1 will be presented and discussed in the following section.

1.2 Research Objectives and Issues

The objective of the present work is to construct a model of daydreaming and related phenomena using the computer as a tool. We seek to apply the techniques of artificial intelligence to the problem, to flesh out these techniques in terms of particular knowledge structures and constructs, to extend these techniques where they fall short, and to develop new techniques. We seek to gain new psychological insights, to raise new psychological issues, to generate new hypotheses, and to suggest possible experiments for testing these hypotheses. In sum, our goal is to investigate daydreaming from the standpoint of computer modeling, and in so doing to advance the fields of both artificial intelligence and psychology.

The fundamental hypothesis of our work is that daydreaming is useful for humans, in particular, that it serves the following important functions: (a) learning from imagined experiences, (b) creative problem solving, and (c) useful interaction with emotions. In this section, we discuss these functions, how they are useful in a computer system, and, specifically, how they are demonstrated in DAYDREAMER. We present the associated research problems addressed by the present work.

1.2.1   Learning from Imagined Experiences

Previous work in artificial intelligence has addressed learning from *real* experiences: DeJong (1981), Carbonell (1983) and Kolodner, Simpson, and Sycara-Cyranski (1985) have investigated how behavior can be improved as a result of previous experience; Schank (1982) and Dyer (1983a) have pointed out the importance of learning from failure experiences. But in order to learn from a given situation, why should one have to wait for that situation to come up in the real world? It should be possible to learn from *imagined* experiences in addition to real ones. In fact, humans frequently daydream about hypothetical future experiences (Pope, 1978; J. L. Singer, 1975, pp. 118-119, 1966; Klinger, 1971, pp. 47-48; Rapaport, 1951, pp. 718-719; G. H. Green, 1923, pp. 16, 180). We propose that this activity enables learning—or the useful modification of future behavior—in several ways.

First, *daydreaming enables the formation of intentions to perform certain actions in the future.* For example, in the following daydream, DAYDREAMER thinks of a way she could find out Harrison Ford's unlisted telephone number in order to contact him again:

RECOVERY1

> I have to ask him out. I have to know his unlisted telephone number.
> I remember the time Alex knew Rich's unlisted telephone number by
> logging into the TRW credit database. I ask Alex to look up Harrison
> Ford's unlisted telephone number in the TRW credit database.

The next time DAYDREAMER sees Alex, she will ask him to help her find out the actor's telephone number.

Second, *daydreaming enables decision-making by (a) exploring alternative plans for achieving a goal, and then (b) forming an intention to carry out the best such plan.* McDougall (1923) proposes a similar function for imagination (of which he considers daydreaming to be one form): "The essential and primary function of imagination is to carry out the process of trial and error on the imaginary plane, to depict each situation and the consequences of each step of action, before the action is accomplished or even begun" (p. 292).

Third, *daydreaming enables preparation in the face of various contingencies through (a) the exploration of alternative situations that might arise in trying to achieve a goal, and (b) the formation of intentions to perform certain actions should those situations arise.* For example, before a job interview one may daydream about possible questions of the interviewer and various responses to those questions.

Fourth, *daydreaming—when one is otherwise unoccupied with a task— enables discovery of negative consequences of carrying out a future intention that might not otherwise have been recognized.* Then, *once a negative consequence of a future intention has been discovered, the future intention may be adjusted as appropriate, and tested through further daydreaming.* Thus daydreaming provides an adaptive safety mechanism to avoid undesirable courses of action. For example:

NEWSPAPER

I want to be entertained. I feel hopeful. I have to read the newspaper. I have to grab the newspaper.

I go outside. I get wet. I feel displeased. I fail at keeping dry.

What if I put on a raincoat? I go outside. I grab the newspaper. I read the newspaper. ...

I want to avoid getting wet. External action: **I put on a raincoat. I go outside. I grab the newspaper. I read the newspaper.**

DAYDREAMER has the intention to go outside to get the newspaper. She then imagines carrying out this intention. Since it is raining, she gets wet.[1] As a result of this negative consequence, she adjusts her intention and tests it out through another daydream—she imagines putting on a raincoat before going outside.

Why is this learning? Before the daydream, DAYDREAMER knew only that (a) a plan to be at some location is to go to that location, and (b) one gets wet if one is at some location and not wearing a raincoat while it is raining at that location. After the daydream, the program also knows that a plan to be at some location while it is raining at that location is to put on a raincoat and then go to that location. In this case, the negative consequence of the plan happens to be obvious to most people (but not to the program). In other cases, however, the negative consequences may not be so obvious.

Fifth, *by spending free time exploring possible future situations in the world, daydreaming enables the discovery of negative consequences of a situation that*

---

[1]The example of getting the newspaper while it is raining is borrowed from Wilensky (1983).

might not otherwise have been noticed. Consequently, *daydreaming enables one to take steps to prevent those negative consequences well in advance of the future situation.* Alternatively, *daydreaming enables one to form the intention to take certain steps as soon as the future situation arises, in order to prevent those negative consequences from occurring.* For example:

REPERCUSSIONS1

Input: *There is an earthquake in Mexico City.*

What if there is an earthquake in Los Angeles? My apartment collapses. My possessions are destroyed. I am killed. I feel very worried.

I have to go to the doorway. I have to get insurance.

What if there is an earthquake in Los Angeles? I go to the doorway. I get hit by a falling plant. I get hurt. My possessions are destroyed. I feel very worried.

I have to move the plant away from the doorway. I have to buy insurance from some company.

External action: **I go to State Farm. I pay Sally.**

Input: *Sally gives me the insurance.*

I am insured. I feel very pleased.

    . . .

DAYDREAMER imagines there is an earthquake, her apartment collapses, and she is killed. In order to prevent this negative consequence of an earthquake, she forms the intention to run to the doorway as soon as the earthquake begins. She imagines carrying out this intention in another daydream. This time, however, she is hit by a plant which is hung above the doorway. So she decides to move the plant well before a possible future earthquake. Without exploring the possibility of running to the doorway in a daydream, the need to move the plant might never have been recognized.

Some Olympic gymnasts and divers prepare for performances by mentally "running through" those performances in advance (Suinn, 1984). Thus, in humans, rehearsing future actions through daydreaming may increase the accessibility of planned actions and the skill with which they can be performed. (J. R. Anderson, 1983, pp. 172-173, however, suggests that short distractions—daydreams—about something *other* than a task may be useful to that task: He hypothesizes that bringing an item out of short-term memory and then back in again results in a strengthening of the long-term memory of that item; therefore, short distractions which remove an item from short-term memory and then restore that item are useful in increasing retention of facts.)

Humans daydream not only about hypothetical future experiences, but about hypothetical *past* experiences as well (Klinger, 1971, p. 301; Neisser,

1982c, p. 14). After an experience, one often imagines alternative sequences of events which might have led to different outcomes. We propose that *daydreaming about imaginary variations of a past experience enables learning as a result of that experience.* By considering these alternative scenarios, one may learn better courses of action for use in future similar situations—or one may reinforce the original course of action. For example, after LOVERS1, DAYDREAMER generates the following daydream:

REVERSAL1

I feel embarrassed. What if I had put on nicer clothes? I would have gone to the Nuart Theater. I would have asked him out. He would have accepted. I feel regretful.

From this daydream, DAYDREAMER learns to wear nicer clothes when going out, if she is looking to form a new relationship, as demonstrated in the following:

LOVERS2

I want to be going out with someone. I feel hopeful. I have to be acquainted with someone. I have to go to the Nuart Theater. I want to avoid someone having a negative attitude toward me. I put on nicer clothes. I go to the Nuart Theater.

Input: *Robert Redford is at the Nuart Theater.*

... External action: **I ask him out.**

Input: *He turns you down.* ...

It is therefore proposed that learning, rather than being a one-shot event in response to an external world experience, is an ongoing process carried out in part during daydreaming. An external experience triggers any number of daydreams from which one may learn and continue to learn if that experience is later *reinterpreted* in light of new experiences and knowledge.

Daydream Production

Are daydreams random? We assume not—at least not entirely—and therefore DAYDREAMER must deal with the following issues:

- *The triggering of daydreams*: What instigates a daydream in the first place?

- *The generation of daydreams*: Once triggered, how are the events of a daydream generated?

- *The direction of daydreams*: Since at any point in a daydream there may be many directions for the daydream to proceed, how is one of those directions selected?

Representation of the Interpersonal Domain

In order to produce daydreams, DAYDREAMER must be able to represent a number of concepts in the domain of interpersonal relations, for example: "going out with," "breaks up with," "turn him down," "get even with," and so on.

The program must have knowledge about what is required to initiate, maintain, and terminate relationships. For example, in LOVERS1, DAYDREAMER attempts to form a relationship with Harrison Ford through appropriate actions such as introducing herself to him, asking him out, and so on. In RATIONAL-IZATION1, DAYDREAMER recognizes that when the movie star leaves to shoot a film in Egypt, their relationship is "in trouble."

The program must be able to represent attitudes such as "interested in" and to assess and modify attitudes. For example, in LOVERS1, DAYDREAMER infers that the movie star does not "think much of" her; she later tells the star she likes his movies in order to get him to have a more positive attitude toward her.

Goals and Needs

Humans have many needs, from biological needs for oxygen, water, food, sex, and shelter, to higher-level needs, such as for self-esteem, companionship, friendship, love, and sometimes also for material goods, money, achievement, a career, and so on. Typically, when a need is unsatisfied (or when a satisfied state is threatened), a goal is activated to satisfy that need (or preserve the satisfied state). For example, in LOVERS1, DAYDREAMER activates goals to form a relationship and to be entertained.

What collection of needs and goals should DAYDREAMER have? Is there a basic set of needs from which all other needs derive? The answer is not simple since even basic needs interact with each other. For example, satisfaction of the need for money may enable satisfaction of the shelter need; satisfaction of the need for friendship may contribute to satisfaction of the need for self-esteem.

Episodic Memory

Past personal and secondhand experiences are often the topic of daydreaming. In order for DAYDREAMER to daydream about such experiences, it will have to have something like a human *episodic memory* (Tulving, 1972). We must therefore address:

- *The representation of episodes*: The program must first be given an appropriate representation for previous experiences.

- *The storage of episodes*: The program should then be able to enter new experiences into memory, such as LOVERS1.

- *The retrieval of episodes*: The program should then be able to retrieve previous experiences appropriate to the current daydream or situation.

- *The application of episodes*: The program should then be able to make use of a retrieved episode, to incorporate the episode into the events of a daydream.

In RATIONALIZATION1, for example, DAYDREAMER is reminded of a previous secondhand experience—the time Harrison Ford "had a job with Paramount Pictures in Cairo." This experience is employed in the daydream being generated.

Learning Through Daydreaming

After a daydream, what remains in memory? How does what remains in memory modify subsequent behavior? We must address the following issues in *machine learning* via daydreaming:

- *Influence of daydreaming on future external behavior*: How are intentions formed in daydreaming later recalled and acted upon? How are daydreams stored in memory and later applied in generating external behavior? In REVERSAL1, for example, DAYDREAMER imagines she wore nicer clothes when going to the movies, and that this led to a success. Later, in LOVERS2, DAYDREAMER remembers to wear better clothes. How are daydreams evaluated and selected? How are realistic daydreams distinguished from unrealistic ones? For example, as a result of RATIONALIZATION1, should or will DAYDREAMER avoid movie stars in the future?

- *Influence of daydreaming on future daydreaming*: How are daydreams stored in memory and later applied in generating new daydreams? For example, suppose DAYDREAMER goes to work and her boss fires her. DAYDREAMER is then reminded of REVENGE1 which assists in the production of the following daydream:

REVENGE3

I remember the time I got even with Harrison Ford for turning me down by studying to be an actor, being a star even more famous than he is, him calling me up, him asking me out, and me turning him down. Say I am a powerful executive. Agatha offers me a job. I turn down her offer. I get even at her for firing me. I feel pleased.

1.2.2 Creative Problem Solving

Consider the traditional paradigm for problem solving in the field of artificial intelligence: Given a collection of operators, an initial state, and a description of a goal state, carry out a process of search in order to find a sequence of operators for transforming the initial state into a goal state. Programs based on this approach, such as GPS (Newell, Shaw, & Simon, 1957; Ernst & Newell, 1969) and STRIPS (Fikes & Nilsson, 1971) are given a problem to solve, perform the search in order to generate a solution or solutions, and then terminate. The generation of solutions to problems in human daydreaming, on the other hand, has three important properties which are not captured by this view of problem solving.

First, *each time one daydreams about a problem, different information may be available that will enable a different and possibly better solution.* That is, new solutions can be generated in light of new experiences. We propose that daydreaming occurs in the context of a memory which is constantly being updated—called a "dynamic memory" by Schank (1982). Each time a problem is examined, it may be moved one step closer to its solution; daydreaming thus enables the *ongoing revision* of previous solutions. For example, in RECOVERY3, the scheming to find out Harrison Ford's telephone number is advanced by one step. Now DAYDREAMER needs to find out where the actor went to college. A classic example of how the production of creative products involves incremental revision is provided by Beethoven's progressive modification of the first eight measures of the second movement of his *Third Symphony* (see, for example, Dallin, 1974, pp. 4-5).

Second, *daydreaming often explores possibilities which are unrealistic or fanciful at first glance, but which can sometimes be incrementally modified into realistic, useful solutions to problems.* In particular, scenarios are pursued in daydreaming which might not have otherwise been pursued either because of physical, social, cultural, or other constraints, or simply because their utility to some problem is not immediately evident. Humans daydream about possibilities such as: What would it be like if I were a laboratory mouse? What if I were on Mars? What if I found out that I was dreaming right now? What if I could fly? Daydreaming is thus a natural instance of such explicit techniques for improving creativity as "brainstorming" (Osborn, 1953) and "lateral thinking" (de Bono, 1970), which encourage one to generate wild ideas even if they seem unrelated to the problem being solved. DAYDREAMER, for example, produces the following daydream:

RECOVERY2

I have to ask him out. I have to call him. I have to know his telephone number. He has to tell me his telephone number. I have to know where he lives. Suppose he tells someone else his telephone number. What do you know! This person has to tell me his telephone number. He has to tell this person his telephone number. He has to want to be going out with this person. He has to believe that this person is attractive. He believes that Karen is attractive. He is interested in her. He breaks up with his girlfriend. He wants to be going out with her. He wants her to know his telephone number. She tells him she would like to know his telephone number. He tells her his telephone number. I have to tell her that I want to know his telephone number. I call her. I tell her that I want to know his telephone number. She tells me his telephone number. I call him. I ask him out.

Here DAYDREAMER wishes to find out Harrison Ford's telephone number. Instead of Harrison Ford telling DAYDREAMER his number, it is imagined that he tells someone else his number. Whether or not this possibility will help DAYDREAMER find out the actor's telephone number is not clear at first. However, this possibility then proceeds to be worked into a potential solution: DAYDREAMER will get her attractive and rich (and happily married) friend Karen to ask Harrison Ford for his telephone number, and then give it to her.

Third, *while daydreaming about one of several ongoing problems, one sometimes stumbles into a solution to another one of those problems.* That is, *serendipity*—the accidental recognition and exploitation of relationships among problems—often occurs during, and as a result of, daydreaming. For example:

RECOVERY3

I want to be entertained. I feel interested in being entertained.

External action: **I go outside. I grab the mail.**

I have the UCLA Alumni directory.

Input: *Carol Burnett went to UCLA.*

Input: *Carol's telephone number is in the UCLA Alumni directory.*

What do you know!

External action: **I read her telephone number in the UCLA Alumni directory.**

I succeed at being entertained. I feel amused. I have to read Harrison Ford's telephone number in the Alumni directory. I have to know where he went to college. I remember the time I knew where Brooke Shields went to college by reading that she went to Princeton University in People magazine. I have to read where he went to college in People magazine. ...

DAYDREAMER receives an alumni directory from UCLA which happens to contain the number of Carol Burnett. Since DAYDREAMER still has an active goal to contact Harrison Ford, she realizes that this experience is applicable to finding out his telephone number: DAYDREAMER now has to obtain a copy of the alumni directory from Harrison Ford's college (if any). Even if in fact there is no such directory, this may lead to other ideas such as finding the number of his agent in the guild directory.

Sometimes a reminding and serendipity are simultaneously triggered by noticing a physical object in one's environment, as in the following:

COMPUTER-SERENDIPITY

Input: *Computer.*

What do you know! I remember the time Harold and I broke the ice by me being a member of the computer dating service, and by him being a member of the computer dating service. I knew his telephone number by the dating service employee telling me Harold's telephone number. I have to be a member of the dating service. He has to be a member of the dating service. I have to pay the dating service employee.

LAMPSHADE-SERENDIPITY

Input: *Lampshade.*

What do you know! I remember the time Karen thought highly of the comedian because she thought he was funny. She thought he was funny because he wore a lampshade.

...I have to wear a lampshade. ...

In each case, a physical object provided as input leads to recall of an episode related to a currently active goal of DAYDREAMER: to form a new relationship. The episode is then applied to the goal. Although either the goal or the physical object alone may be insufficient to retrieve an appropriate episode, the two in conjunction *are* sufficient. The solution resulting from such a combination is *serendipitous* because the retrieved episode is not directly related to the active problem, and was not previously known as a possible solution to that problem. For example, the episode retrieved in LAMPSHADE-SERENDIPITY does not involve a goal to form a relationship with someone. Rather, it involves the goal to be entertained at a nightclub. Nonetheless, it is recognized as applicable, albeit indirectly, to the goal to form a relationship with someone.

Despite the fundamental importance of creativity to human intelligence, this unique ability of humans has rarely been investigated by artificial intelligence researchers.[2] This is understandable in light of the difficulty of the problem— creativity and the creative process are indeed quite complex. However, there are certain needless limitations of most present-day artificial intelligence programs which make creativity difficult or impossible: They are unable to consider bizarre possibilities and they are unable to exploit accidents. Several researchers

---

[2]Some notable exceptions are AM (Lenat, 1976), BACON (Langley, Bradshaw, & Simon, 1983), and TALE-SPIN (Meehan, 1976), considered in Chapter 5.

(see, for example, Sacerdoti, 1974, 1977; Sussman, 1975) have a cremental refinement of slightly incorrect or incomplete solution... have not addressed the generation and application of highly fanciful sequences of events such as those found in human daydreaming. Wilensky (1983) has proposed the use of "meta-planning" for dealing with multiple problems. Still, this mechanism does not address how a solution to one problem may accidentally suggest a solution to another problem.

Modeling Everyday Creativity

Our task here is not to account for the creative products of great artists, writers, scientists, and so on. Rather, we seek to address *everyday creativity*—so called for the following reasons: First, all humans are creative at some time or another (J. L. Adams, 1974; Wertheimer, 1945). As Polya (1945) puts it: "A great discovery solves a great problem but there is a grain of discovery in the solution of any problem." (p. v) Second, the problems of interest in daydreaming are the everyday topics of interpersonal relations, employment, and the like (J. L. Singer, 1975). Third, creative solutions occur on a daily basis in daydreaming (J. L. Singer, 1981; Klinger, 1971, pp. 217-221; Wallas, 1926, pp. 104-105; Varendonck, 1921, pp. 213-215) and are thus not restricted to exceptional individuals. Of course, the mechanisms which we will propose for everyday creativity may also be involved in more profound forms of creativity.

The following research issues must be addressed:

- the *generation* of fanciful and realistic solutions

- the incremental *modification* of solutions

- the *use of past experience* in generating new solutions

- the *recognition and exploitation of accidental relationships* among problems

- the *evaluation* of solutions

- the *storage* of solutions

- the *application* of solutions in the real world

Poincaré (1908/1952) and others (Wallas, 1926) have discussed the following phenomenon in creativity: setting aside a problem for a period of time ("incubation") seems to increase the chances for a later insight ("illumination"). How will our theory address this phenomenon?

Perpetual Daydreaming

Humans seem to be able to generate an endless series of novel daydreams. Although it has not reached this point so far, eventually we would like DAYDREAMER to be a sort of computational perpetual motion machine, able to daydream continuously as humans do. How much knowledge do we have to put into our program to get it to daydream endlessly or for a long time?

Of course, humans are constantly taking in new experiences which provide a new source of material for daydreaming. Even a human might run out of things to daydream about if not provided with input; subjects participating in sensory deprivation experiments report visual hallucinations (Bexton, Heron, & Scott, 1954) rather than normal daydreams. Will a daydreaming computer also require new input? If not provided with input, would the program eventually reach a limit point at which time it repeats the same daydreams over and over?

It is easy to construct a program which generates superficially different daydreams. One can imagine any number of schemes for generating daydreams which are trivial variations on previous daydreams, such as changing the names of all the characters in a previous daydream. However, constructing a program which continuously generates truly novel daydreams is very difficult. The daydreams of such a computer program must influence future daydreams in ways which are unpredictable to the programmer (since the only way for the programmer to predict an endless sequence of novel daydreams is by running the program itself).

Is it possible for any closed information-processing device to generate new and useful combinations forever? So far, of course, the universe has not exhausted its evolutionary possibilities. Daydreaming humans and computers have the benefit of outside input to enhance their daydreaming. But is it always a benefit? In what ways are creative processes aided or hindered by outside input?

1.2.3 Emotions

Previous artificial intelligence research in problem solving (see, for example, Fikes & Nilsson, 1971; Wilensky, 1983) has ignored the role of emotions. How might emotions be useful in a problem-solving program? Certainly it would be a disadvantage, say, for a theorem prover to become so angry about not being able to prove a theorem one way that it gave up trying other ways. But future intelligent problem-solving systems will not be directed toward single goals specified in advance by the user. Consider the example of a household robot: Such a system may have a number of ongoing goals, such as keeping its users happy and safe, answering the door, answering the phone, cleaning the house, preparing meals, and so on. The system will have to activate goals on its own, in response to the environment. When several goals are active simultaneously, the system will have to decide which to work on first. *Emotions are one mechanism for selecting from among multiple goals: (a) Emotions determine the focus of attention in processing, and (b) various events during processing in turn in-fluence emotions.* For example, when a problem-solving system anticipates a future failure, a negative emotion is produced. This negative emotion in turn motivates a goal to avoid the failure. This goal will compete for processing time with other goals motivated by other emotions. Emotions provide a partial alternative to other proposed mechanisms for coping with multiple goals, such as "universal subgoaling" (Laird, 1984) and "meta-planning" (Wilensky, 1983).

Previous computer models of human emotions have addressed the following: the emotional responses of a paranoid patient to a psychiatrist's questions and the influence of emotional state on the patient's subsequent verbal behavior (Colby, 1975); the emotional responses of a person to real-world situations and their influence on later behavior (Pfeifer, 1982); and the comprehension by a reader of the emotional reactions of story characters (Dyer, 1983b). These models all assume that a single emotional response results from an event in the world. However, in humans, an entire *sequence* of daydreams and corresponding emotions may result from an event, perhaps at last resting on a final emotion. One must therefore account not only for the relationship between emotions and events in the external world, but also between emotions and *internal* events or daydreams. Our work addresses how emotions result from imagined as well as real events.

Daydreaming interacts with emotions in a useful way: In humans, daydream-ing is often concerned with the reduction or elimination of negative emotions resulting from a past failure (Varendonck, 1921, pp. 249-250; J. L. Singer, 1975; Izard, 1977, pp. 339, 398). Thus, *daydreaming enables one to feel better.* In effect, failures result in a form of cognitive dissonance (Festinger, 1957) which must somehow be reduced. DAYDREAMER models this human function of day-dreaming: For example, in RATIONALIZATION1 the failure is rationalized and the disappointment reduced through the generation of a daydream demonstrat-ing that going out with Harrison Ford would be undesirable. In REVENGE1, DAYDREAMER achieves revenge against the movie star, resulting in a positive emotion. In ROVING1, DAYDREAMER shifts the topic to a more pleasant episode.

Of course, sometimes an increase, rather than a decrease, of negative emo-tion results from daydreaming about a failure. This is also modeled in DAY-DREAMER: For example, in REVERSAL1, DAYDREAMER regrets not having worn nicer clothes.

Modeling Emotions

A potential problem arises in the computer modeling of emotions: Emotions are subjective experiences *felt* by humans whereas computers are presumably not able to feel. An assumption of the present work is that it is possible, in principle, to model abstractly the interaction of emotions and (external and internal) behavior without actually getting the computer to feel, whatever that might mean. (The philosophical issues involved here are discussed in greater detail in Chapter 10.)

Given this assumption, the following research issues must be addressed:

- *Representation of emotions*: How do we represent emotions such as the "embarrassed," "angry," and "pleased" which appear in the daydreams reproduced above?

- *Initiation and modification of emotions in response to real and imagined (daydreamed) events*: What rules govern the generation of emotional responses in daydreaming? In RATIONALIZATION1, for example, a positive emotion results from what would normally be considered a negative experience by DAYDREAMER—losing her job. Why is this?

- *Influence of emotions on daydreaming and behavior*: How do emotions, such as the "angry" emotion of LOVERS1, modify processing? If this emotion resulted in no modification of the behavior of DAYDREAMER, then there would be no reason to have it.

- *Emotion regulation*: What strategies do humans employ for generating daydreams such as RATIONALIZATION1 which enable one to feel better? Why is it necessary to feel better in the first place—for humans, and possibly even for computers?

1.3   Foundations of DAYDREAMER

So far we have seen that daydreaming is a pervasive aspect of human experience and fundamentally related to human learning, creative problem solving, and emotions. But is not daydreaming then too all-encompassing, too difficult a research problem to undertake?

In this section, we present (a) the simplifying assumptions which have been imposed in order to facilitate the difficult task of constructing a theory of daydreaming and associated computer program, and (b) a definition of the phenomenon of daydreaming to be addressed by our work.

1.3.1   Excluded Phenomena

Some daydreaming consists of an interior monologue (J. L. Singer, 1975) in which one comments to oneself on the current situation or attempts to solve a personal problem by silently talking to oneself. Although the output of DAYDREAMER is in English, our theory and program do not address verbal daydreaming. Rather, the program generates verbal protocols of nonverbal, conceptual, and emotional daydreaming. Modeling of the generation of protocols itself is not of primary concern to us in the present work. English output is employed to make the stream of thought easily accessible to the human observer of the program. The English generator of DAYDREAMER functions merely as an output filter—it has no influence upon the content of daydreaming, unlike concurrent verbalization in humans which most likely does modify and slow down daydreaming (Pope, 1978). Instead of an English generator, DAYDREAMER might have employed an output module to produce graphical animations of its daydreams. For example, the representation for transfer of location (**PTRANS**) would result in a little person walking from one place to another on the screen; locations such as home, work, and the Nuart Theater would be represented graphically; and so on. Such output graphics would have had no impact on daydreaming.

Our theory does not address mental imagery or the quasisensory experiences which are often a part of daydreaming (J. L. Singer & Antrobus, 1972). Is this simplification justified? We must distinguish the *experience* of imagery from imagery *representations*: The experience of imagery escapes objective definition as does any inherently subjective phenomenon (Dennett, 1978, pp. 174-189; Nagel, 1974). Imagery representations, however, may be defined as knowledge structures which enable operations such as three-dimensional rotation (Shepard & Metzler, 1971) or the inference of spatial relations among objects from English sentences (Waltz & Boggess, 1979). Imagery representations may be essential to certain kinds of daydreaming: an artist or architect, for example, has detailed visual representations (Arnheim, 1974) which may enable daydreaming about artistic or architectural objects, just as a composer may have detailed musical representations (L. B. Meyer, 1956).

Imagery representations do not appear essential in the domain of interest to DAYDREAMER, that of interpersonal relations. We choose to ignore the effects of the visualization of real or imaginary events—whatever those might turn out to be—and concentrate on the *conceptual* and abstract *emotional* content of daydreaming. Although the daydream protocols we have consulted in constructing our theory sometimes refer to visual images, imagery representations have not been necessary to account for such daydreams. Varendonck (1921) came to the conclusion that "the thought associations, which are rendered in words when we succeed in becoming conscious of our fancy, are the principal part of the phantasy, the visual images only the illustrations" (p. 74, emphasis removed).

DAYDREAMER does not account in detail for daydreams (or external experiences) involving conversations. Nor does our theory account for daydreams involving arguments.

Despite the inherently phenomenological nature of daydreaming, DAYDREAMER is not a detailed theory of the phenomenology of inner experience. Thus our theory does not account for the subjective feeling states of the stream of consciousness, including the altered state of consciousness which often accompanies daydreaming. This is a point of departure for addressing the larger problem: As we construct more and more detailed conceptual representations, we may find that we are increasingly able to account for diverse aspects of subjective experience. Another largely subjective aspect which is not addressed is the distinction between directed, voluntary, intentional thought and undirected, involuntary, unintentional thought.

The theory does not address metacognition (Flavell, 1979) or the knowledge and thoughts people have about their own thoughts and thought processes. Although one may observe one's daydreams and consciously develop new strategies for daydreaming, this phenomenon is not addressed in the current work.

The theory does not account for development of daydreaming in childhood (J. L. Singer, 1975, pp. 123-148; Klinger, 1971; G. H. Green, 1923). However, we *are* concerned with learning through daydreaming.

The theory does not account for nonconscious thoughts and the influence of nonconscious thoughts on the conscious stream of thought.

For concreteness, DAYDREAMER is arbitrarily chosen to be a young heterosexual female. This makes little difference in the program since we make no attempt to model age, gender differences, or differences in sexual preference (although it does make a difference in the concrete details of the daydreams which the program produces).

1.3.2   Definition of Daydreaming

In order to construct a computational theory of a particular phenomenon, it is necessary to have a precise definition of that phenomenon. People have different ideas about what the words and expressions "daydreaming," "daydream," "fantasy," "reverie," "stream of consciousness," and "stream of thought" refer to. In this section, we examine previous attempts to define human daydreaming and related phenomena, and then we present our own definitions. (A detailed review of past work in daydreaming is provided in Chapter 9.)

J. L. Singer (1975) describes daydreaming as a shift of attention from the external environment or task being performed toward an internal sequence of events, memories, or images of future events which have varying degrees of likelihood of occurring. Daydreaming often ends with an awakening sensation (Varendonck, 1921, pp. 154-165; G. H. Green, 1923, p. 25)—either as a result of some environmental stimulus or a sudden awareness of being lost in thought. However, as Klinger (1971) notes, daydreaming may also be freely intermixed with task-related thought—in such cases there is no clear distraction followed by daydreaming followed by awakening.

In response to earlier conceptions of daydreaming as *stimulus-independent* and *task-irrelevant* thought (J. L. Singer, 1966; Antrobus, J. L. Singer, & Greenberg, 1966), Klinger (1978) points out that *undirected* thought—another property identified with daydreaming—may be stimulus-dependent and, conversely, directed thought may be stimulus-independent. Klinger therefore proposes that thought be characterized along the following four conceptually independent dimensions: undirected vs. directed, stimulus-independent vs. stimulus-dependent, fanciful vs. realistic, and integrated vs. degenerate.

Directedness refers to the degree to which thought is under deliberate, voluntary control. In directed thought, one has the impression that one is consciously steering the stream of thought, whereas in undirected thought the stream of thought seems to steer itself. Stimulus dependence refers to the degree to which thought is directly related to the external environment or current task situation. Realism refers to the likelihood or plausibility of depicted events or situations. Integrated vs. degenerate distinguishes regular thoughts from more bizarre, dreamlike, or nonsensical thoughts which sometimes intrude into waking thought. Degenerate thoughts are similar to the *hypnagogic* thoughts present at the onset of sleep (Maury, 1878; S. Freud, 1900/1965; Silberer, 1909/1951; Foulkes, 1966).

Although the classical daydream is undirected, stimulus-independent, fanciful, and integrated, daydreaming is capable of having any values along the above dimensions. It would serve no purpose to rule out, say, a directed, stimulus-dependent, and fanciful stream of thought, or an undirected, stimulus-dependent and realistic stream of thought, as instances of daydreaming.

Daydreaming may be decomposed into *segments* consisting of a coherent sequence of events or having distinct thematic content (Klinger, 1971). Daydreaming shifts from segment to segment (Pope, 1978); segments vary in length and need not run to their natural completion—each segment transition is in effect a distraction from one segment to the other. Hereafter, when we speak of a *daydream* or a *fantasy*, we mean such a segment.

Daydreaming, therefore, is not as homogeneous in content over time as some might think. Rather, daydreaming is similar to stream of consciousness thought (James, 1890a). However, the stream of consciousness is composed of a rich and varied collection of subjective phenomena including sensory perceptions, imagery, feelings, memories, drug-induced and meditative states, and so on (Ornstein, 1973; Tart, 1969; Pope & J. L. Singer, 1978b). For the purposes of the present work, we must limit our investigation to some well-defined subset of the contents of conscious thought. We choose a subset consisting of the following frequent components of daydreaming:

- emotional states

- recalled past events

- imagined variations on past events

- imagined future events

We rule out degenerate thoughts, imagery, and other consciousness phenomena from consideration (in accordance with the previous discussion). We also rule out metacognition. Since daydreaming is an event performed internally, an objective definition of daydreaming must refer to external behavior, in particular, to verbal protocols. (A detailed argument for this point is presented in Chapter 10.)

Accordingly, *human daydreaming is defined as any sequence of thoughts reported in a verbal protocol, where thoughts are composed of self attitudes, goals, and emotions, beliefs about the thoughts of others, beliefs about world states and events, hypothetical past, present, or future thoughts of varying degrees of realism, and memories of past thoughts.* This is a recursive definition: "Hypothetical thoughts," for example, refers to hypothetical self attitudes, goals, and emotions, hypothetical beliefs about the thoughts of others, and hypothetical beliefs about world states and events; these cases correspond to daydreaming about imaginary past or future events. For simplicity, we rule out hypothetical hypothetical thoughts (i.e., hypothetical thoughts about hypothetical thoughts) and hypothetical memories (although such thoughts are possible in humans). However, memories of past hypothetical thoughts are not ruled out: This is the case of recalling prior imagined events. But memories of memories are also ruled out for simplicity.

Our definition incorporates neither beliefs about self thought processes nor the notion of directed (vs. undirected) thought—two requirements for metacognition, not explored in the current work. Also, note that no mention is made of the dependence or independence of daydreaming upon current world states and events—the stimulus-dependence dimension of Klinger.

Daydreaming is restricted to verbal protocols for convenience only. We certainly would not wish to claim that a person or animal unable to communicate using language does not daydream. From other behaviors it may be possible to infer beliefs about world states and events, thoughts about hypothetical events, and so on. In fact, Griffin (1981) presents evidence which suggests that animals may daydream. He writes:

> [A] type of mental experience that may occur in non-human animals is belief that something has happened or will happen in the future. ...[A]nimals might experience thoughts roughly described in the following words: ...A cottontail rabbit, "If I run into this briar patch, that big, threatening animal won't catch and hurt me." ...(p. 15)

William James (quoted by Jung, 1916) writes:

> Our thought consists for the great part of a series of images, one of which produces the other; *a sort of passive dream-state of which the higher animals are also capable.* ...(p. 21, emphasis in original)

However, McDougall (1923) writes:

> Animals, if they are confronted by a problem, solve it, if at all, *ambulando,* in the course of action; they do not sit down and think out a plan .... Such planning, such purposive deliberation, would be the principal condition of the natural man's superiority to the animals. (p. 208)

Although the phrase "stream of consciousness" is normally a broad expression which includes all the subjective phenomena of consciousness (and thus encompasses daydreaming as we have defined it), we define *stream of consciousness* thought here as one kind of daydreaming, in particular, as any daydreaming which contains only segments of relatively short duration. The term "scenario" is used to mean a sequence of thoughts as defined above.

Some readers might use the word "daydreaming" to refer to a narrower or otherwise different range of mental phenomena. Why did we choose this word to describe the subject of our research? We might have instead called it, for example, "spontaneous fanciful and realistic past and future scenario exploration with emotional responses and recall of past experiences." On the other hand, "daydreaming" is short, memorable, and has been used by psychologists to mean just the phenomenon we are interested in.

1.4 Daydream Protocol Analysis

How can we obtain data on the content of the stream of consciousness? Daydreaming is unlike most other human activities which artificial intelligence researchers attempt to model, since it is a behavior which is not visible externally. The fact that there is no apparent "I/O" while a person is daydreaming presents several problems for us: How do we know when a person is daydreaming? Given that a person is daydreaming, how can we find out what that person is daydreaming about? Under what circumstances are people able to remember their daydreams? How can we hope to record a phenomenon as varied as the stream of consciousness? In this section, we review methods for gathering data about and analyzing daydreaming which have been used in the past.

1.4.1 Retrospective Reports

Retrospective reports involve the description of a previous stream of thought as it is recalled from memory. Varendonck (1921) obtained many retrospective reports of his own daydreams. He would start from the end of a daydream and work backward—in a similar fashion to when two people attempt to discover how a particular topic was reached in a conversation:

> I try to retrace, step by step, all the ideas which have succeeded one another on the screen of my fore-consciousness, but not at random. Usually I start from the last link (which I at once write down) and try to recapitulate the last but one, and so on .... The whole process requires some practice .... (p. 29)

Foulkes (1985, p. 83) similarly observes that the last portion of a night dream is a more effective cue for retrieval of the remainder of the dream than other content cues.

Here is a portion of one of Varendonck's retrospective daydream reports:

> Shall I get my book and read, or shall I continue to think about Miss
> X.? But what is the good of thinking about her? And what should
> I say to my children to explain why I prefer her to Miss Y., to whom
> they seem to incline? (Here my imagination reproduced before my
> mental eye an occurrence which took place the day before yesterday:
> Miss Y. is in my home with two members of her family; we see them
> out, and at the front door we all repeat the same remarks. This
> recollection is interrupted by the thought:) Miss X. looks younger
> than Miss Y. But the latter seems so attached to her father that I
> should have to detach her from him before she could really become
> attached to me. I might let her read some works on psycho-analysis.
> That might help. But how could I make her realize that I am not
> any more the simple teacher of fifteen years ago? For I remember
> that she once said in my presence that she would never marry an
> insignificant man. I could let her read one of my publications. I
> might start by sending her an order written on one of my visiting-
> cards that bears my academic title. By the way, I must not forget
> to send a card of congratulation to my friend V. on the occasion of
> his election as a member of the Flemish Academy. ... (Varendonck,
> 1921, p. 79, emphasis removed)

We assume that daydreaming consists of the continual addition (and removal) of items into short-term memory. A "memory trace" is a (possibly
partial) series of snapshots of the state of short-term memory. Retrospective reports may be generated immediately, while the memory trace of the daydream
is still in short-term memory, or later by retrieving the trace from long-term
memory (M. K. Johnson & Raye, 1981; Ericsson & Simon, 1984, pp. 10-20).
Retrieval of daydream traces from long-term memory, however, is subject to at
least two problems. First, if the daydream is recalled it will be subject to the
same kinds of confusions, distortions, and omissions which other memories are
subject to (F. C. Bartlett, 1932; Neisser, 1967, 1982b; Loftus, 1975).

Second, it may be difficult to recall the daydream at all. Smirnov (1973)
performed studies in which subjects were asked to report on their stream of
thought during particular episodes (such as walking to the office) which had
occurred an hour or two earlier. Most of the consciousness material which subjects were able to recall was directly related to the activity being performed at
the time. However, subjects felt that other material had been forgotten. Why
might daydreams be forgotten? S. Freud (1900/1965) proposes that the forgetting of dreams—and presumably also daydreams—is caused by the resistance
created by repression. A similar view is adopted by G. H. Green. Ericsson
and Simon (1984, p. 151), however, propose the following account of Smirnov's
results: When the content of internal experience is unrelated to the task being
performed, memory connections are not formed between the internal experience

events and the task events (especially when the task is a routine one, such as driving to work) and therefore the subject will be unable to recall those internal experiences when the task is used as a retrieval cue. They propose (p. 19) that retrospective reports be obtained immediately after a given thought stream, while it is still unnecessary to give specific retrieval cues—presumably, appropriate retrieval cues are already present in short-term memory. Varendonck (1921) similarly writes: "my memory is very deficient as regards my day-dreams, and for this reason I retrace and register them immediately if I think they may be useful." (p. 214) (However, at other times [p. 327] Varendonck claims to have no trouble recalling daydreams.) One way of generating retrieval cues long after the fact is to ask subjects whether they can recall ever experiencing a particular daydream (for example, "have you ever imagined walking on the ceiling?"). One method of gathering retrospective reports, the 400-item Imaginal Processes Inventory (IPI) of J. L. Singer and Antrobus (1972), operates in just this fashion.

1.4.2 Think-Aloud Protocols

Retrospective reports rely on memory and so are not always accurate: similar internal experiences may be confused. However, they do have the advantage of not interfering with the process itself. In the *think-aloud protocols*, first employed extensively by researchers interested in problem solving (Bloom & Broder, 1950), the subject verbalizes the stream of thought as it is taking place. Several investigators (Pope, 1978; Klinger, 1971; Antrobus & J. L. Singer, 1964) have used think-aloud protocols in examining spontaneous stream of consciousness thought. The experiments of Pope (1978) have suggested that think-aloud protocols may inhibit the subject or slow down and thus modify the stream of thought. Specifically, he found that the duration of thought segments relating to a particular theme or topic lasted about 30 seconds when thinking aloud, whereas when the subject was not thinking aloud, thought segments—which were indicated by pressing a button—lasted only about 5 or 6 seconds.

Think-aloud protocols derive from the method of *free association* used by S. Freud (1900/1965) in interpreting the night dreams of patients. In this method, patients were instructed to verbalize any thoughts coming to mind in connection with a given element in a dream, no matter how irrelevant, absurd, meaningless, or unpleasant to the patient. The purpose of free association was to uncover, and eventually remove, a pathological idea underlying the patient's problems. Reik (1948) gives a transcript of his own free associations, of which we reproduce a portion below in order to show the similarity between what has been called free association by psychoanalysts and what we are here calling daydreaming and stream of consciousness thought:

> What are my thoughts at this moment? I see the pussy willows
> on my bookcase ...a prehistoric vase ...spring, youth, old age
> ...regrets ...the books ...the *Encyclopedia of Ethics and Religion*

... the book I did not finish ... My eyes wander to the door ... A pho-
tograph of Arthur Schnitzler on the wall ... my son Arthur ... his
future ... the lamp on the table ... What a patient had said about
the lamp once when it was without a shade ... the table ... it was
not there a few years ago ... my wife bought it ... I did not want
to spend the money at first ... she bought it nevertheless ... (pp.
28-29, ellipses in original)

Reik does not specify exactly how this report was generated; it was either a
"think-in-writing" protocol or a transcription of a think-aloud protocol recorded
on a dictaphone. He does make it fairly clear that retrospective elaboration and
modification (for the purposes of publication) were performed.

Rapaport (1951, p. 452) reports using the method of recording (presumably
by writing down) the contents of his hypnagogic reveries continuously, interrupted only by temporarily dozing off.

1.4.3   Thought Sampling and Event Recording

One variation on the immediate retrospective report is the method of *thought
sampling* used by Klinger (1978). In this method, subjects would carry a beeper
with them during the day. When the beeper sounded at a random time, subjects would be asked to describe or rate their most recent thoughts. Descriptions
would be provided in the form of a narrative retrospective report; subjects would
also complete a questionnaire in order to rate their most recent thoughts along
the following variables: duration of most recent thought segment, duration of
next to most recent thought segment, vagueness versus specificity of imagery,
amount of directed thought, amount of undirected thought, amount of detail in
imagery, number of things that seemed to be going on simultaneously, visualness
of imagery, auditoriness of imagery, attentiveness to environmental events, degree of recall of environmental events, degree to which imagery felt controllable,
degree of trust in accuracy of memory of latest experience, usualness of latest
experience, distortedness of latest experience, and time of life associated with
experience.

In the method of *event recording* discussed by Klinger, subjects are asked to
report whenever a particular kind of event occurs in the stream of consciousness. Thus, for example, subjects can be asked to note whenever they have a
daydream of revenge. Pope (1978) used a button-press method to indicate a
new daydreaming topic.

1.5 Contents of the Book

The remainder of this book is organized as follows:

Chapters 2 through 7 present and discuss our computer model of human daydreaming. The major contributions of each chapter are as follows:

	Artificial intelligence	*Psychology*
Chapter 2	Emotion-driven planning	Algorithms for modeling daydream production
Chapter 3		Interaction of emotions and daydreaming
		Daydreaming taxonomy and strategies
Chapter 4	Strategies for learning	Memory for daydreams
	Perpetual planning	
Chapter 5	Serendipity-based planning and learning	Algorithms for transfer in problem solving
Chapter 6	Representation of interpersonal planning knowledge	
Chapter 7	Detailed construction of program	

Chapter 8 presents a critical comparison of techniques for episode storage and retrieval, including discrimination nets, spreading activation, and connectionist nets.

Chapter 9 presents a comprehensive review of previous psychological work on daydreaming and the related phenomenon of night dreaming.

Chapter 10 examines some of the underpinnings, both philosophical and scientific, of the present work.

Chapter 11 presents some potential future applications of DAYDREAMER and daydreaming, suggestions for future work, and conclusions.

Appendix A presents annotated traces of the operation of DAYDREAMER.

Appendix B presents the generator used to convert program representations into English.

Chapter 2

Architecture of DAYDREAMER

How do we build a daydreaming machine? What existing artificial intelligence techniques can be employed? How must these techniques be extended? What new techniques need to be developed?

This chapter presents the basic architecture of DAYDREAMER: We present an overview of the program components, propose that daydreaming can be modeled by the process of planning for multiple goals, introduce the mechanism of *emotion-driven planning*, and present the control structure and procedures of the program.

2.1 Overview of Program

DAYDREAMER contains the following processing mechanisms:

- *Emotion-driven control*

- *Analogical planner*

- *Serendipity mechanism*

- *Mutation mechanism*

- *Parser*

- *Generator*

The program contains the following static data structures:

- *Personal goals*

- *Daydreaming goals*

- *Planning rules*

- *Inference rules*

The program contains the following dynamic data structures:

- *Episodic memory*

- *Emotions*

- *Concerns*

- *Planning contexts*

Personal goals are the basic goals of the program (such as to maintain self-esteem, form friendships, and so on). Daydreaming goals are an additional set of goals which model human daydreaming activities and enable the program to improve its ability to achieve its personal goals.

Concerns are processes or tasks on behalf of top-level goals. At any moment a number of concerns are active—some are for personal goals, others are for daydreaming goals. One or more emotions are associated with each concern. The emotion-driven control mechanism determines which concern goal to work on at any moment, based on these associated emotions.

The analogical planner employs planning and inference rules in order to generate plans for achieving a given top-level goal. It also generates plans through the retrieval and analogical application of plans previously stored automatically in episodic memory. (Some episodes are also added manually by the programmer.)

The analogical planner reads from and writes to planning contexts: A short-term memory for the maintenance of alternative hypothetical worlds. When the program is started, the first planning context is loaded with the initial state of the external world.

The serendipity mechanism finds previously overlooked plans in response to input states, retrieved episodes, and random possibilities generated by the mutation mechanism.

English input is converted into representational data structures by the parser and placed into the current planning context. Information added to the current planning context is converted into English and produced as output by the generator. (While the recursive descent generator is able to produce a fairly natural-sounding English sentence given any representation in the program's repertoire, the parser is implemented simply by mapping entire canned sentences into canned representations—parsing and generation of English is not a major focus of this work.)

2.2 Daydreaming as Planning

In daydreaming, one sometimes reviews existing, past world events. Other times, one constructs new, imaginary world events. That is, daydreaming is *generative* (Jakobson, 1978; Chomsky, 1965).

In DAYDREAMER, *the generation of imaginary sequences of events is accomplished through the process of planning*[1] (see, for example, Newell, Shaw, & Simon, 1957; Fikes & Nilsson, 1971; Wilensky, 1983). Planning attempts to achieve a *top-level goal* by breaking it into *subgoals*, breaking each of those subgoals into further subgoals, and so on, until subgoals are reached that are already satisfied or can be satisfied by performing an action. An action may in turn require further subgoals (often called preconditions) to be achieved, which may in turn require other actions to be performed.

In this section, we review past work in psychology and artificial intelligence which attempts to find a suitable set of top-level goals for modeling humans. We then present a set of such goals chosen for DAYDREAMER. Finally we discuss how knowledge for achieving these goals is represented in the program as planning and inference rules.

2.2.1 Previous Personal Goal Sets

The objectives which a human strives to achieve or maintain—such as health, shelter, safety, friendship, love, self-esteem, satisfaction of hunger, thirst, and sexual needs, and so on—are called *personal goals* in DAYDREAMER. Personal goals—in the past also variously known as "ends," "aims," "objectives," "desires," "needs," "wishes," "impulses," "instincts," and so on—are assumed to be the basic determinants of human behavior.

Researchers have always grappled with the problem of finding a "fundamental" or "primitive" set of personal goals. Of developing one of his theories, S. Freud (1920/1961) writes:

> No knowledge would have been more valuable as a foundation for true psychological science than an approximate grasp of the common characteristics and possible distinctive features of the instincts. But in no region of psychology were we groping more in the dark. Everyone assumed the existence of as many instincts or 'basic instincts' as he chose, and juggled with them like the ancient Greek natural philosophers with their four elements—earth, air, fire, and water. (p.45)

For example, McDougall (1923) proposed a set of instincts consisting of escape, combat, repulsion, appeal, mating, curiosity, submission, self-assertion, food-

---

[1]Italicized assertions regarding the processes and data entities of DAYDREAMER are to be taken also as assertions of our theory of human daydreaming—that is, as hypotheses about human daydreaming subject to empirical validation.

seeking, acquisition, construction, laughter, the gregarious instinct, and the parental instinct.

Instead of postulating a large number of instincts, Freud chose merely to classify instincts as sexual or ego instincts. Sexual instincts are concerned with preservation of the species through reproduction, while ego instincts are concerned with self-preservation (including satisfaction of hunger, thirst, protection from danger, and so on). Sexual instincts are governed by the "pleasure principle," while ego instincts are governed by the "reality principle."

Maslow (1954) argues that the endeavor of constructing lists of drives or needs should be given up: such lists imply equality of the various drives, ignore their dynamic nature, and treat drives as isolated entities, when in fact they interact with each other. Moreover, he argues, the particular collection of drives which one chooses depends on the degree of specificity of analysis. Maslow thus proposes a *hierarchy* of needs, with lower-level needs having priority over higher-level ones: At the lowest level are physiological needs such as sex, health, and satisfaction of hunger and thirst. One level higher are safety needs such as those for security, comfort, protection, and freedom from fear. Next, belongingness and love are needs satisfied—for example, through friends, spouse, and family. At the next level are esteem needs, divided into needs for self-respect and needs for the esteem of others. At the highest level is the need for self-actualization, or fulfillment of the true nature or potential of the particular individual in creative or other activities.

Schank and Abelson (1977) propose a taxonomy of goals for use in story comprehension; they are concerned with elaborating a set of goals which a reader of a story assumes a character to have, when no explicit mention of goals is made. In their scheme, goals are classified according to whether they are to be repeatedly satisfied (S-HUNGER, S-SEX, S-SLEEP), are pursued for enjoyment (E-TRAVEL, E-ENTERTAINMENT), are to be achieved (A-POSSESSIONS, A-GOOD-JOB), are states to be preserved (P-HEALTH), or are instrumental to other goals (D-KNOW, D-PROX).

2.2.2   A Set of Personal Goals

The personal goal set of DAYDREAMER roughly approximates the personal goals of humans; however, several common human goals (such as safety, shelter, sleep, oxygen, and water) are omitted for simplicity. The following personal goals correspond to (physical or mental) *states* to be achieved and maintained; they are called *achievement goals*:

SELF-ESTEEM Achieve and maintain a positive self attitude

SOCIAL ESTEEM Achieve and maintain a positive attitude of another person toward self

LOVERS Achieve and maintain a romantic relationship with another person

FRIENDS Achieve and maintain a friendship with another person

EMPLOYMENT Achieve and maintain employment

The following personal goals require repeated satisfaction; they are called *cyclic goals*:

FOOD Satisfy a recurring need for food

SEX Satisfy a recurring need for sexual gratification

LOVE-GIVING Satisfy a recurring need for giving love

LOVE-RECEIVING Satisfy a recurring need for receiving love

COMPANIONSHIP Satisfy a recurring need for companionship

ENTERTAINMENT Satisfy a recurring need for entertainment

MONEY Satisfy a recurring need for money

POSSESSIONS Satisfy a recurring need for more possessions

Cyclic goals correspond to *need states* whose level of satisfaction is continually declining. Once the level of satisfaction of a need state falls below a certain level, a personal goal to satisfy that need state is activated. Performing an appropriate action then restores the level of satisfaction of that need state to a certain level. For example, when the level of satisfaction of the need for food falls below a certain level (i.e., one becomes hungry), one eats. This restores the level of satisfaction of the need for food to a high level. With the passage of time, the level again falls below a certain level, one eats again, and so on.

Personal goals are not isolated entities; rather, they interact dynamically in various ways: The satisfaction of one personal goal may continually enable the satisfaction of another personal goal (directly or indirectly); this situation is called *goal subsumption* (Schank & Abelson, 1977). For example, as long as one continues to have a job, the need for money will continue to be satisfied; thus **EMPLOYMENT** subsumes **MONEY**. In a similar fashion, **LOVERS** subsumes **SEX**, **LOVE-GIVING**, **LOVE-RECEIVING**, **COMPAN-IONSHIP**, and **ENTERTAINMENT**; **FRIENDS** subsumes **COMPANIONSHIP** and **ENTERTAINMENT**; **MONEY** subsumes **POSSESSIONS** and **ENTERTAIN-MENT**. All achievement and cyclic goals enable satisfaction, to some extent, of **SELF-ESTEEM** and **SOCIAL ESTEEM**.

When a need state falls below a certain level of satisfaction, DAYDREAMER takes a different action depending on whether its corresponding subsumption state is satisfied: As shown in Figure 2.1, if the subsumption state is not satisfied, a goal to achieve that subsumption state is initiated. Otherwise, a goal to achieve the need itself is initiated. So, for example, if one does not have a job and runs low on money, one initiates the goal to get a job. If one has a job and

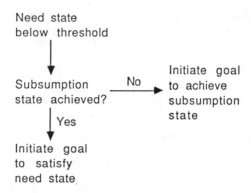

Figure 2.1: Needs with Subsumption States

runs low on money, one initiates the goal to go to work. Running low on money is therefore what drives the program to maintain employment.

It might be argued that the goal set presented here is incorrect (e.g., is the goal for money really a basic goal?), or incomplete (e.g., it omits the goal to acquire information through reading). But it must be realized that any goal set in an artificial intelligence program is arbitrary—it is always relative to what the function of the program is. Nor is it possible to specify a final set of human goals from which all other goals follow, given our unique ability to create idiosyncratic goals. As Lewin (1926/1951) put it: "The extraordinary liberty which man has to intend any, even nonsensical actions—that is, his freedom to create in himself quasi-needs—is amazing" (p. 136). Schank and Abelson (1977) also note:

> If one were to set out to list all the different kinds of things that could be desired, there would be no end to it. One may be told and be prepared to believe that someone likes banging his head or wants to line his patio with Martian rocks. ... (p. 112)

2.2.3   Planning and Inference Rules

In order to achieve personal goals, DAYDREAMER contains knowledge about persons, physical objects, locations, possession of physical objects, locations of objects, environmental conditions, personal attributes (such as occupation, external appearance, social class), mental states (such as beliefs, goals, attitudes, emotions), interpersonal relationships (such as friends, lovers, employee-employer), actions (such as changing location, communicating information, eating), and activities (such as seeing a movie, eating at a restaurant, going on a date). We refer to this realm of world entities generally as the *interpersonal domain*.

Representation of the interpersonal domain in DAYDREAMER is based on previous work by Dyer (1983a) and Schank and Abelson (1977) on representing relationships and stereotypical sequences of events in the everyday world, and Heider (1958) on representing sentiments or generalized attitudes and how they relate to interpersonal relations. Our contribution is the addition of planning knowledge for forming, maintaining, and terminating relationships, scripts for asking someone out, going out on a date, and so on.

Domain knowledge is represented in the program as *rules*. There are two kinds of rules—*planning rules* and *inference rules*. Planning rules specify alternative methods for breaking down a goal into subgoals, while inference rules specify unavoidable consequences of various states or events.

A rule consists of an *antecedent pattern* (indicated by *IF*) and a *consequent pattern* (indicated by *THEN*). Here are two sample rules from DAYDREAMER:

IF	person has **POS-ATTITUDE** toward something that self has **POS-ATTITUDE** toward
THEN	**POS-ATTITUDE** toward person

IF	**ACTIVE-GOAL** to have **POS-ATTITUDE** toward person
THEN	**ACTIVE-GOAL** for person to have **POS-ATTITUDE** toward something that self has **POS-ATTITUDE** toward

The first rule is an inference rule which states that if one likes something that one believes another person likes, one will like that person. The second rule is a planning rule which states that if one has a goal to like someone, the subgoal may be created for one to like something that one believes the other person likes. Rules often occur in complementary pairs such as this. In fact, such pairs are represented in DAYDREAMER as a single combined planning and inference rule. Hereafter, combined planning and inference rules will only be given in their planning form.

Rules are applied through *unification* (Charniak, Riesbeck, & McDermott, 1980; Robinson, 1965) and *instantiation*: If the antecedent pattern of a rule unifies with an existing representational data structure, a new representational data structure is constructed by instantiating the consequent pattern of the rule with the variable bindings resulting from the unification.

If several planning rules unify with a given subgoal, each rule is applied in its own separate world model or planning *context*. Thus, *planning rules give rise to alternative states of a hypothetical world*.

Although the above rules are stated in terms of "self," they may be applied to other entities than DAYDREAMER via *other planning*. Using this mechanism, DAYDREAMER is able to represent the mental states of others, make inferences about the mental states of others (simulate the viewpoint of others), and perform planning in order to influence the mental states of others.

In addition to the contribution of other daydreaming mechanisms, the fanciful scenarios characteristic of daydreaming result in part from the use of *plausible* planning and inference rules—rules whose applicability in the real world is less

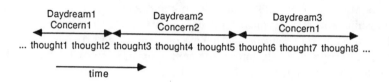

Figure 2.2: Interleaving of Concerns

than certain. For example, DAYDREAMER has the plausible inference rule that if someone smiles at her, that person is romantically interested in her. This can lead to a fanciful daydream about a relationship in response to a mere friendly smile.

The particular rules employed by DAYDREAMER in generating daydreams in the interpersonal domain are discussed in Chapter 6.

2.3   Daydreaming as Multiple Goal Planning

Most artificial intelligence programs (a) take a request from a human user, (b) handle the request, and (c) terminate: For example, a medical expert system such as MYCIN (Shortliffe, 1976) is asked to recommend a course of treatment given a description of symptoms, or a natural language understanding system such as BORIS (Dyer, 1983a) is asked to read and answer questions about a given segment of text.

In contrast, DAYDREAMER has a collection of goals which are initiated and processed as the program sees fit. Instead of terminating once a goal is achieved, DAYDREAMER switches to another one of its goals.

Daydream protocols may be divided into segments which relate to a particular "current concern," or goal not yet achieved, of the daydreamer (Klinger, 1971). Therefore, in DAYDREAMER, a task or activity on behalf of an active top-level goal is called a *concern*.

At any time, several concerns may be active. The program operates by selecting a concern and performing some planning for that concern, then selecting another concern and performing some planning for *that* concern, and so on repeatedly. Planning activity results in the production of representational data structures; these are the "thoughts" of the program which are converted into English and produced as output. In DAYDREAMER, then, *daydreaming consists of the production of a sequence of representations through a process of planning which shifts from concern to concern.* Each subsequence of contiguous representations produced on behalf of a given concern may be called a "daydream" (see Figure 2.2).

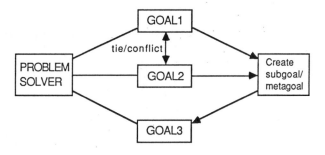

Figure 2.3: Meta-Planning/Universal Subgoaling

2.4 Daydreaming as Emotion-Driven Planning

How does the program decide which of its concerns to process at any given moment, and which others to defer? Some overall guiding mechanism is needed to direct attention to the concern deemed most important at any moment in time.

Wilensky (1983) proposed the mechanism of *meta-planning* and the use of *meta-themes* such as **DON'T WASTE RESOURCES** for dealing with multiple goals. Laird (1984) proposed the use of a goal selection subgoal within the framework of *universal subgoaling*. In both cases, general problem solving is invoked in order to cope with multiple goals: as shown in Figure 2.3, a tie or conflict between two goals is resolved by initiating a third goal.

These approaches suffer from two problems: First, in practice, there is often no time to invoke general meta-planning or subgoaling when making behavioral decisions in an external environment. Rather, a computationally inexpensive set of heuristics is required. Second, from a psychological standpoint, thoughts about "what to think about next" are relatively infrequent in the protocols of human daydreaming we have examined—whereas meta-planning and universal subgoaling could very well end up in a deep recursion in determining what to do next.

As an alternative, we propose the mechanism of *emotion-driven planning* in which *emotions* provide the primary motivation for daydreaming: First, *in order to daydream at all, emotions must be present.* Second, *emotions determine*

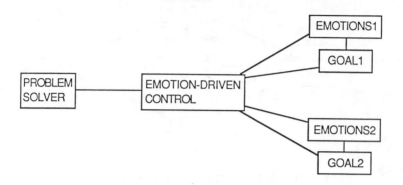

Figure 2.4: Emotion-Driven Planning

which concern to process at any particular time.[2] [3]

One or more emotions are associated with each concern, as shown in Figure 2.4. The *emotional motivation* of a concern (also called its *dynamic importance*) is the absolute value of the sum of the *strengths* of each of the emotions associated with that concern. The primary control algorithm of DAYDREAMER is then as follows:

1. Select concern with highest emotional motivation.

2. Perform one unit of planning for that concern.

3. Go to step 1.

While a concern is the most highly motivated, planning activity proceeds towards this concern. If at some point another concern becomes more highly motivated, then planning activity switches to this other concern. That is, *a shift from one concern to another occurs when the motivation of the latter is greater than the motivation of the former.* The processing of concerns depends upon the association of emotions with concerns and the setting and modification of the strengths of those emotions.

---

[2]It should be noted that meta-planning handles problems which emotion-driven planning is not designed to handle: For example, suppose the program has concerns to maintain a friendship with Karen, satisfy a need for food, and rationalize being turned down for a job. Meta-planning would enable the generation of a plan in which the daydreamer has dinner with Karen in order to achieve two of her concerns (satisfy need for food and maintain friendship) simultaneously. If Karen also happens to be good at coming up with rationalizations for failures, the three concerns of the daydreamer could be achieved in a single plan. Achievement of multiple concerns can only occur accidentally in DAYDREAMER (through serendipity and fortuitous subgoal success). A complete system would have to have both meta-planning for long-term planning, and emotion-driven planning for short-term decisions about what to do next and what to daydream about next.

[3]The view that emotions serve as motivation in humans will be argued in Chapter 3.

In DAYDREAMER, emotions are activated, deactivated, associated with concerns, and disassociated with concerns in various situations:

When a concern is initiated, a new emotion is created and associated with the concern. The strength of the emotion is set according to the *intrinsic importance* (a measure of importance assigned by the programmer) of the goal. For example, in DAYDREAMER the friendship goal has a greater intrinsic importance than the entertainment goal. Thus, all other things being equal, the program will pursue a friendship goal before attending to an entertainment goal.

Our notion of intrinsic importance is similar to the notion of relative goal importance Carbonell (1980) developed for modeling personality traits and political ideologies. The emotion representation created upon concern initiation is intended to model the emotion of "interest-excitement" which is hypothesized to provide a source of motivation in humans (Tomkins, 1962; Izard, 1977). This emotion accompanies a motivated task and is sustained through successful performance of this task. Rothenberg (1979) is most likely referring to the same emotion when he writes: "While engaging in the creative process, the creator is stimulated and aroused" (p. 46).

When a concern terminates (with a success or failure), a new positive or negative response emotion is created. The strength of this new emotion is set according to the dynamic importance of the concern—that is, the absolute value of the sum of the strengths of the emotions associated with the concern. Such negative-response emotions may initiate and become associated with new concerns. Any motivating emotions associated with the concern not connected to other concerns are deactivated.

The strengths of response emotions decay over time and such emotions are disassociated with any concerns and deactivated when their strength falls below a certain threshold. A concern whose motivation falls below a certain threshold is terminated.

When a real or imagined personal goal success or failure occurs as a side effect of processing to achieve some concern, a response emotion is created and associated with that concern. Its strength is proportional to the intrinsic importance of the goal. This provides an additional force of motivation for a concern when unexpected personal goal successes are generated, and a reduction in motivation when unexpected goal failures are generated.

When a subgoal of a concern succeeds while working on another concern, or when a potential plan for achieving the top-level goal of some concern is accidentally discovered by the serendipity mechanism, a new positive emotion is created and associated with the concern. Its strength is some fraction of the strength of the motivating emotions of the concern.

Here is an example which will serve to clarify the interaction between emotions and concerns: Suppose two concerns—*concern$_1$* and *concern$_2$*—are initiated. Motivating emotions are created and associated with the two concerns as shown in Figure 2.5. Their strengths are equal to the intrinsic importances of the concerns; in this case, *concern$_1$* has a greater intrinsic importance than

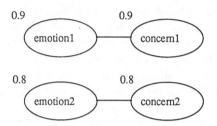

Figure 2.5: Emotions and Concerns 1

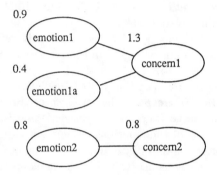

Figure 2.6: Emotions and Concerns 2

$concern_2$ and so its emotion has a greater strength. Therefore, $concern_1$ will be processed first.

Now suppose that a side-effect personal goal success occurs while processing $concern_1$. A new emotion is created and associated with that concern, as shown in Figure 2.6. The strength of the new emotion is proportional to the intrinsic importance of the personal goal which succeeded. $Concern_1$ now has two emotions connected to it; its dynamic importance is now even more than it was before. Therefore, $concern_1$ will again be processed rather than $concern_2$.

While processing $concern_1$, suppose an accidental success of a subgoal of $concern_2$ occurs. This creates a positive emotion (corresponding to the feeling of surprise in humans) which is associated with $concern_2$, as shown in Figure 2.7. The strength of this emotion is such that the dynamic importance of $concern_2$ has now surpassed that of $concern_1$. For the first time, $concern_2$ will be processed instead of $concern_1$.

Now suppose that $concern_2$ terminates with a failure. A negative response emotion is created; its strength is equal to the dynamic importance of the concern before it terminated. This new emotion in turn activates a new concern,

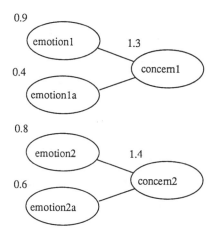

Figure 2.7: Emotions and Concerns 3

as shown in Figure 2.8. Since the dynamic importance of *concern₃* is greater than that of *concern₁*, it will be processed.

Next suppose a new concern is initiated and an associated emotion activated, as shown in Figure 2.9. The dynamic importance of this new *concern₄* (which so far is equal to its intrinsic importance) is greater than the dynamic importance of both *concern₃* and *concern₂*, so it will be processed.

Figure 2.10 summarizes this example: concerns were processed in the following sequence: *concern₁*, *concern₁*, *concern₂*, *concern₃*, *concern₄*. Of course, in an actual DAYDREAMER run, many steps are performed for each concern before an event occurs causing the program to switch to another concern. Decay of emotion strengths has also been ignored in this example.

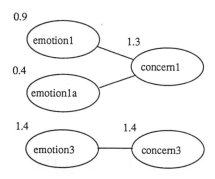

Figure 2.8: Emotions and Concerns 4

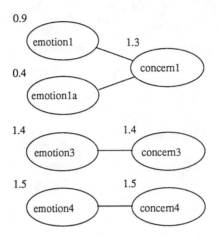

Figure 2.9: Emotions and Concerns 5

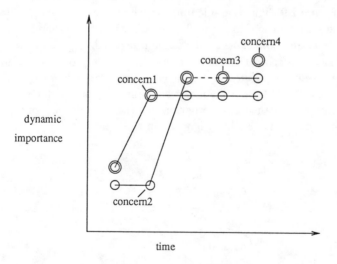

Figure 2.10: Emotions and Concerns

The primary control mechanism of DAYDREAMER is similar to the AM (Lenat, 1976) agenda mechanism, except that whereas AM tasks are assigned a resource limit after which processing switches to another task, in DAY-DREAMER execution of a concern continues until it is interrupted by a higher priority concern (for example, as a result of the activation of a new top-level goal or upon detection of a serendipity). Another difference is that within a DAYDREAMER concern, certain heuristics are used to direct the search process within that concern. Thus DAYDREAMER has agendas within agendas.

The idea of multiple concerns motivated by emotions in DAYDREAMER is related to Minsky's (1977) computational theory of the mind as a society of intercommunicating and conflicting entities. He describes a child playing with blocks: Internally, the WRECKER in the child wants to destroy the tower being built by the BUILDER. Meanwhile, the I'M-GETTING-HUNGRY entity is growing in strength. As the control of the BUILDER weakens, the child destroys the tower and gets up to go home and eat.

Emotional motivations associated with concerns in DAYDREAMER are similar in some ways to priorities associated with processes in a computer operating system (see, for example, Thompson, 1978, pp. 1936-1937). However, operating systems rapidly *interleave* tasks, giving more time to higher-priority processes and less time to lower-priority processes. Humans do not process concerns in a "round-robin" fashion because of the overhead required for bringing a new concern into short-term memory (J. R. Anderson, 1983). Humans tend to stay with one concern until interrupted for a particular reason by another, as does DAYDREAMER.

2.5 Procedures for Daydreaming

DAYDREAMER consists of a number of procedures, shown in Figure 2.11. This section presents the procedures and explains the gist of what they do. Further details and examples will be presented later on.

These procedures make use of the following existing techniques in artificial intelligence:

- planning (Fikes & Nilsson, 1971),

- planning for multiple goals (Wilensky, 1983),

- analogical planning (Carbonell, 1983; Fikes, P. E. Hart, & Nilsson, 1972), and

- episodic memory (Kolodner, 1984; Schank, 1982).

The novel aspects of these procedures are the following:

- Emotion-driven planning: a mechanism based on emotions for determining which goal to focus on at any moment in planning.

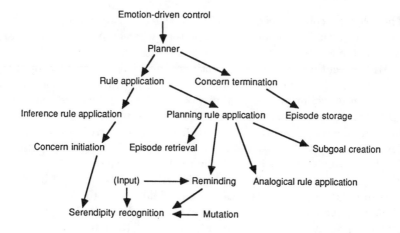

Figure 2.11: DAYDREAMER Procedures

- Serendipity-based planning/learning: a mechanism for finding previously overlooked plans in response to input states, recalled episodes, and generated mutations of events. Once an overlooked plan has been found through serendipity, it is stored in memory, so the program does not have to stumble upon the same solution in the future.

- Integration: The various techniques, old and new, have been integrated into a complete system.

- Application to daydreaming: The techniques are applied to the problem of modeling human daydreaming.

Emotion-driven control is the top-level procedure of the program. It consists of the following steps:

1. Invoke inference rule application to initiate concerns.

2. Select the concern with the highest dynamic importance.

3. Invoke planner on that concern.

4. Go to step 2.

It is intended for the program to continue to daydream for as long as it is turned on. Thus the above procedure consists of an endless processing loop.

The *planner* procedure is as follows:

1. Invoke rule application.

2. If top-level goal succeeded, invoke concern termination with "success."

3. Otherwise, if no subgoals were created in step 1:

   (a) Attempt to backtrack.

   (b) If all possibilities have been exhausted, invoke concern termination with "failure."

Once a *single* plan has been generated for achieving a top-level goal, the concern terminates successfully. A concern terminates unsuccessfully when *all* planning routes have failed.

Rule application is as follows:

1. For each active subgoal whose subgoals have all succeeded or whose objective is satisfied:

   (a) If subgoal is for another concern, create a positive emotion and associate it with the other concern.

   (b) If objective of subgoal is an action, assert it.

   (c) Change subgoal from active to succeeded.

2. Invoke inference rule application.

3. Invoke planning rule application.

A subgoal succeeds when all of its subgoals have succeeded. In addition, a subgoal can succeed:

- as a result of an input state or action, or

- as a side effect of planning for another subgoal.

This other subgoal can be on behalf of the current concern or another concern entirely. In the latter case, a positive emotion is generated and associated with the other concern, making it more likely for that concern to be processed in the future.

Concern termination is as follows:

1. Destroy otherwise unconnected emotions associated with concern.

2. Create a new positive or negative response emotion, whose strength is equal to the dynamic importance of the top-level goal.

3. If concern terminated with "success," invoke episode storage on the generated plan.

4. Destroy concern.

A positive emotion is created upon success, and a negative emotion is created upon failure.

Episode storage stores a plan in episodic memory indexed under the top-level goal, subgoals, and surface-level features (including persons, physical objects, and locations).

Inference rule application is as follows:

1. For each applicable inference rule:

   (a) Assert the consequent of the rule.

   (b) If consequent is a personal goal success or failure, generate an emotional response and associate it with the current concern.

   (c) If consequent is a new goal, invoke concern initiation.

Note that inference rule application is what initiates concerns in the first place. In general, concerns are initiated when a *need state* (such as for money) in the program is unsatisfied.

Concern initiation is as follows:

1. Create a new concern.

2. If an emotion was an antecedent of the inference rule which resulted in the creation of this concern, associate that emotion with the concern.

3. Otherwise, create a new motivating emotion and associate it with this concern. The strength of the new emotion is proportional to the intrinsic importance of the personal goal.

From step 2 above, one can see that a new concern resulting from the success or failure of an old concern assumes a dynamic importance equal to that of the old concern.

Planning rule application is as follows:

1. For each active subgoal of the current concern having no subgoals:

   (a) Invoke episode retrieval using subgoal objective and other active indices.

   (b) If episode retrieved, invoke reminding and analogical rule application procedures.

   (c) Otherwise, for each accessible planning rule which is applicable to subgoal, in a new context, create subgoals as specified by the rule.

Episode retrieval retrieves a planning experience from memory given a set of indices. An experience is retrieved only if a certain number of its indices are provided.

The *reminding* procedure is as follows:

1. Activate other indices of the retrieved episode E (limiting the number of active indices).

2. Invoke episode retrieval on the active indices.

3. If another episode E' is retrieved, invoke reminding procedure recursively on E'.

4. Invoke serendipity recognition on E.

This procedure can produce an associative stream of recalled experiences.

Analogical rule application uses a retrieved plan to suggest which rule to apply to the current subgoal, its resulting subgoals, and so on. This procedure reduces search in planning. If a suggested rule does not apply in the current situation, the program reverts to trying all accessible, applicable rules.

Serendipity recognition is as follows:

1. Perform an intersection search through the space of planning rule connections from the current top-level goal to a salient concept.

2. If a path is found:

   (a) Create a positive emotion and associate it with the current concern.

   (b) Verify the potential plan given by the path through recursive invocation of the planner.

The intersection search through rule connections is a way of discovering that a salient concept is related to the current top-level goal—in particular, that the concept is of potential use in achieving the goal. The salient concept can be:

- a concept provided as input to DAYDREAMER

- certain previously inaccessible rules contained in episodes that have just been retrieved

- a concept generated by mutation

As in the case of a fortuitous subgoal success, the generated positive emotion makes it more likely that the concern will be processed over other concerns in the future.

The idea underlying the serendipity mechanism is as follows: All of the rules in the program applicable to a given goal may not be accessible when attempting to achieve that goal. Some rules may have been improperly or incompletely indexed when they were first acquired. However, when a rule is brought into focus, the system should check whether that rule is applicable to any currently active goal and, if so, immediately apply that rule to that goal. This is exactly what the serendipity mechanism does.

Mutation is as follows:

1. For each unsatisfied subgoal whose objective is an action, for each mutation of the action generated by permuting objects, generalizing objects, or
   changing the action type, invoke serendipity recognition.

Mutation is invoked in DAYDREAMER as a last resort when all plans for achieving a goal have failed.

   In the chapters to come, the above mechanisms will be fleshed out and the
particular knowledge structures operated upon by these mechanisms will be
presented and discussed.

Chapter 3

Emotions and Daydreaming

In 1890, William James proposed that emotions are the awareness of the bodily changes resulting from the perception of some fact. According to James, "we feel sorry because we cry" and "angry because we strike" (p. 450). Based on evidence from experiments conducted on animals, Cannon (1927) criticized this theory on the grounds, among others, that the perception of the bodily states in question is too vague to account for the variety and detail of human emotions. Schachter and J. E. Singer (1962) demonstrated experimentally that physiologically similar states induced by injecting epinephrine into a human subject could be experienced as either positive and negative emotional states depending on the *cognitive appraisal* of the situation by the subject.

Recent work in the computer modeling of emotions (Colby, 1975; Pfeifer, 1982; Dyer, 1983b) follows Schachter and J. E. Singer (1962) in assuming that emotional states can be described in cognitive terms. However, this work addresses only emotions resulting from real-world events.

In this chapter, we extend previous work by addressing how emotions are influenced by daydreaming, and how daydreaming is influenced by emotions. First, we address emotional responses to daydreamed events. Second, we discuss the representation of emotions. Third, we introduce a set of *daydreaming goals* or strategies for daydreaming. Fourth, we discuss the daydreaming goals of rationalization, revenge, and roving which are activated in response to emotions and which serve a useful *emotion regulation* function. Fifth, we speculate on how highly idiosyncratic emotional states might be represented in humans. Sixth, we review previous related work.

3.1   Emotional Responses in Daydreaming

Emotions relate to goal outcomes (Dyer, 1983b; Weiner, 1982; Abelson, 1981): generally speaking, a goal success results in a positive emotion while a goal failure results in a negative emotion. Is this also true of goal successes and failures that are *imagined* during daydreaming? That is, can one become so involved in a daydream that one experiences an emotional reaction to an event which has not even happened in the real world?

Retrospective reports of daydreaming—including narrative reports (McNeil, 1981) and responses to items of the Imaginal Processes Inventory (J. L. Singer & Antrobus, 1963)—indicate that positive and negative emotional responses in fact do occur in daydreaming. Beck (1967) reports a patient who "experienced intense humiliation as though the fantasied event was occurring in reality." (p. 329)

Although emotions are often produced in daydreaming (J. L. Singer, 1978; Pope & J. L. Singer, 1978a; Varendonck, 1921), imagined goal successes and failures do not always have the same emotional consequences as real ones. For example, an imagined goal success can result in a negative, rather than positive, emotion under certain circumstances. In the examples which follow, we explore the additional complexity of emotional responses in daydreaming.

First of all, it is possible to experience a real goal failure or success during daydreaming by making new inferences or through reinterpretation of a past experience:

> I realize that the professor actually told us that the passing grade on the psychology midterm was 50, not 60, so I actually passed the test!

> I realize that I left my car in a Tuesday morning street cleaning zone, and so there is a parking ticket waiting for me.

There is nothing new here: A goal success in the first example results in a positive emotion and a goal failure in the second results in a negative emotion. The emotional reactions to a goal outcome detected during daydreaming are the same as those resulting from a goal outcome in a real-world experience. One might object that if the goal outcome no longer matters to the daydreamer, because, for example, the reinterpreted experience happened so long ago, or a more recent goal outcome has superseded the old one, then the emotional response may not occur, or may occur with reduced strength. But then this is no longer considered a real goal outcome.

What about imagined goal outcomes? Consider the following:

> I imagine that I get the top grade on my upcoming midterm.

> I imagine that the problems on my upcoming psychology midterm are so difficult that I have no idea how to answer them. I fail the test.

In the first example, one might feel *pleasure* or *hope*, while in the second, one might feel *displeasure* or *worry*.

Thus far, successes have corresponded to positive emotions while failures have corresponded to negative emotions, as one might normally expect. However, consider the following example:

> I wish that I had done the sample questions at the end of the chapter. Then I would have known how to do the second question on the test and I would have passed.

In this case, passing the test, which is normally considered a goal success, results in a negative emotion. This is because it results from an alternative past action which was not performed. The word *regret* is often used for the negative emotion in such situations. Similarly, an imagined goal failure resulting from an alternative which was not performed results in a positive emotion:

> I imagine cheating on the test my copying my neighbor's answers. But I would probably have gotten caught.

Here one would be glad about not having cheated: Imagining that one would have gotten caught, normally a goal failure, here results in a positive emotion often called *relief*. Thus, *the sign (negative or positive) of the emotion resulting from an imagined goal outcome is negated if that outcome results from an alternative past.*

How strong are the emotions resulting from an imagined situation? Is the strength of the emotion the same (possibly with opposite sign) as for a real goal? The strength of the emotion is certainly proportional to the importance of the goal (Dyer, 1983b). For example:

> I imagine that if I didn't fail my psychology midterm, I wouldn't have had to eat lunch an hour late in order to talk to the professor.

Here, eating lunch on time is not a particularly important goal, and so the resulting negative emotion is not particularly strong.

Now consider the following:

> I imagine that I actually had all the knowledge of the psychology professor. Then I would have aced the exam.

In this case, acing the exam does not result in a very negative emotion. This is because the scenario is unrealistic. As a result, we see that *the strength of the emotion is in proportion to the realism of the scenario as well as the importance of the goal.* Some experimental evidence for such an effect is provided by Kahneman and Tversky (1982) who presented subjects with two scenarios: missing a plane which had just departed because it was delayed, and missing a plane which had departed on time long ago. To 96 percent of the subjects, the former was rated as a more upsetting scenario than the latter. In our terms,

it is more realistic to have caught the plane in the former scenario than in the
latter.

The following situations have now been examined: real goal outcomes, imagined future goal outcomes, and imagined goal outcomes resulting from alternative pasts. We have seen that in the case of alternative pasts, the sign of
emotions is negated. We have also seen that the strength of emotions is proportional to the realism of the scenario and the importance of the goal. We
therefore hypothesize the following *emotional response* rule for daydreaming:

> For a goal outcome g with realism r, the strength of the resulting
> emotion is proportional to $r \cdot importance(g)$ and the sign of the re-
> sulting emotion is the same as that of g unless the scenario represents
> an alternative past in which case the sign of the resulting emotion
> is opposite to that of g.

Psychological experiments are needed in order to test the above rule: One
approach would be to compare subjective ratings provided by subjects for scenario realism and goal importance with those for emotion strength. However,
there are several potential difficulties: How do we elicit the necessary daydreams
in subjects? If stories are used in lieu of daydreams, how do we know that similar
emotional responses occur in natural daydreaming?

The above rule requires that DAYDREAMER have the capabilities of:

- maintaining whether or not a given scenario is an alternative version of
  the past

- assigning importance to goals

- assessing the realism of a scenario

The first capability is easily added to the program: Whenever a scenario takes
a past situation as its starting point, it is marked as an alternative past scenario. The goal importance is determined according to the dynamic importance
of a goal; this in turn is based upon (a) its intrinsic importance set upon goal
activation, and (b) modifications to the importance performed during later daydreaming. Scenario realism is assessed based upon the *plausibilities* (Shortliffe
& Buchanan, 1975; Duda, P. E. Hart, & Nilsson, 1976) of the rules employed in
generating a given scenario.

DAYDREAMER maintains a running assessment of the realism of a given scenario. However, a human may defer evaluation of the realism of a daydream—or
may never evaluate its realism. In certain *hallucinatory* daydreaming states, one
may believe that the events of the daydream are actually happening. Foulkes
and Fleisher (1975) report that a surprising 19 percent of the subjects interrupted while in a state of relaxed wakefulness described their thoughts as hallucinatory. If a person has a hallucinatory daydream, that is, if a person fails to
make assessments about the realism of a scenario and whether the daydream is

an alternative past scenario, emotional responses will be calculated according to default assessments that the scenario is realistic and not an alternative. Therefore, the strength of the emotion will be greater without a realism judgment and the sign of the emotion will be inverted if the scenario is not assessed as an alternative version of the past. Thus, for example, after barely avoiding a severe car accident, one may first experience pangs from imagining what it would have been like had the accident occurred, and then feel relief after realizing that the accident only occurred in an alternative version of the past. In general, the emotional response to a situation depends on the cognitive appraisal of that situation (Schachter & J. E. Singer, 1962; Mandler, 1975). For example, as the causal attributions for a failure or success change over time, so do the corresponding emotions (Weiner, 1980b, p. 369).

Under what conditions does one make or fail to make such assessments in daydreaming? Additional research is needed in this area. It is possible that failure to make realism and alternative assessments is the default in certain daydreaming states, with such assessments being produced only if required to reduce a sufficiently strong negative emotion resulting from a daydream without such assessments. For simplicity, the current version of DAYDREAMER continually maintains such assessments.

3.2 Emotion Representation

The representation of emotions in DAYDREAMER is based on Dyer's (1983b) previous representation of emotions for use in narrative comprehension. We extend his previous work, addressing the problems of: (a) how daydreaming influences emotions, and (b) how emotions influence daydreaming and behavior. Emotions are represented in DAYDREAMER in terms of the following components:

- *Sign*: positive or negative.

- *Strength*: non-negative real number specifying the magnitude of the emotion.

- *To (optional)*: person toward whom the emotion is directed (as in *anger*).

- *From goal (optional)*: failed or succeeded goal from which the emotion resulted as a response.

- *To goal (optional)*: active goal for which the emotion serves as motivation.

- *Altern*: flag specifying whether or not the emotion resulted from an imagined alternative past scenario.

In the case of the *to goal*, the emotion may have either caused the activation of the goal (as in the case of daydreaming goals), or may have been activated

Word	Toward	Altern?	To goal	From goal
pleasure		no		succeeded goal
relief		yes		failed goal
amusement		no		succeeded **ENTERTAINMENT**
satiation		no		succeeded **FOOD**
pride		no		succeeded **SELF-ESTEEM**
poise		no		succeeded **SOCIAL ESTEEM**
interest			active goal	
surprise			active goal	
hope		no	active goal	succeeded goal
gratitude	person	no		succeeded goal

Table 3.1: Positive Emotions in DAYDREAMER

Word	Toward	Altern?	To goal	From goal
displeasure		no		failed goal
regret		yes		succeeded goal
shame		no		failed **SELF-ESTEEM**
embarrassment		no		failed **SOCIAL ESTEEM**
rejection		no		failed pos rel goal
heartbreak		no		failed existing **LOVERS**
worry		no	active goal	failed goal
anger	person	no		failed goal
humiliation	person	no		failed **SOCIAL ESTEEM**

Table 3.2: Negative Emotions in DAYDREAMER

simultaneously with the goal. Various terms for emotions in English may be represented using the above scheme, as shown in Table 3.1 (positive emotions) and Table 3.2 (negative emotions). (The active goal may not be a daydreaming goal in the case of *interest.*)

Two terms are used to refer broadly to positive and negative emotions:

- *Pleasure,* also called satisfaction, joy, happiness, and so on refers to a positive emotion.

- *Displeasure,* also called dissatisfaction, unhappiness, disappointment, sadness, and so on, refers to a negative emotion.

Two terms refer to goal resolutions in imagined alternative past scenarios:

- *Relief* refers to any positive emotion resulting from an imagined past goal failure.

- *Regret* refers to any negative emotion resulting from an imagined past goal success.

Several terms refer to the specific kind of goal which resulted in the emotion:

- *Amusement* refers to a positive emotion resulting from a succeeded **ENTERTAINMENT** goal.

- *Satiation* refers to a succeeded **FOOD** goal.

- *Pride* refers to a succeeded **SELF-ESTEEM** goal.

- *Poise* refers to a succeeded **SOCIAL ESTEEM** goal.

- *Shame* refers to a negative emotion resulting from a failed **SELF-ESTEEM** goal.

- *Embarrassment* refers to a failed **SOCIAL ESTEEM** goal.

- *Rejection* refers to a failed goal for a positive interpersonal relationship with someone.

- *Heartbreak* refers to a failure of an existing **LOVERS** relationship.

Other emotion terms refer to emotions connected to (and providing the motivation for the achievement of) active goals:

- *Interest* refers to any positive emotion connected to an active goal (this emotion is similar to the "interest-excitement" emotion of Izard, 1977; Tomkins, 1962).

- *Hope* refers to a positive emotion which resulted from an imagined goal success and is connected to an active goal.

- *Worry* refers to a negative emotion which resulted from an imagined goal failure and is connected to an active goal.

- *Surprise* is generated upon fortuitous success of a subgoal or upon serendipitous discovery of a new plan for achieving an active top-level goal (see Chapter 5). This emotion is associated with the appropriate active top-level goal, and provides an additional force of motivation to perform processing activity toward that goal.

Finally, a group of emotion terms are employed when the emotion is directed toward someone:

- *Gratitude* refers to a positive emotion which is directed toward a person and connected to a succeeded goal.

- *Anger* refers to a negative emotion which is directed toward a person and connected to a failed goal.

- *Humiliation* refers to a negative emotion which resulted from a failed **SELF-ESTEEM** goal and is directed toward a person (thus one might conceive of humiliation as a combination of anger and embarrassment).

Emotions directed toward someone are generated when a top-level goal failure occurs as a result of some input action of another person (as in LOVERS1).

In DAYDREAMER, these tables are used by the English generator in order to generate an appropriate English word given an emotion represented in the program. The generator selects a word according to:

- whether or not the emotion is directed toward someone

- whether or not the goal success or failure (if any) is an imagined alternative

- the kind of active top-level goal (or concern) associated with the emotion (if any)

- the kind of succeeded or failed goal associated with the emotion (if any)

Each table is ordered from most broad to most specific; the most specific term is selected by the English generator for any given emotion.

The words which we have chosen to describe emotions in DAYDREAMER are only intended to be suggestive—there is no absolutely correct word for a given situation. For example, "shame," "humiliation," and "embarrassment" are interchangeable according to most emotion theorists (Tomkins, 1962; Izard, 1977), whereas they are distinguished in the generator of DAYDREAMER. The program does not "think in natural language." Rather, the conceptual representations generated by the program are converted into natural language for understanding by humans (and as a protocol of the program's daydreams). The English generator of DAYDREAMER is discussed in detail in Appendix B.

The words for emotions themselves, of course, have no functional role in the behavior of DAYDREAMER. That is why we are not concerned with finding a finite set of "primary" emotions (see, for example, Tomkins, 1962; Izard, 1977). However, the *representation* of the emotion affects the behavior of the program, since certain emotions activate certain daydreaming goals. For example, if a negative emotion is directed toward someone, then a daydream of revenge may be generated. The particular goal failure connected to this emotion will determine what type of revenge is sought. The particular values of the strengths of emotions determine the choice among multiple concerns. Emotions thus provide one way of pruning (for better or for worse) the large space of daydreaming possibilities.

3.3 Daydreaming Goals

Daydreams are triggered in DAYDREAMER by a set of goals above and beyond the personal goals called *daydreaming goals*. *Daydreaming goals are heuristics for how to exploit surplus processing time in a useful way.* They initiate processing whose purpose is to improve the ability of the system to satisfy its personal goals in the present and in the future. They are thus related to Wilensky's (1983) "metagoals."

Through analysis of a variety of daydream protocols, we have abstracted the following collection of daydreaming goals:

RATIONALIZATION Modification of the interpretation of a previous personal goal failure in order to reduce the negative emotional state resulting from that failure.

ROVING Shifting of attention from a personal goal failure to a positive episode or imagined scenario in order to reduce the negative emotional state resulting from that failure.

REVENGE Generation of imaginary scenarios of retaliation after another person thwarts a personal goal of the daydreamer.

REVERSAL Generation of imaginary alternative past or possible future scenarios in which a real past or imagined future personal goal failure is prevented, in order to learn from real and imagined failures.

RECOVERY Generation of possible scenarios for achieving a personal goal in the future which failed previously.

REHEARSAL Generation of possible future scenarios for achieving an active personal goal.

REPERCUSSIONS Exploration of the consequences of a hypothetical future situation.

Once activated, a given daydreaming goal is achieved by the planning mechanism through the generation of one or more imaginary past or future sequences of events. DAYDREAMER has a number of strategies for achieving each daydreaming goal; these strategies are represented as planning and inference rules. *Daydreaming goals are activated by emotions.* Moreover, *the emotion which activates a given daydreaming goal becomes the motivation associated with that goal.* Thus daydreaming goals which result from emotions of higher strength will be processed before goals resulting from emotions of lower strength.

In general, we have the following feedback loop: Emotions are activated in response to real-world situations. These emotions then motivate daydreaming goals which influence future daydreaming. Daydreaming in turn modifies the emotional state of the program and initiates new emotions. These new

Figure 3.1: Goal Taxonomy

emotions influence the future course of daydreaming through the motivation of other daydreaming goals. New emotions are generated, which activate further daydreaming goals, and so on.

As shown in Figure 3.1, top-level goals are divided up into daydreaming goals and personal goals. Daydreaming goals are further divided into two categories: *emotional daydreaming goals* (**RATIONALIZATION**, **ROVING**, and **REVENGE**) and *learning daydreaming goals* (**REVERSAL**, **RECOVERY**, **REHEARSAL**, and **REPERCUSSIONS**). The function of the first category of daydreaming goals is modification of emotional state, while the function of the second category is learning.

The learning daydreaming goals have an impact on emotional state as well. For example, daydreaming about **REVERSAL** of a prior failure may enable one to feel less upset about that failure if there is no **REVERSAL** which the daydreamer could reasonably have been expected to perform in the past situation; daydreaming about **RECOVERY** from a failure enables one to feel less upset about that failure; **REHEARSAL** for a future event enables one to be less (and sometimes more) worried about that event. Thus, *learning daydreaming goals have emotional consequences.*

The emotional daydreaming goals have an impact on learning, since any scenario generated through daydreaming has the potential for future application. For example, a daydream of revenge may be carried out in reality if the daydreamer is sufficiently angry at a person or has been sufficiently wronged and believes that retaliation would prevent future harm from the person. Modification of emotional state might itself be considered a form of learning since emotional state affects future processing. Thus, *emotional daydreaming goals provide another source of daydream scenarios for potential use in the future.*

Daydreaming goal	Activated by	Primary function
RATIONALIZATION	neg	emotional
ROVING	neg	emotional
REVENGE	neg directed	emotional
REVERSAL	neg	learning
RECOVERY	neg	learning
REHEARSAL	pos	learning
REPERCUSSIONS		learning

Table 3.3: Daydreaming Goals

Daydreaming goals are activated when certain emotions are of sufficient strength: The **RATIONALIZATION**, **ROVING**, **REVERSAL**, and **RECOVERY** daydreaming goals are activated in response to negative emotions (such as embarrassment, rejection, and anger). The **REVENGE** daydreaming goal is activated in response to negative emotions directed toward another person or persons (such as anger). The **REHEARSAL** daydreaming goal is activated in response to positive emotion of *interest* associated with an active personal goal. The **REPERCUSSIONS** daydreaming goal is not activated by emotions, but by a collection of heuristics. Daydreaming goals are summarized in Table 3.3. The emotional daydreaming goals are discussed in detail in Chapter 3, while the learning daydreaming goals are discussed in detail in Chapter 4. A concern is called a *personal goal concern* if its top-level goal is a personal goal, and a *daydreaming goal concern* if its top-level goal is a daydreaming goal.

Are daydreaming goals to rationalize, rehearse, and so on, cognitive entities of which humans are aware? Do humans pursue daydreaming goals nondeliberately or deliberately, involuntarily or voluntarily, automatically or with effort? Are daydreaming goals conscious, learned, metacognitive (Flavell, 1979) strategies for self-regulation, or are they innate and nonconscious? Humans are sometimes aware of daydreaming goals, and sometimes not: Although each of the daydreaming goals may be pursued deliberately, there are times when each of these goals is pursued nondeliberately. On the two ends of the spectrum, **ROVING** is frequently performed as a deliberate strategy (this strategy is similar to the various "guided daydreaming" and related therapy techniques, see, for example, Desoille, 1966; Leuner, 1969), while **REVENGE** is pursued spontaneously (Izard, 1977, p. 335; Weiner, 1980a) and sometimes against the wishes of the daydreamer. In the present work, we do not attempt to distinguish deliberate from nondeliberate applications of daydreaming goals.

The problem of whether daydreaming goals are learned or innate is orthogonal to the problem of whether daydreaming goals are voluntary or involuntary: learned processes which at first require effort may become automatic through practice. There is evidence that certain strategies for daydreaming may be learned starting at an early age (see, for example, Mischel, 1979; G. H. Green, 1923); other evidence (Darwin, 1872; Izard, 1977) points to the innate nature of certain external and internal behaviors. For simplicity, DAYDREAMER models an adult daydreamer assumed to have a fixed, hard-wired set of daydreaming goals. The program is not capable of learning this set of daydreaming goals, nor is it able to extend this set. However, DAYDREAMER does have the capacity to learn new strategies, or plans, for achieving existing daydreaming goals.

3.4   Emotional Feedback System

In DAYDREAMER, the emotional daydreaming goals model human daydreaming in response to a negative emotion resulting from a real goal failure. Specifically, **RATIONALIZATION**, **ROVING**, and **REVENGE** generate scenarios which result in positive emotions; these positive emotions offset the original negative emotions. What, then, is the purpose of reducing negative emotions?

Negative emotions serve the important function of motivating the program to take appropriate actions in response to whatever real or imagined negative situation caused those emotions: Via the **REVERSAL** daydreaming goal, the program attempts to prevent the occurrence in the future of failures similar to a real or imagined goal failure. Negative emotions thus indicate to the program that there is a threat, and that the program should take actions to eliminate this threat.

If negative emotions are useful, then why is it necessary for the emotional daydreaming goals to reduce negative emotions? The reason is that excessive negative emotions, and negative emotions lasting longer than necessary to motivate appropriate behaviors, have a negative impact. In humans, some of the effects of a long-term negative emotional state (such as depression and other negative moods) are decreased interpersonal attraction (Gouaux, 1971), lowered expectations of success (Loeb, Beck, & Diggory, 1971), increased recall of negative memories (Bower & Cohen, 1982; Clark & Isen, 1982), and low self-esteem, loss of satisfaction from activities, and loss of motivation (Beck, 1967, pp. 10-43). Negative emotional states also lead to negative behavior, which often leads to further negative feedback from the external world: If, for example, the daydreamer is angry at a person and retaliates against that person, the person may retaliate against the daydreamer, causing further anger, and so on.

In DAYDREAMER, emotional states, both positive and negative, tend to *strengthen* themselves for the following reasons: First, since episodes are indexed in DAYDREAMER by emotions, *episodes congruent with the current emotional state are retrieved*. In addition, *emotions associated with a retrieved episode are partially reactivated*. As a result, the current emotional state is strengthened. Second, retrieved episodes are employed analogically in the generation of daydream scenarios (see Chapter 4). Thus, *retrieved episodes congruent with the current emotional state result in analogous goal outcome scenarios, resulting in congruent emotional responses.*

Nonetheless, the emotional state of the program does not continue to increase in a positive or negative direction: emotion *decay* counteracts the above strengthening effects. The decay of emotions, of course, must not be too steep, or the motivating function of emotions would be lost and the program would fail to generate any behavior. The net effect of strengthening and decay is that *a positive or negative emotional state tends to perpetuate.*

In DAYDREAMER, negative emotional states are undesirable because: (a) negative emotional states indicate that the system is failing to achieve its personal goals, (b) negative emotional states lead to the recall of negative episodes resulting in the analogical generation of negative daydreams, and (c) negative emotional states tend to perpetuate themselves. Thus, *negative emotional states result in negative daydreams and, ultimately, behavior which hinders the achievement of personal goals.* Positive emotional states, on the other hand, are desirable, because they indicate that the program is achieving its personal goals. Thus we would like the program to strive for a positive emotional state. *Emotional daydreaming goals provide a way of breaking a cycle of negative emotions and bringing the program into a more positive emotional state.*

A diagram of the full emotional feedback system of DAYDREAMER is shown in Figure 3.2. Negative emotions activate various daydreaming goals, several of which ultimately result in the generation of positive emotions (**RATIONAL-IZATION**, **REVENGE**, and **ROVING**) and others which may result in positive or negative emotions (**REVERSAL** and **RECOVERY**). Positive emotions activate **REHEARSAL** daydreaming goals and motivate external behavior—both of which may result in positive or negative emotions. Other sources, the activation of a new concern, and the occurrence of a serendipity (accidental discovery of a solution to a problem) may result in positive emotions. (For simplicity, this diagram does not indicate that episodes can be retrieved for reasons other than active positive or negative emotions, that negative daydreams might still be generated in response to a positive episode and vice versa, and that, if unsuccessful, **RATIONALIZATION** might actually result in a negative emotion. It also ignores the negation of the sign of emotional responses resulting from alternative past scenarios.)

Instead of **RATIONALIZATION**, why could we not simply inject the program with a dose of positive emotion whenever it is unhappy? This is, after all, what **ROVING** does. Why does rationalization have to be "reasonable"? In DAYDREAMER, *emotions associated with an episode are modified through rationalization of a goal failure associated with the episode.* Bower and Cohen (1982) call this emotional "reappraisal" of an episode. The next time the episode is retrieved, the modified emotion will be activated. Thus, *rationalization reduces the likelihood of negative emotional states in the long term*, in addition to providing short-term reduction of negative emotions. The consistent application of rationalization can get the program out of a *depressed* state (if we use the term depressed to mean both a short- as well as long-term negative emotional

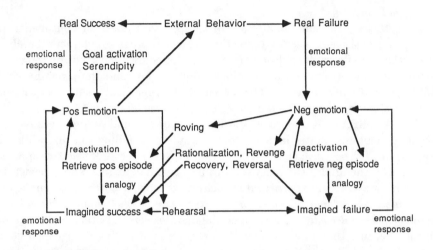

Figure 3.2: The Emotional Feedback System

state). (See Beck, 1967, for a related discussion of a "circular feedback model" of depression.)

Why should we bother implementing these human "features," the negative side effects of negative emotions? Perhaps a program would be much better off if episodic recall were not sensitive to emotions, if anger did not generate retaliation, and so on. In humans, these apparent artifacts are fundamental aspects of the emotional system and its adaptive purpose. Pervasiveness of emotional state is necessary to prevent one from becoming distracted from the problem at hand. By pervasiveness, we mean that emotions are *generalized* to situations other than that which caused those emotions—for example, after being insulted by one's boss, one may later take it out on a friend, one may hug the nearest person after winning the state lottery, and so on. The motivational function of negative emotions would be lost if one could simply direct attention to another topic while avoiding any cognitive impact of the negative emotion from the original topic. It is because negative emotions interfere with normal operation of the entire person that they serve as a fundamental driving force toward adaptive behavior. Negative emotions must have a global effect if they are to retain their utility as a motivator. A similar argument could be made for computer programs such as DAYDREAMER, where evolution consists of modifications by the programmer enabling the system to function adaptively.

There is another function of the emotional daydreaming goals. In humans, emotional well-being contributes to biological well-being. Various relaxation and imagery techniques, such as progressive relaxation, guided affective imagery, and biofeedback appear to induce beneficial physiological changes (see, for example, Luthe, 1969; J. L. Singer & Pope, 1978). These effects, however, are impossible

Figure 3.3: The Emotional Daydreaming Goals

to simulate in the nonbiological DAYDREAMER.

In the following sections, we discuss specific strategies for **RATIONALIZA-TION**, **REVENGE**, and **ROVING**. These strategies, and their impact on emotions, are summarized in Figure 3.3.

3.5 Rationalization

When one is unhappy, one's daydreams often turn towards consolations or thoughts designed to alleviate the unhappiness—one tries to make oneself feel better, to *rationalize* a goal failure in some way. The daydreams RATIONAL-IZATION1 and RATIONALIZATION2 (see page 4) are examples. We define rationalization as any process for adjusting one's interpretation of a situation in order to reduce the negative emotional consequences of that situations. In effect, rationalization provides a way of reducing the cognitive dissonance (Festinger, 1957) resulting from a goal failure.

Rationalization is a frequent component of daydreaming: Varendonck (1921) noticed the rationalization aspect of daydreaming in his hypothesis-rejoinder pairs: "Every thought in the second list implies a soothing process, a consolation, a reparation for the unpleasant impression left on the mind by the recollections in the first list" (pp. 249-250). Izard (1977) reports that 18.2 of the subjects presented with photographs representing the emotion of shame described the thoughts resulting from that emotion as "rationalize, escape from feeling." (p. 398) Hofstadter (1985, pp. 257-259) gives examples of creative variations on a theme called "subjunctive instant replays"; in our terminology these are alternative past scenarios which function as rationalizations (or as their opposite).

The **RATIONALIZATION** daydreaming goal is activated by a negative emotion of sufficient strength:

IF	**NEG-EMOTION** of sufficient strength resulting from a **FAILED-GOAL**
THEN	**ACTIVE-GOAL** for **RATIONALIZATION** of failure

Humans will also sometimes rationalize an anticipated *future* goal failure: that is, an emotion of *worry* about a future failure may be reduced through rationalization of that potential failure should it occur. One may also employ rationalization to reduce a negative emotional state when a particular personal goal cannot be attempted or achieved in real life because of negative social or other consequences. These cases are ignored in the current version of DAYDREAMER for simplicity.

Once **RATIONALIZATION** is active, how is it achieved? That is, how are rationalization daydreams produced? In this section, we describe four methods for rationalization: *mixed blessing, hidden blessing, external attribution,* and *minimization.*

3.5.1   Rationalization by Mixed Blessing

In rationalization by mixed blessing, one explores a hypothetical situation in which the goal succeeded instead of failed, and attempts to find negative consequences of that situation. That is, *rationalization by mixed blessing involves generation of a scenario in which the imagined success of the goal which failed in reality leads to an imagined goal failure.* This plan for rationalization, employed in RATIONALIZATION1, is expressed by the following rule:[1]

IF	**ACTIVE-GOAL** for **RATIONALIZATION** of failure
THEN	**ACTIVE-GOAL** for success to **LEADTO** failure

IF	**NEG-EMOTION** associated with failure less than a certain strength or **POS-EMOTION** associated with failure
THEN	**RATIONALIZATION** of failure

The strength of the emotion associated with the goal failure is then modified via a mechanism first presented in Chapter 2: During planning, *any emotion generated as a result of a side-effect personal goal outcome is associated with the current concern, and the strength of such an emotion is diverted to the primary motivating emotion of that concern.*[2] This mechanism rewards concerns

---

[1]Note that mixed blessing scenarios in DAYDREAMER are instances of Lehnert's (1982) MIXED BLESSING plot unit.

[2]Since, in general, more than one emotion can provide the emotion for a given concern, the *primary motivating emotion* of a concern is the emotion which first resulted in the creation of the concern (such as a negative emotion resulting from goal failure), or which was first activated upon creation of the concern (such as a motivating positive *interest* emotion created upon personal goal activation).

generating additional positive emotions in planning to achieve a top-level goal and discourages concerns generating additional negative emotions. Now recall that:

- a goal failure generated in an alternative past scenario results in the *positive* emotion of *relief* whose strength is proportional to the realism of the scenario and the importance of the goal which failed (while a goal success results in the *negative* emotion of *regret*)

- the negative emotion associated with the goal failure is the same as the motivating emotion associated with the active **RATIONALIZATION** goal

Consequently, goal successes and failures which result during planning will, respectively, increase or decrease the strength of the negative emotion. Thus if an imagined goal failure is generated, the original goal failure will be rationalized in proportion to the strength of the new positive emotion of *relief* resulting from the goal failure. However, if an imagined goal success is generated, the opposite of a rationalization will result: A new negative emotion of *regret* will be activated and the strength of the original negative emotion will be increased.

3.5.2 Rationalization by Hidden Blessing

When someone is unhappy because of some negative situation, event, or outcome, we often tell that person to "look on the bright side"—to shift attention to the positive aspects of the situation in order to feel better. This is also expressed in the proverb "Every cloud has a silver lining." Employed in RATIONALIZATION2, *rationalization by hidden blessing involves generation of a scenario in which a real goal failure leads to a real or imagined goal success:*[3]

IF	**ACTIVE-GOAL** for **RATIONALIZATION** of failure
THEN	**ACTIVE-GOAL** for failure to **LEADTO** success

Since this form of rationalization involves generation of a present or future— rather than alternative past—scenario, goal failures which are generated in the course of planning result in negative emotions and goal successes result in a positive emotions. Thus resulting goal successes will decrease the strength of the original negative emotion and goal failures will increase the strength of the original negative emotion. Thus when a goal success is generated, the original goal failure will be rationalized in proportion to the strength of the new positive emotion resulting from the goal success. Of course, if a goal failure is generated, DAYDREAMER is further away from a rationalization than when it started: A new negative emotion will be activated and the strength of the original negative emotion will be increased.

---

[3]Note that hidden blessing scenarios in DAYDREAMER are instances of both Lehnert's (1982) HIDDEN BLESSING plot unit and Dyer's (1983a) thematic abstraction unit TAU-HIDDEN-BLESSING.

3.5.3   Rationalization by External Attribution

Another method for rationalization is suggested by attribution theory (Heider, 1958; Kelley, 1967). An *attribution* is the cause or causes that one attributes to a past success or failure. Attributions have an impact on emotional state. In this framework, rationalization may be viewed as the process of finding an attribution with the most positive emotional consequences or with the least amount of "cognitive dissonance" (Festinger, 1957).

For failures, finding an *external attribution*—attributing the failure to another person, lack of luck, environmental factors, lack of ability, lack of effort, or fatigue—will often reduce a negative emotion resulting from blaming oneself for a failure. Thus, *external attribution is another possible strategy for rationalization.*

External attribution is not currently implemented as a separate strategy in DAYDREAMER but rather is accomplished as follows:

- The **REVERSAL** goal attempts to generate an alternative past scenario which might have prevented the failure (as described in detail in Chapter 4).

- The alternative scenario leads to the same goal failure.

- A positive emotion of *relief* results (because this is an alternative past scenario), reducing the strength of the negative emotion associated with the original goal failure.

If the above occurs for each alternative past scenario, the following external attribution may be made: "No matter what I might have done, the failure would still have resulted." Of course, if the **REVERSAL** is successful, the negative emotion of *regret* results. This situation occurs in REVERSAL1.

Another mechanism in DAYDREAMER relates to external attribution: *A negative emotion resulting from a personal goal failure is directed toward another person when an input action of that person is the cause for the failure.* However, this does not result in a reduction of the strength of the negative emotion, but rather a shift of the kind of emotion from *displeasure* to *anger* (which is then handled by the **REVENGE** daydreaming goal.)

3.5.4   Rationalization by Minimization

Suppose that, in LOVERS1, the movie star mentions the name of a famous director and DAYDREAMER responds that she has never heard of that director, and she is embarrassed as a result. Here are several rationalizations that might result from such an experience:

He knows that most people have never heard of that director anyway.

He might have just thought I was playing dumb to engage him in conversation.

He wouldn't dismiss someone simply because she had never heard of that director.

I don't think much of him anyway.

How might rationalizations such as the above be generated? *Rationalization by minimization involves reducing the likelihood of a goal failure and correspondingly the strength of the resulting negative emotion through the generation of reasons why antecedents of that goal failure are not necessarily true* (and sometimes merely by asserting the negation of those antecedents). (This process is named after a similar process discussed by Plutchik, 1980.)

An antecedent for embarrassment is a negative judgment by another person of the self. A negative judgment might result if a person expects everyone to have certain knowledge and believes another person does not have that knowledge. Thus the first rationalization above may be produced: The movie star in fact does not expect everyone to know about the particular director. The second rationalization results from a negation of the other half of the rule: The movie star in fact does not believe the daydreamer does not know about the director, and an alternative belief of the star regarding the reason for the behavior of the daydreamer is generated. The third rationalization above amounts to a negation of the negative judgment rule itself. The fourth rationalization negates another antecedent for embarrassment: having a positive attitude toward the person one is embarrassed in front of.

In general, for each fact which led to the negative emotion, the daydreamer attempts to generate a reasoning chain which shows the negation of that fact. If the daydreamer is successful in showing the negation of the fact with a given strength or degree of likelihood s', then the strength of the original non-negated fact s is adjusted to be $s/(s + s')$. In some cases, an attempt to minimize will backfire since worse consequences may result from the process.

DAYDREAMER produces two minimizations in the daydream RATIONAL-IZATION3 (see page 5). In the current version of DAYDREAMER, minimization may negate attitudes without justification. The impact of the new attitude on other attitudes of the program is then evaluated through the application of inference rules. However, there is currently no way for the program to reject an attitude which would result in cognitive dissonance (Heider, 1958; Festinger, 1957) when evaluated with respect to the program's overall belief system, as is done in Abelson's (1963) program.

3.6 Revenge

When one is angry at a person because that person has caused a failure for the self, one often daydreams about retaliation against the person or getting back at that person. An example is the daydream REVENGE1 (see page 4).

Human daydreaming is often concerned with revenge: Items of the Imaginal Processes Inventory (J. L. Singer & Antrobus, 1972) refer to ways of "getting even," "rubbing it in," "attaining revenge," and so on. Izard (1977) reports that when subjects (approximately 130 college students) were presented with photographs representing the emotion of anger, 43.9 percent of the subjects described the thoughts that follow from experiencing that emotion as "of revenge, attacking others, destruction" (p. 335).

The **REVENGE** daydreaming goal is activated by a directed negative emotion (such as *anger* and *humiliation*) of sufficient strength:

IF	**NEG-EMOTION** toward person resulting from a **FAILED-GOAL**
THEN	**ACTIVE-GOAL** to gain **REVENGE** against person

In this section, we investigate the purpose of revenge daydreams, and discuss some strategies for generating them.

3.6.1   The Purpose of Revenge Daydreams

Achieving revenge in a daydream results in a positive emotion of satisfaction. Should we then conclude that the function of revenge daydreaming is to make one feel better after experiencing a goal failure caused by another? It seems that revenge daydreams sometimes have the effect of intensifying, rather than reducing, the original anger. If this is true, we must question the utility of revenge daydreaming.

It has sometimes been proposed that daydreaming serves a *catharsis* function (Breuer & S. Freud, 1895/1937)—that through daydreaming one may partially discharge aggression in order to prevent its often damaging expression in real life. Some studies (Feshbach, 1956) have, in fact, found a reduction in aggression after engaging in fantasy activity. However, other studies have failed to support this hypothesis: In an experiment conducted by Paton (1972), subjects were insulted while performing a task. Afterwards, one group was shown pictures containing aggressive material, while another was shown nonaggressive pictures. These two groups were then given a chance to daydream and to report their daydreams to the experimenter. A third group had no opportunity to daydream. Those who had a chance to engage in fantasies exhibited less anger after being insulted than those who did not. However, those who had had aggressive fantasies exhibited less reduction of anger than those who had had more neutral fantasies. J. L. Singer (1975) proposes that reduction of anger through daydreaming is accomplished primarily through shift of the focus of attention away from the anger toward more pleasant fantasies, rather than through a catharsis effect. This would suggest that reduction of anger is accomplished more successfully through a strategy such as our **ROVING** daydreaming goal than through **REVENGE**.

What is the function of the emotion of anger which gives rise to revenge? Anger is a safety mechanism which ensures that we will try to resist being wronged or "stepped on" by others. It motivates us to fight back against those who harm us—to stand up for ourselves—and as such is an adaptive emotion (Izard, 1977).

Revenge daydreaming thus has the potential to generate realistic plans for retaliatory actions necessary in some situations. Through daydreaming, one may consider the consequences of retaliation in order to determine whether or not such actions are desirable. While a particular unrealistic scenario (such as REVENGE1) might not be worthwhile to pursue, a variation or outgrowth of such a scenario may turn out to be useful in recovery or adaptive retaliations to prevent recurrences. Thus revenge daydreaming is another way of generating scenarios which may or may not lead to productive use.

We conclude tentatively that the purpose of revenge daydreaming is to generate scenarios of potential future use for both retaliation and other purposes, and to provide a short-term positive emotional benefit. However, if an organism gets carried away with revenge daydreaming the results can be destructive. If the organism devoted more and more energy to bigger and better retaliations it would lose sight of reality; it would be obsessed. This effect would occur if revenge satisfaction intensified the original anger, causing further daydreams of revenge, resulting in even greater anger, and so on.

3.6.2 Achievement of Revenge

In order to get even with a person for thwarting a personal goal of the self, one must cause a personal goal failure for that person. The importance of the personal goal should be approximately the same as the self personal goal thwarted by the person. Two specific strategies are used in DAYDREAMER to accomplish such goal failures: *turning the tables* and *physical harm.*

Revenge by turning the tables involves causing the failure of a similar goal to the goal of the self thwarted by the person. A specific rule accomplishes this form of revenge for failed **FRIENDS, LOVERS,** and **EMPLOYMENT** relationships:

IF	**ACTIVE-GOAL** to gain **REVENGE** against person for causing self a failed **POS-RELATIONSHIP** goal
THEN	**ACTIVE-GOAL** for person to have failure of same **POS-RELATIONSHIP**

Revenge by physical harm involves generation of a scenario in which the person who caused a self goal failure is physically damaged. Such scenarios are generated by the following rule:

IF	**ACTIVE-GOAL** for **REVENGE** against person
THEN	**ACTIVE-GOAL** for person to have **FAILED-GOAL** of **NOT** being **HURT**

This rule results in the following daydream:

REVENGE4

I want to get even with my boss for firing me. I beat her up. I feel
pleased.

3.7   Roving

One method of feeling better when one is upset is to ignore the negative feelings
and divert attention to a more positive topic. An example is the daydream
ROVING1 (see page 5).

This strategy, which we call *roving*, is often employed in daydreaming: According to J. L. Singer (1975), positive daydreaming is frequently used as a form
of self-entertainment, or as a diversion or form of escape from negative emotions
and thoughts. Izard (1977) reports that 15.0 percent of subjects (approximately
130 college students) shown photographs representing the emotion of disgust described the following thoughts as "of others, trying to forget, escape situation"
(p. 339).

The **ROVING** daydreaming goal is activated by any negative emotion of
sufficient strength:

IF	**NEG-EMOTION** of sufficient strength resulting
	from a **FAILED-GOAL**
THEN	**ACTIVE-GOAL** for **ROVING**

Employed in ROVING1, *a strategy for roving is to recall any particularly
pleasant episode*—real or daydreamed:

IF	**ACTIVE-GOAL** for **ROVING**
THEN	recall pleasant episode

Beck (1967, p. 330) reports he was able to alleviate patients' feelings of inad-
equacy by instructing them to recall past successes. (However, recalling a past
success can also have a negative effect if the daydreamer starts wishing that the
past success could happen again.) Another plan is to generate a wish-fulfillment
daydream, or a daydream in which an active top-level goal is achieved. (However, in DAYDREAMER this is accomplished via the **REHEARSAL** daydreaming
goal.)

Although **ROVING** in response to a negative emotion associated with a goal
failure reduces or eliminates the negative emotional state of the program, it
does not enable the program to deal directly with the goal failure by recovering
from the failure or determining how to avoid future similar goal failures. Rather,
these functions are provided by the **REVERSAL** and **REHEARSAL** daydreaming
goals, respectively. Nonetheless, **ROVING** may be important to reduce the

negative emotional state of the program to a point where it can then perform such constructive activities.

When a negative emotion of sufficient strength is activated, several daydreaming goals are activated in response and that emotion becomes the motivating emotion for each of those goals. Thus no one daydreaming goal has priority over the other. (A discussion of how daydreaming goals might be prioritized is presented elsewhere [E. T. Mueller, 1987b].)

3.8 Emotions as Motivation

We view emotions as the primary motivation for external and internal (or daydreaming) behavior. How is this view justified? Varendonck (1921) has previously offered the similar view that daydreaming "is a mental process in which the ideation is directed by our affects according to the pleasure-and-pain principle." (p. 255) Klinger (1971, pp. 202-206) also views affect as instigators and guides of daydreaming and other behavior. The concept of emotions as motivation derives from Darwin (1872), who investigated the expressions, movements, postures, and gestures which emotions give rise to, and the value of these emotional reactions for the welfare of the organism—for example, fear prepares the organism for attack of an unknown object. Tomkins (1962) argues that emotions are the primary motivational system in humans and that biological drives only have motivational impact when accompanied and amplified by emotions. Izard (1977) and Plutchik (1980) take similar views. Other theorists (Lazarus, 1968; Mandler, 1975), however, argue against the concept of emotions as motivation. Because our view is somewhat controversial, we examine the arguments for and against the view in some detail.

Lazarus (1968, pp. 178-189) presents the following arguments against the view of emotions as motivation:

- This concept distracts our attention from studying emotions as real states or responses (because emotions become subsumed under concepts of motivation and drive).

- This concept distracts us from the causes of emotions (because emotions tend to be defined in terms of what adaptive actions follow the emotion).

- This concept tends to encourage one-dimensional emotional constructs, such as fear and anxiety.

- This concept tends to separate out components of the total emotional event in a misleading way (such as adaptive acts which are "disconnected" from the emotion itself and made a consequence of emotion).

- This concept is not predictive of adaptive response, since additional information beyond the emotions themselves is needed to predict the specific response.

We respond to these arguments as follows: Does viewing emotions as motivation distract us from the true nature of the emotions? Although in the present work we have not concentrated upon the phenomenology of emotions, other researchers who view emotions as motivation have explored the nature of the emotions in detail, including facial expressions and the subjective experience of emotions (Izard, 1977).

Does the view of emotions as motivation distract our attention from the conditions which result in certain emotions? On the contrary, in order for emotions to function as motivation, it is necessary to examine the situations in which certain emotions are activated. Thus in DAYDREAMER, an emotion of *worry* is activated when a future goal failure is imagined, and this emotion then functions as motivation to plan to prevent the imagined failure.

Lazarus (1968, p. 183) claims that in viewing emotions as motivation one easily forgets that emotions are an intervening variable, and one may end up stating that emotions, by themselves, cause behavior. Although we do not claim that emotions by themselves cause behavior, we do claim that emotions motivate behavior. What, then, do we really mean by motivation? Motivation has the following specific definitions in DAYDREAMER:

- *The activation of certain concerns in response to certain emotions*: Of course, various conditions lead to the activation of those emotions. Can we then eliminate the intervening variable of emotions? A measure of simplicity would be lost if we did: Since a given emotion can activate more than one kind of concern, the complete conditions for the activation of each concern activated by the same conditions would have to be repeated for each such concern. On the other hand, different emotions can activate the same kind of concern. Again, the specific conditions for activation of a certain kind of concern would have to be added to that concern.

- *Selection from among multiple courses of action*: At any time, the concern whose motivating emotions have the highest strength is performed. The strength of an emotion is determined initially by the conditions which activated the emotion, and is subject to modification under various other conditions.

Emotions are therefore a useful abstraction in DAYDREAMER for capturing the various conditions for activation of concerns and selection from among multiple concerns. Emotions provide generalizations of the current situation—each active emotion contains only the information necessary to select appropriate kinds of concerns for activation and processing. Of course we still have the following problem: Are the emotion constructs which we have postulated the same as what we call "emotions" in humans? Resolution of this problem, however, requires no less than solutions to the mind-body problem and the problem of the nature of scientific explanation. Whether or not our emotion constructs correspond to human emotions, the use of such constructs in intelligent computer systems may be justified on processing grounds alone: Some mechanism is necessary to direct processing when it is possible for more than one concern to be active at any given moment. Emotions provide such a mechanism. Furthermore, emotions provide a useful abstraction enabling the activation of adaptive activities—such as those provided by daydreaming goals—in appropriate situations.

Does viewing emotions as motivation imply that our emotional constructs will be narrow? Not at all—we distinguish 19 emotions in DAYDREAMER.

Have we separated out the components of emotion in a misleading way? As argued above, emotions provide useful abstractions. In order to identify appropriate abstractions, we have divided the problem into two components: How and when are certain emotions activated and modified? What is the influence of active emotions upon behavior? Simply because the antecedents and consequents of emotions are described separately does not mean that they are not "part" of the total emotions.

Does viewing emotions as motivation imply that emotions are insufficient to predict specific responses? This is not the case: In DAYDREAMER, it is necessary to retain any and all information within emotions that will be necessary to activate specific concerns. Of course, processing activity on behalf of that concern may be influenced not only by the emotions motivating that activity, but by the entire system state. It is unreasonable to require that the performance of a concern depend upon its motivating emotion and nothing else.

There is some experimental evidence for the view of emotions as motivation. For example, Weiner (1980a) conducted experiments to test the hypothesis that emotions such as pity and anger—rather than the abstract attributions of uncontrollability and controllability which are related to those respective emotions—are the direct motivators of actions such as help and lack of help, respectively. In these experiments, subjects were given various stories involving interpersonal requests for assistance, and asked to assess the controllability or uncontrollability of the need, to rate the level of pity or anger, and to rate the likelihood of giving assistance. The hypothesis was supported: It was found that the relationship between pity and help, and between anger and withdrawal, was not altered when the perceived controllability was statistically held constant. Furthermore, when the emotions were held constant, the relationship between controllability and assistance was greatly reduced.

It is logically possible that the feeling of an emotion (that which to humans *is* the emotion) itself does not result in a given behavior. That is, perhaps the feeling is an epiphenomenon which accompanies, but does not cause, a certain behavior. However, humans are aware of the feeling of emotions and, as a result, this feeling may have a causal impact. But can the feeling of emotions be described as motivational? Such feelings may be considered as motivations at least in the sense that through metacognitive strategies one may decide to perform a certain behavior as a result of feeling a certain emotional state: "I feel depressed. I don't want to feel depressed. I should do something about this feeling." But once such a decision is made, what carries out the decision?

Presumably, the hardware of the brain, which is already in motion. At some point one must postulate a construct which determines what the brain will do next. Whether or not this construct is identical with emotions is an open question. We assume for the time being that it is.

3.9   Idiosyncratic Feeling States

Salaman (1970) describes a phenomenon which she calls "involuntary memories"—the unexpected and sudden recall of a past experience accompanied by strong emotions and the sensation that one is living in an instant of the past. There is an overwhelming feeling that the past moment is being experienced exactly as it was originally experienced. For Salaman, involuntary memories appeared to be linked to a shock or disturbance which had otherwise been inaccessible to conscious thought.

A large number of such memories started coming back to Salaman when she was in her early fifties and writing about her childhood. She gives examples of involuntary memories described by other autobiographers and novelists: Chateaubriand, Proust, and De Quincey. Why are not more people aware of such memories? Salaman believes her involuntary memories were a result of her continued interest in past memories and attempt to recreate them in her writing. If one does not attend to these elusive experiences, she says, one may be aware merely of a slight change in mood. Involuntary memories are an interesting topic for future psychological and philosophical investigation: How accurate are these memories both in terms of their abstract contents and their phenomenological contents? Are prior feeling states reproduced authentically? What do we mean by "authentic reproduction"? As Hofstadter (Hofstadter & Dennett, 1981, p. 413) points out, knowing what it truly felt like to be "me" seconds, days, months, or years ago suffers from the same philosophical problems as knowing what it feels like to be someone else (Nagel, 1974).

Although involuntary memories may be rare occurrences, most people are familiar with the profound memory recall effects (James, 1890a, p. 556) produced by certain smells, tastes, sounds (especially music and voices), seasonal changes, locations, and other stimuli. These triggers sometimes activate strong unique feeling states associated with a particular past personal experience or era. For example, try using a particular brand of soap which you have not used for years. This will most likely bring back a whole set of feelings associated with the time when you were using that soap.

McKellar (1957, pp. 51-72) also discusses several types of subjective experiences which occur in the general population: the "falling experience" before falling asleep at night (reported by 75.87 percent of 182 subjects in a questionnaire study), déjà vu (reported by 69.23 percent), hypnagogic imagery (a kind of dreaming involving imagery which occurs just before falling asleep, reported by 63.18 percent), hypnopompic imagery (a kind of dreaming involving imagery which occurs while waking up, reported by 21.42 percent), synesthesia (a stimulus in one sensory modality evoking an image in another modality, reported by 21.42 percent), diagram forms (visual imagery of a spatial arrangement corresponding to numbers, days of the week, months of the year, and so on, reported by 7.69 percent), and color associations (associating colors with, say, days of the week, reported by 20.41 percent of 191 subjects).

Do such feeling states influence behavior or are they merely epiphenomenal? Certainly, such states affect mood, which in turn affects behavior. One may also comment to a friend on having experienced such a state. But what is the unique contribution of a particular feeling state? What *are* these feeling states? How might they be represented in a computer program? Might we be able to build such states out of a set of "primary" feelings or emotions (see, for example, Izard, 1977; Tomkins, 1962) mixed in various combinations? For example, could a certain idiosyncratic feeling state be represented as 60 percent interest-excitement, 20 percent guilt, and 20 percent shame? Although such emotional mixtures describe an infinite set of distinct emotional states, we doubt that these mixtures would be able to capture the richness, variety, and idiosyncrasy of personal feeling states (Salaman, 1970; James, 1890b, pp. 454, 468).

Although the true nature and influence on behavior of such feeling states is a topic for future research, we hypothesize that these subjective states result from a sufficiently complex conceptual representation. DAYDREAMER, for instance, does not have a set of fundamental emotions, since emotions are represented as arbitrary data structures. That is, although a particular data structure may be designated as an emotion, what is really important is the relationship of this data structure to other data structures and how these data structures are employed in processing. English words such as *embarrassment* are employed merely for convenience in understanding the operation of the program: The English generator contains a set of templates interpreting certain emotional data structures connected in a certain way to other data structures as certain emotions. One such connection is the association between a personal goal outcome and the emotional data structure—for example, a negative emotion resulting from a failure of the **SOCIAL ESTEEM** personal goal is generated as *embarrassment*. But suppose some new, idiosyncratic personal goal (such as recharging by plugging into a wall outlet) is acquired by the program. How, then, should the corresponding emotion be named? Perhaps there is no equivalent emotion in a human!

Although the subjective aspect of feelings is indeed a difficult philosophical paradox (Nagel, 1974; Dennett, 1978), we propose that the behavioral impact of what we call complex feeling states (including verbal descriptions of these states) results from a very complex conceptual representation—that *complex feeling states are in fact identical to complex conceptual representations*. L. B. Meyer (1956) takes a similar view in his analysis of emotional and intellectual responses to (nonsymbolic) music. Zajonc (1980), however, argues that affective judgments can occur to stimuli before cognitive understandings of those stimuli are formed, and thus that affect and cognition are partially independent. We would reply that although emotions may indeed occur prior to other, more "cold" conscious cognitions, some cognitive process must still have caused those emotions—perhaps a nonconscious one.

3.10   Previous Related Work in Emotions

Several researchers have previously argued the importance of emotions in theories of cognition: Neisser (1963) wrote that emotions are "conspicuously absent from existing or contemplated computer programs" (p. 195) and that "[h]uman thinking begins in an intimate association with emotions and feelings which is never entirely lost." (p. 195) Norman (1981) included emotions in his list of 12 issues for cognitive science. Sloman and Croucher (1981) argued the necessity of emotions in robots.

Since the psychological literature on emotions is quite large (see, for example, Mandler, 1975 and Izard, 1977 for overviews), in this section we review only the work in psychology and artificial intelligence most relevant to our current research, specifically, the (to our knowledge) unimplemented emotion models of Simon, Abelson, Weiner, and Bower and Cohen, and the implemented models of Colby, Dyer, and Pfeifer.

3.10.1   Simon's Interruption Mechanism

An early system which handles multiple goals is described by Simon (1967). When the system notices a high-priority need, processing toward the current goal is interrupted so that processing may be performed to handle the new goal. Simon equates this interruption mechanism with emotional behavior and the corresponding subjective feelings in humans.

3.10.2   Colby's PARRY

Colby (1973, 1975, 1981) constructed a computer simulation, called PARRY, of a paranoid patient participating in a psychiatric interview. PARRY is similar to DAYDREAMER in that it uses emotional states to modify and motivate behaviors. In PARRY the behavior is overt verbal behavior, while in DAYDREAMER the behavior consists of both overt behavior and daydreams. PARRY's strategy of blaming others in order to cope with shame-humiliation is similar to the external attribution strategy of DAYDREAMER (which is used by nonparanoids in rationalizing certain failures during daydreaming).

3.10.3 Abelson's Affect Taxonomy

Abelson (1981) proposes a model of affect in which affective responses occur when two construals of a situation, derived from real events as well as imaginings, are incompatible. A variety of emotion terms from "relief" to "mortification" can be captured by this scheme.

3.10.4 Weiner's Causal Attribution Model of Emotions

Weiner (1982) proposes a model of emotions based on the attributed causes of a positive or negative outcome. He distinguishes three causal dimensions: "locus of causality," which refers to whether the outcome is perceived as caused by self factors (such as intelligence, attractiveness, personality, and other self attributes) or by environmental factors (such as help from others, the objective difficulty of a task, and so on); "stability," which refers to whether the causing factor is perceived to be enduring (e.g., blindness is considered an enduring factor whereas luck may not be); and "controllability," which refers to whether the outcome is perceived to have been controllable by the person deemed the locus of causality. This scheme also captures a variety of emotion terms.

3.10.5 The Emotion Model of Bower and Cohen

Bower and Cohen (1982) propose a model of emotional responses based on "cognitive interpretation" (C-I) rules and "emotional interpretation" (E-I) rules. The emotional state of the system is represented as a (fixed-length) vector of emotion activation levels; each element corresponds to a type of emotion such as "fear," "anger," "happiness," "sadness," "disgust," and so on. The right-hand side of an E-I rule specifies how each of these emotions is to be adjusted (or not adjusted). For example, one E-I rule might state that if "someone acts in a negative way and the factors causing the act are internal, stable, uncontrollable," then pity is incremented by 0.7 and anger is incremented by 0.2.

The emotion representation of Bower and Cohen (1982), unlike that of DAY-DREAMER, assumes a fixed set of "primary" emotions (Izard, 1977; Tomkins, 1962, 1963). Emotions are not "typed" in DAYDREAMER (except by the English generator)—this allows for the possibility that new types of emotions may arise by virtue of being associated with new types of goals (or goals unique to certain systems). In addition, multiple instances of the same type of emotion may exist simultaneously in DAYDREAMER—for example, an *anger* emotion directed toward one person and an *anger* emotion directed toward another person may be active at the same time. Our philosophy is that emotions should be retained in their most specific form and that any emotional generalizations should be performed by rules which make use of those emotions. Both schemes enable the representation of mixed or conflicting emotions.

3.10.6   Dyer's BORIS Affect Model

The representation of emotions in DAYDREAMER derives from the AFFECT representations employed in the BORIS narrative comprehension program (Dyer, 1983a). Using this scheme, a variety of words used to describe the emotions of story characters may be represented.

The BORIS affect model is concerned with the reader's abstract understanding of the emotional states of story characters rather than the modeling of emotions themselves and the influence on processing of being in certain emotional states. However, in constructing DAYDREAMER, we have demonstrated that the BORIS affect model is in fact applicable to such a problem.

3.10.7   Pfeifer's FEELER

Pfeifer (1982) implemented a model of emotional behavior called FEELER. FEELER performs simple goal-based processing; the implementation of FEELER consists of 27 production rules and a declarative representation of the planning knowledge for taking a plane trip.

Emotions are generated in FEELER by interrupts and completion of a plan or subplan. FEELER models the influence of emotions on behavior in the following ways: First, emotions prime memories emotionally congruent with those emotions and thus increase the probability of the retrieval of those memories through spreading activation (similar mechanisms have been proposed by Bower & Cohen, 1982, and Clark & Isen, 1982). Second, certain production rules are sensitive to emotional state. For example, one rule states that if one is angry (at anyone) and someone makes a request to the self, anger is generated toward that person. Another rule states that if one is angry at a person, the goal to harm that person is activated (this rule is similar to the rule in DAYDREAMER which activates the **REVENGE** daydreaming goal in response to *anger*).

Although emotions decay in FEELER, other factors lead to the sustenance of both positive and negative emotions: First, when a memory emotionally congruent with the current emotional state is retrieved, the emotions associated with that episode are reactivated, and thus the existing emotional state is reinforced. Second, certain rules (such as generating anger toward a person when already angry at someone else) perpetuate emotional states. Pfeifer suggests that positive states may be maintained in the long run by having more rules for positive emotions than for negative emotions. Unlike DAYDREAMER, FEELER has no strategies (such as **RATIONALIZATION** and **ROVING**) for reducing negative emotional states and maintaining positive emotional states. However, Pfeifer does describe one unimplemented defense mechanism (A. Freud, 1937/1946) which shifts anger into self-pity.

The control mechanism for selecting among multiple goals, planner, and planning knowledge of FEELER is quite limited. In addition, FEELER does not address the issues of daydreaming, creativity, and learning addressed in the present work. Nonetheless, FEELER is quite close in spirit to the emotion component of DAYDREAMER and the construction of DAYDREAMER has benefited from this previous work.

3.11 Work Related to Daydreaming Goals

This section reviews previous work related to daydreaming goals and their activation by emotions.

3.11.1 Defense Mechanisms

"Defense mechanisms" (A. Freud, 1937/1946) are the unconscious processes employed by the "ego" as a protection against dangerous "id" impulses. Defense mechanisms either distort impulses before they enter consciousness or prevent them from reaching consciousness altogether. It is hypothesized (A. Freud, 1937/1946; S. Freud, 1926/1936) that defense mechanisms are set in motion by anxiety and serve to avoid or transform that anxiety.

Plutchik (1980) proposes a mapping from specific emotions to specific defense mechanisms. For example, anger results in displacement which shifts the object of the anger from a dangerous one to a less dangerous one, disgust with the self results in projection which shifts blame to someone else, and sadness results in sublimation or compensation for a loss.

The defense mechanism of rationalization was first introduced by Jones (1908). This mechanism is not discussed by A. Freud (1937/1946) and is rarely mentioned by S. Freud.

Defense mechanisms are similar to daydreaming goals in that both result from emotions and in turn eliminate or modify those emotions. For example: **ROVING** reduces a negative emotion by displacing negative thoughts from consciousness and substituting more positive ones; this daydreaming goal thus resembles repression and denial in that it prevents or defers dealing with these negative thoughts. **RATIONALIZATION** reduces a negative emotion associated with a past failure through the generation of a rationalization daydream. **REVERSAL** reduces a negative emotion associated with an imagined future failure through the generation of a realistic scenario in which that failure is avoided. However, daydreaming goals are generally adaptive whereas defense mechanisms are generally maladaptive.

3.11.2 McDougall's Emotions and Instincts

McDougall (1923) proposes a correspondence between emotions and instincts analogous to our relationship between emotions and daydreaming goals (except that in his view the emotions accompany, rather than initiate, the corresponding instinct). For example, fear corresponds to the instinct of escape, anger to the instinct of combat, disgust to the instinct of repulsion, lust to the instinct of pairing, tender emotion to the parental instinct, and so on.

3.11.3   Festinger's Reduction of Cognitive Dissonance

Festinger (1957) presents 12 methods for reducing cognitive dissonance which we may view as rationalization strategies. Among them are: increasing the desirability of a chosen alternative (similar to our hidden blessing strategy if the "chosen" alternative is taken to be the goal outcome), decreasing the desirability of the unchosen alternative (similar to our mixed blessing strategy), perceiving certain characteristics of the chosen and unchosen alternatives as equivalent, decreasing the importance of aspects of the decision (similar to our minimization), and rejecting those who disagree (which arises as a special case of minimization). Although Festinger speaks of dissonance as being "psychologically uncomfortable" (p. 3), cognitive dissonance reduction is considered to be cognitively driven rather than emotion driven (unlike rationalization in DAYDREAMER).

3.11.4   Abelson's Simulation of Hot Cognition

Abelson (1963) implemented a program to produce rationalizations. The program simulates the cognitive processes of a person who is confronted with a new proposition, either self-generated or taken as input. Propositions in the program refer to a causal relationship between two elements: the "antecedent" and the "consequent" of the proposition. Each element has an "evaluation" associated with it which captures the degree to which that element is considered to be "good" or "bad." A proposition is "balanced" if the evaluative signs of the antecedent and consequent are the same. Otherwise the proposition is "imbalanced." When a new proposition is received, the program must decide what to do with it. If the input proposition is balanced, it is simply added to the program's database of propositions. However, if the input proposition is imbalanced, it is then subject to denial and rationalization.

The program attempts three rationalization strategies in sequence until a successful rationalization is produced. The "Reinterpret final goal" strategy involves retrieving a proposition from the database whose antecedent is the same as the consequent of the input proposition, and whose consequent's evaluation is greater in magnitude than, and opposite in sign to, that of the consequent of the input proposition. That is, one may rationalize the fact that something "good" results in something "bad" if it can be shown that the "bad" thing results in something more "good" than the "bad" thing is "bad." This is similar to rationalization by hidden blessing in DAYDREAMER, except that instead of dynamically generating an arbitrary length imagined scenario of varying degrees of realism, the process is limited in Abelson's program to the application of one "inference," which must already be instantiated in the database. The "Accidental by-product" and "Find the prime mover" strategies explain the imbalanced proposition as an "accident" by retrieving a proposition from the database whose antecedent's evaluation had the same sign as that of the consequent of the input proposition, and whose consequent is the same as the consequent of the input proposition or the antecedent of the input proposition, respectively for each strategy. These two strategies are similar to our rationalization by external attribution. Again these processes are limited to a single "inference."

3.11.5 Colby's Neurosis Model

Colby (1963) implemented a computer model of neurosis which includes rationalization and 10 other defense mechanisms. The program contains a collection of beliefs, expressed as subject-verb-object propositions. Each belief has associated with it a "charge" representing the degree of importance of that belief. When a belief is "expressed" (produced as output), it loses most of its charge. The system's objective is to express its more charged beliefs, in order to reduce the charge on those beliefs. However, a belief may be expressed directly only if it is not in conflict with other "imperative" or "superego" beliefs about what is or is not acceptable. For example, *I like Debra* might conflict with the imperative belief *I must not like movie stars*. Unacceptable beliefs are only expressible once they have been rendered acceptable by one of eight "transforms." For example, using the transform of "deflection," *I like Debra* might be transformed into *I like Sarah*.

Rationalization in the neurosis simulation refers to the justification of a newly-created belief, while in DAYDREAMER it is defined as any process whereby a situation may be rendered more emotionally acceptable. In both systems, several strategies are used to produce rationalizations.

There are a number of similarities and differences between aspects of Colby's neurosis simulation and the process of rationalization in DAYDREAMER. In both cases, defensive strategies render an undesirable belief more acceptable. In the neurosis simulation the undesirable belief is one which violates a moral or personal imperative, while in DAYDREAMER it is the belief that an important personal goal failure has occurred. Successful application of the defense mechanism involves distortion of the unacceptable belief into an acceptable one in the neurosis simulation, and reduction of the importance of the goal failure in DAYDREAMER. Emotional state is regulated upon successful defense in both cases: In the neurosis simulation the danger level is brought under control while in DAYDREAMER the strength of a negative emotion is brought below a threshold.

In the neurosis simulation (Colby, 1963), unacceptable beliefs are taken to be unavailable to conscious thought, while in DAYDREAMER, goal failures *are* considered to enter into consciousness after which a conscious rationalization process occurs. Thus different phenomena are being modeled: Colby seeks to explain the neurotic patient's distorted beliefs and the anxieties which are of unknown origin to the patient, while we seek here to explain the generation of the conscious emotions and resulting daydreams of a normal person recalling or experiencing a goal failure. Also, the defense mechanisms of Colby are highly distortive, causing loss of information and the generation of incorrect information, whereas rationalization in DAYDREAMER is more of the flavor of subtle reinterpretation. The mechanisms in Colby's program are maladaptive, while rationalization in DAYDREAMER is an adaptive mechanism (if not used in excess). Still, both investigations are relevant to each other since most defense mechanisms may be applied both consciously and nonconsciously (see, for example, Plutchik, 1980) and there is a fine line between adaptive and maladaptive coping strategies.

Colby notes that if a program simulating neurosis could be constructed, various psychotherapeutic alternatives for removing maladaptive defense mechanisms or beliefs could be tested on the program itself in reduced time.

Suppes and H. Warren (1975) later formalized a collection of 44 or more defensive transformations on propositions which are a generalization of Colby's transforms (not including his more complex mechanisms of isolation, denial, and rationalization). However, they are concerned not with the modeling of cognitive processes, but merely with the generation and classification of a large collection of possible defense mechanisms starting from more basic transformations.

3.11.6   Janis and Mann's Bolstering Strategies

Janis and Mann (1977) discuss six bolstering strategies which are related to our plans for rationalization: exaggerating favorable consequences (related to our rationalization by hidden blessing and mixed blessing), minimizing unfavorable consequences (related to our minimization, hidden blessing, and mixed blessing), denying aversive feelings (related to our ROVING daydreaming goal), exaggerating remoteness of action commitment, minimizing social surveillance (similar to our minimization where the goal is social esteem), and minimizing personal responsibility (similar to our rationalization by external attribution).

3.11.7   Summary and Comparison of Strategies

The "emotion regulation" strategies of Festinger (1957), Abelson (1963), and Janis and Mann (1977) are summarized and compared to strategies employed in DAYDREAMER in Table 3.4. The correspondences, however, are only approximate. To our knowledge, only the strategies of Abelson (1963) have previously been implemented in a computer program. In this case, they are achieved through the application of a single, literal inference rule; parameter substitution is not performed, much less the generation of a daydream scenario.

The remaining strategy of Janis and Mann, exaggerating remoteness of action commitment, could be implemented in DAYDREAMER through the genera-

	Hidden blessing	Mixed blessing	External attribution	Minimization	Roving
Festinger (1957)	Increase desirability of chosen alternative	Decrease desirability of unchosen alternative		Decrease importance, Reject those who disagree	
Abelson (1963)	Reinterpret final goal		Accidental by-product, Find prime mover		
Janis & Mann (1977)	Exaggerate favorable consequences	Minimize unfavorable consequences	Minimize personal responsibility	Minimize social surveillance	Deny aversive feelings

Table 3.4: Comparison of Emotion Regulation Strategies

tion of future daydreams involving various alternative actions. Some of the other strategies of Festinger could be added, although they are really all just instances of his more general principle of cognitive dissonance reduction. Also, some of the remaining strategies involve external behavior rather than daydreams. Freudian defense mechanisms such as those employed by Colby (not shown in the table) could be added to a more pathological version of DAYDREAMER.

Chapter 4

Learning through Daydreaming

Why do humans spend time daydreaming about hypothetical past and future situations? There must be something that is gained. Humans must be retaining or *learning* something in the process of daydreaming. Daydreaming should result in similar benefits for computer programs: Instead of leaving a computer sitting idle when it has no task to perform, it should be daydreaming in order to improve the usefulness and efficiency of its future behavior.[1] Not every daydream need be useful. In fact, *even if only 10 percent (or less) of all generated daydreams eventually prove to be of use, a daydreaming system still has an advantage over systems which merely sit idle when unoccupied with tasks.*

Previous researchers have addressed learning from *real* experiences (Lebowitz, 1980; DeJong, 1981; Schank, 1982; Carbonell, 1983; Kolodner, 1984). We address learning from *imagined* experiences or *thought experiments*: If a situation arises which is similar to one explored in a previous daydream, the program should be equipped to handle that situation better and more efficiently. For example, when the program daydreams about alternative actions which might have prevented a past failure, those actions may be recalled and applied in case a similar situation comes up. Similarly, if possible future situations and responses to those situations are daydreamed and stored in memory, they may be used if those situations should arise.

In this chapter, we address how DAYDREAMER learns—improves its future

---

[1]The idea of exploiting free processing time has been around in one form or another for years. In an early version of the UNIX operating system, a lowest-priority background process applied any idle CPU time toward the task of calculating the first million digits of the mathematical constant e (Ritchie & Thompson, 1974). The PARRY program (Colby, 1975), a computer simulation of a paranoid patient, continued processing while waiting for the next input of the psychiatrist by using its own output as input (K. M. Colby, personal communication, March, 1987).

behavior—through the exploration of hypothetical situations. We consider: (a) the generation of alternative past scenarios, (b) the generation of possible future scenarios, (c) the use of such scenarios in improving future behavior, (d) the formation and later recall of *intentions* to perform certain future actions.

First, we present the mechanisms of *analogical planning* and *episodic memory* which provide the basic framework for learning in DAYDREAMER. Second, we explore what is meant by learning through daydreaming. Third, we discuss several *daydreaming goals* which initiate and guide learning activity. Fourth, we present strategies for daydream evaluation and selection—mechanisms for *decision-making* in daydreaming. Fifth, we discuss mechanisms for the formation and recall of *intentions*. Finally, we review related previous work in artificial intelligence.

4.1   Episodic Memory and Analogical Planning

In order for people to learn through daydreaming, this activity must somehow modify memory. Tulving (1972, 1983) proposed a distinction between two types of human memory—episodic and semantic.[2] Episodic memory contains concrete, unique autobiographical memories, while semantic memory contains abstract, generalized world knowledge. DAYDREAMER contains an episodic memory which includes both real and *imagined* episodes. That is, *daydreams generated in the past are stored in episodic memory along with real experiences* (gained in performance mode or taken as input). In addition to personal experiences, "real" episodes include *secondhand* experiences which do not contain DAYDREAMER as a character. Currently, secondhand experiences are added to the program as hand-coded episodes. DAYDREAMER also contains a semantic memory which consists of its collection of planning and inference rules.

Adding daydreams to episodic memory enables learning via the following sequence: *(a) a daydream—hypothetical past or future scenario of potential use in the future—is first generated; (b) next, the daydream is evaluated and possibly stored in episodic memory; and (c) if the daydream is later retrieved from episodic memory, it may then be applied in generating new internal or external behavior.*

What evidence is there that humans store daydreams in episodic memory? Just as real experiences may be forgotten (Linton, 1982), so may daydreams (Smirnov, 1973). (Of course, it is possible that the experience still exists somewhere in long-term memory and could be recalled if one were able to find the right retrieval index.) Nonetheless, humans often do remember their daydreams—sometimes in great detail: Subjects are able to recall and report daydreams seconds after they occur (Klinger, 1978) and days, months, and

---

[2]Tulving (1972) introduced the term episodic memory; the term semantic memory had been used previously by Quillian (1968). See Tulving (1983, pp. 18-21) for a review of the history of the episodic-semantic distinction in psychology and philosophy.

years afterwards (J. L. Singer & Antrobus, 1963, 1972). Retrospective reports of daydreaming are often quite elaborate (McNeil, 1981; Varendonck, 1921). People sometimes confuse real and imaginary experiences (M. K. Johnson & Raye, 1981). People frequently construct future plans (J. L. Singer, 1975) or intentions (Lewin, 1926/1951) during daydreaming which are later carried out. Varendonck (1921) reports near-complete recall of his daydreamed plans for writing: "When I am composing letters I often afterwards write them almost exactly as I worded them in my phantasy" (p. 327).

How are real experiences distinguished from daydreamed ones in episodic memory? Very subtle memory modifications may occur during daydreaming: Neisser's (1982a) analysis of the testimony of John Dean shows that he often remembered conversations in terms of his own fantasy about how those conversations *should* have been, rather than how they actually were. M. K. Johnson and Raye (1981) propose that humans distinguish memories of real and imagined experiences based on differences in how those experiences are represented in memory as well as through general reasoning processes. They hypothesize and present empirical evidence that externally generated memories have a greater amount of spatial and temporal contextual information, sensory attributes, and semantic detail, while internally generated memories have associated with them a greater amount of information about cognitive operations. In DAYDREAMER, episodes are simply tagged as to whether they are real or imaginary.

4.1.1 Representation of Episodes

Real and imaginary episodes are represented as planning trees produced by the planner. Figure 4.1 shows the planning tree for the REVENGE1 (see page 4) daydream. A planning tree consists of a tree of goals, with the root of the tree (e.g., **REVENGE**) as the top-level goal, and remaining nodes (e.g., **STAR**, **MTRANS**, **VPROX**, and so on) as various subgoals. Leaves of the tree (e.g., **RTRUE**, **KNOW**, and so on) are subgoals which did not require further planning because their objectives were already satisfied in the world model or *context*. **RTRUE** is an objective which is satisfied in any context. Actions are never leaves because all actions have preconditions in DAYDREAMER. Each nonleaf goal (e.g., **REVENGE**) in the tree has a two-way connection to the rule (e.g., *Revenge-Plan1*) from which the division of that goal into subgoals was derived.

Recall from Chapter 2 that planning rules consist of an antecedent pattern specifying a goal and a consequent pattern specifying one or more subgoals. A planning rule is applicable to a given active goal if its antecedent unifies with the objective of the goal. The objectives of the subgoals for the active goal are then obtained by instantiating the consequent pattern of the planning rule with the bindings obtained from the unification.

An episode is actually represented as an entire context, which includes the planning tree as well as any states and inference chains not contained within the planning tree.

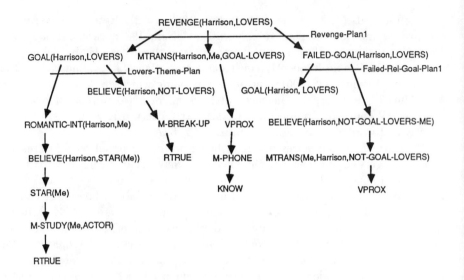

Figure 4.1: Planning Tree for REVENGE1

Episodes arise in three ways in DAYDREAMER: through generation of a daydream, through participation in a performance mode experience (including external input), and by manual entry of a hand-coded episode. In the program, *daydreams, performance mode experiences, and hand-coded episodes are all represented in the same fashion.*

4.1.2   Storage of Episodes

The episodic memory of DAYDREAMER will contain many episodes. An episode will be useful to the current situation when that episode contains a plan for achieving a goal similar to one which is currently active. Thus in DAYDREAMER, *a real or imagined episode is indexed under the planning rule associated with the root goal of the episode.* Since (a) an episode will be retrieved in planning when its rule is selected, and (b) a rule is selected when the antecedent of that rule unifies with (matches) the objective of an active subgoal, *episodes are effectively indexed under goal objective patterns.*

Every single nonleaf subgoal in a planning tree becomes the root of an episode. Thus a top-level planning tree actually becomes many episodes—one for the top-level goal and one for each nonleaf proper descendant of the top-level goal. Effectively, then, episodes are indexed under goal objective patterns corresponding to each of the subgoals achieved in that episode.

Figure 4.2 shows a portion of an episodic memory consisting of REVENGE1 (see page 4), REVENGE2 (see page 91), and REVENGE3 (see page 11). Each

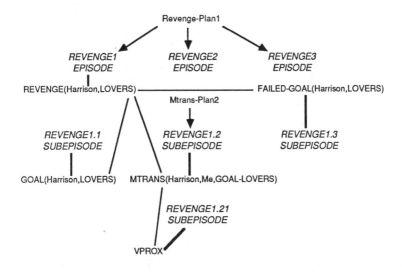

Figure 4.2: Indexing of Episodes by Rules

of these episodes is indexed under the *Revenge-plan* rule (and thus, in effect, under the **REVENGE** goal which is the antecedent of that rule). The top-level **REVENGE** goal is connected to three subgoals, which are the root goals of three *sub-episodes*, REVENGE1.1, REVENGE1.2, and REVENGE1.3. These sub-episodes are also indexed under rules; in the figure, only the index of REVENGE1.2, *Mtrans-plan2*, is shown. The **MTRANS** subgoal in turn has **VPROX** as a subgoal, which is the root goal of the sub-episode REVENGE1.21. The entire planning tree, not shown in the figure, is broken down into sub-episodes in this fashion.

Other researchers have previously noted the importance of goals as memory indices: Reiser (1983) argues that episodes are indexed under causally relevant features; specifically, that episodes are indexed under goals and goal relationships. Schank's (1982) MOPs are indexed under goals.

4.1.3   Analogical Planning

Once imagined and real episodes are stored in episodic memory, how are they later retrieved and applied in an appropriate situation? Episodes, whether imaginary or real, are employed in generating new daydreams or external behavior through the process of *analogical planning*. An existing planning tree for achieving a particular goal, called the *source* goal, is used to guide the generation of a planning tree for achieving a new goal, called the *target* goal.

When planning to achieve a given subgoal, planning rules are first found which apply to that subgoal. A number of episodes are then retrieved using

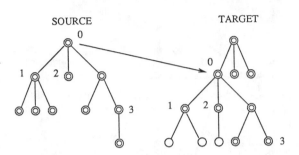

Figure 4.3: Cases in Analogical Planning

those planning rules as indices.   One or more episodes are then selected for application.

The regular planning algorithm involves repeatedly selecting an active subgoal and sprouting plans—activating further subgoals—for that subgoal, with each plan in its own separate context. When an episode is retrieved and selected, the entire planning tree is not employed at once. Instead, the tree serves as a *suggestion* to the planner about which planning rules to employ in sprouting plans for active subgoals. The following cases, shown in Figure 4.3, may occur during this process:

- *Case 0: rule is applicable*: A rule which was applicable to a subgoal in the source situation is applicable in the target situation. Therefore, it is applied.

- *Case 1: rule is inapplicable*: A rule which was applicable to a subgoal in the source situation is not be applicable in the target situation. Regular planning must be invoked in order to achieve this subgoal.

- *Case 2: source tree bottoms out before target*: In planning to achieve the target goal, a subgoal which was satisfied in the source situation (and is thus a leaf of the source episode) is not satisfied in the target situation. Regular planning must be invoked in order to construct a plan for achieving this subgoal in the target situation.

- *Case 3: target tree bottoms out before source*: A subgoal which required planning in the source situation might already be satisfied in the target situation. In this case, no further planning for this subgoal need be performed.

Portions of the source episode which do not apply to the current situation are therefore automatically repaired in the course of planning.

Whenever regular planning is invoked above on a given subgoal, yet another episode may be found applicable to that subgoal; that is, analogical planning may be invoked recursively. As a result, a newly constructed episode may be composed of multiple previous episodes and generic planning rules.

Once an episode for achieving a goal is found, alternative plans for achieving each subgoal—such as those specified by other generic planning rules or episodes—are not employed since the purpose of employing the episode is to reduce the search through such possibilities. Only when a suggested rule proves inapplicable, or if the source episode bottoms out before the target episode, are other possibilities considered.

Here are some specific examples of analogical planning in DAYDREAMER: Since both imaginary and real episodes may be employed in generating both imaginary and real episodes, there are four possibilities. We consider each in turn.

First, a previous daydream may be employed in generating a new daydream. For example, in LOVERS1, DAYDREAMER asks Harrison Ford out on a date and he turns her down. DAYDREAMER then produces REVENGE1. Later when DAYDREAMER is turned down by another movie star, the following daydream is generated by analogy to REVENGE1:

REVENGE2

I remember the time I got even with Harrison Ford for turning me down by studying to be an actor, being a star even more famous than he is, him calling me up, him asking me out, and me turning him down. I study to be an actor. I am a star even more famous than Robert Redford is. He calls me up. He asks me out. I turn him down. I get even with him for turning me down. I feel pleased.

In this case, no repairs or completions were necessary—the entire source episode, as previously shown in Figure 4.1, was employed verbatim (after mapping Harrison Ford to Robert Redford) in the target episode, as shown in Figure 4.4.

In the daydream REVENGE3 (see page 11), however, repairs must be performed. As shown in Figure 4.5, the **GOAL** subgoal of the target episode cannot be achieved by the *Lovers-theme-plan* rule of the source episode (as shown previously in Figure 4.1). Since this rule is inapplicable in the target situation, regular planning is invoked to achieve this subgoal. The rule *Employment-theme-plan* is applied to the subgoal, followed by further rules. Except for this subtree, the remainder of the source episode is employed without further modification (other than parameter substitution).

Second, a previous real episode may be employed in generating a new daydream. Suppose DAYDREAMER now wishes to get in touch with Harrison Ford to try and ask him out again (just in case he changes his mind!). Hand-coded episodes contain plans for finding out the movie star's telephone number which are recalled and employed through analogical planning in RECOVERY1 (see

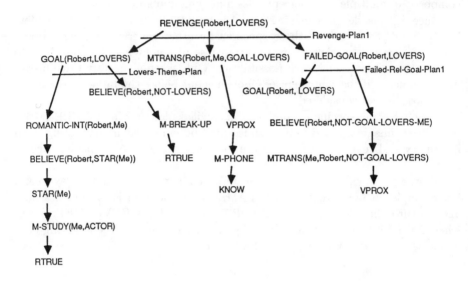

Figure 4.4: Planning Tree for REVENGE2

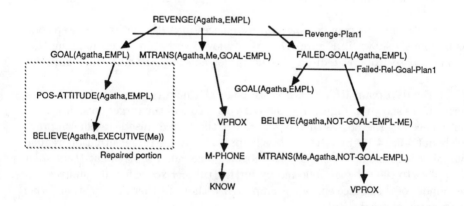

Figure 4.5: Planning Tree for REVENGE3

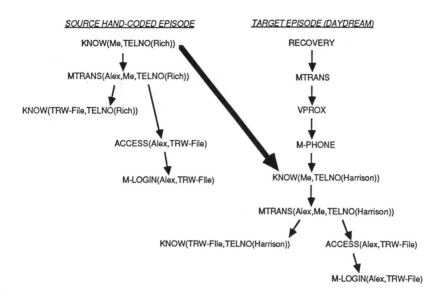

Figure 4.6: Analogical Planning in RECOVERY1

page 6). As shown in Figure 4.6, the source episode becomes a subtree of the target episode (daydream).

The daydream RATIONALIZATION1 (see page 4) provides another example of a (hand-coded) real episode applied to the generation of a daydream—in this case, however, only a small component of the daydream results from analogical application of the episode, as shown in Figure 4.7. The source episode bottoms out at the **RPROX** subgoal, and so regular planning is employed in order to expand this subgoal.

Third, a previous imaginary episode may be employed in generating a real-world (performance mode) episode. For example, consider the following daydream:

REHEARSAL1

I go to his house. We go to a restaurant. We eat dinner. We go to the Fox International Theater. We watch a movie. We go to his house. We kiss. . . .

This daydream is employed through analogical planning in the following performance mode experience:

LOVERS3

External action: **I go to Vicente Foods.**

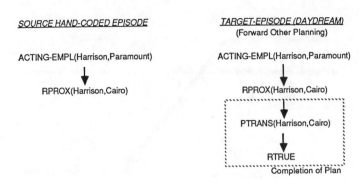

Figure 4.7: Partial Tree for RATIONALIZATION1

Input: *A guy asks you what time it is.*

What do you know!

...

External action: **I tell him I would like to have dinner with him at a restaurant.**

Input: *He tells me his address.*

External action: **I buy groceries. I go home. I eat.**

...

External shared action: **We go to a restaurant. We eat dinner. We go to his house. We kiss.**

...

Fourth, a previous real episode may be employed in generating another real episode. For example, after the successful performance mode experience of LOVERS3, DAYDREAMER could apply this episode by analogy the next time it meets someone.

When the target situation is sufficiently different from the source situation, nontrivial repairs will be made to the source episode. However, if previous episodes are frequently employed in generating new daydreams without repair, then the program will generate what is basically the same daydream over and over again. For example, REVENGE2 is a trivial variation on REVENGE1; future similar experiences will give rise to yet more similar daydreams. There are several ways in which DAYDREAMER attempts to cope with this problem (even if recurrent daydreams do occur in humans [J. L. Singer, 1975, pp. 17-32]):

- Each episode associated with a given rule is moved to the end of the list of episodes associated with the rule whenever it is employed in generating a daydream. When selecting episodes for application, episodes earlier in the list are given priority over episodes later in the list.

- A collection of strategies for creativity, including serendipity, mutation, and environmental input (discussed in Chapter 5), are employed to get the program out of a rut by forcing it to generate or notice new possibilities.

Furthermore, if a daydream generated by analogy to a previous daydream is stored again in episodic memory, there will be an even greater proliferation of almost identical episodes. For this reason, a daydream is not stored in memory if it was derived from a previous episode without any repairs. That is, a new daydream is not stored if it is isomorphic to a previous episode—if it consists of exactly the same tree of rules. Thus REVENGE2 is not stored since it was derived wholly from REVENGE1 without repairs; however, REVENGE3 *is* stored since it was constructed through repair of REVENGE1. Whether or not repairs are performed is monitored during planning, so that a graph comparison does not have to be performed later on.

4.1.4 Issues for Theories of Episodic Memory

Any theory of episodic memory use must address the following questions:

- What information is provided by an episode which is not already available as a result of having found the appropriate knowledge structure for accessing that episode? For example, suppose that episodes are represented in terms of *scripts* (Schank & Abelson, 1977). Scripts may then be used as memory structures for retrieval of an episode involving the same script as the current situation. However, what is the use of retrieving such an episode? Normally, one might wish to employ such an episode in making predictions about the current situation. However, in order to retrieve the episode, one must first determine what script the current situation is an instance of; having done this, the retrieved episode provides no information not already contained in its script. The only reasonable response is that episodes must somehow contain more information than is contained in their indices. It is important to specify exactly what that information is. Scripts were used as an example memory structure. In the offshoot of scripts called MOPs (Schank, 1982), episodes are in fact indexed by deviations of a scene of a script rather than by scripts themselves.

- What information is provided by the episode that is not available elsewhere? For example, if episodes are merely combinations of existing rules from semantic memory, what is gained by accessing the episode? If episode events are fully explained in terms of existing rules, what new knowledge do those events provide? For example, in the explanation-based generalization of Mitchell, Kellar, and Kedar-Cabelli (1986), both the training example and the justified generalization—that is, a generalization of the proof which demonstrates that the example is an instance of a given concept—contain no information that is not derivable from the existing concept definition and the domain theory. However, the justified generalization does satisfy a given criterion of *operationality* (Mostow, 1981)—that the generalization be useful for recognizing examples of the concept efficiently. The requirements that DAYDREAMER increase efficiency and improve control in closed-system learning (as described in the next section) are both examples of operationality criteria.

How do we address these questions in the case of DAYDREAMER? Episodes indeed contain more information than is contained in their indices. In particular, episodes consist of instantiated planning trees enabling search to be reduced. Indices are a heuristic for relevance in DAYDREAMER. Episodes are retrieved when their indices are currently active. Once an episode is retrieved, it may be discarded based on evaluations of *realism* and *desirability* associated with the episode (discussed later in the chapter). If not, the episode will be applied in planning and it may or may not prove useful.

The representation of episodes in DAYDREAMER is based to a large extent on semantic memory—episodes are represented as planning trees which refer to generic rules, which are in turn part of semantic memory. In Figure 4.2, for example, the rules *Revenge-plan1* and *Mtrans-plan2* are part of semantic memory and the remainder is part of episodic memory.

In addition, however, some episodes contain information that is not present elsewhere in any form, that has no basis in terms of other knowledge. That is, episodes may contain causal relationships—inaccessible episodic rules—that are unknown outside that episode, and are thus *not* considered a part of semantic memory.

4.2   Definition of Learning

What does it mean for DAYDREAMER to learn through daydreaming—a process carried out without interaction with the external world? There is a potential problem in claiming that a closed system is able to learn: Any desired state of the system (including processes and data) given the initial state of the system is reachable whether or not the system has "learned"—that is, reached some intermediate state between the initial and desired states. In other words, if you can learn something without external input, then you must really already know it.

First of all, DAYDREAMER is not a closed system. In fact, it accepts the following sources of input which contribute to learning:

- *Input of events and states in performance mode*: This enables the program to gain new experiences automatically through interaction with the external environment. External input is an important source of feedback in learning: Daydreamed plans may be modified if they do not succeed when tested in the real world.[3]

- *Input of hand-coded episodes*: Normally, hand-coded episodes are loaded when the program is started. However, there is no reason why such episodes cannot be added later. This is not a particularly interesting form of learning since it is not carried out automatically. Hand-coded episodes are all loaded during program initialization in the current version of DAY-DREAMER. However, we might eventually wish to compare the behavior of DAYDREAMER given different initial sets of hand-coded episodes, or given new hand-coded episodes after the program has been running for a while. RECOVERY1 (see page 6) is an example of a daydream generated through the application of a hand-coded episode.

- *Input of random physical objects*: This input is used to stir up a stagnant closed system by producing remindings (see Chapter 5). Examples are provided by the daydreams COMPUTER-SERENDIPITY and LAMPSHADE-SERENDIPITY (see page 14).

However, since the majority of the learning carried out by DAYDREAMER does not require external input, we must examine in detail what is meant by closed-system learning. Closed-system daydreaming modifies the state of the system in three ways which contribute to learning:

- *Episode storage*: Daydreams are stored in episodic memory for later recall. A recalled daydream is then employed through the process of analogical planning in generating new daydreams or external behavior.

- *Rule creation*: Inference and planning rules are constructed and saved for possible future use. In particular, new *preservation goal* inference and planning rules are derived by the **REVERSAL** daydreaming goal.

- *Partial concern completion*: A concern is activated, planned up to a point, and then halted until an appropriate time in the future. Such a concern is primed to employ fortuitous subgoal successes or serendipities which may occur later on.

In this section, we concentrate on the problem of what is meant by learning in the case of stored episodes, leaving a discussion of rule creation and partial concern completion to later sections in this chapter.

---

[3]DAYDREAMER also includes a simple mechanism for inducing new planning rules from a sequence of two input states assumed to form a causal chain; this is accomplished through variabilization (converting constants into variables). Although a potentially general mechanism, it is currently employed only in the specific situation of RECOVERY3 (see page 13).

4.2.1   Episodes for Improvement of Future Search

Many daydreams stored as episodes are simply collections of existing generic rules. What new information, then, does such an episode contribute? What has actually been *learned* through storage of the episode?

In generating a plan to achieve a given goal, the goal is broken down into subgoals, which in turn are further broken into subgoals, and so on. Multiple ways of breaking down each subgoal lead to alternative plans, which are evaluated according to their positive and negative consequences (i.e., personal goal successes and failures). If there are several ways of breaking down each subgoal, the search for a good solution can grow large—a combinatorial explosion results. After spending much time searching to solve a problem once, a traditional planner will start from scratch in solving a similar or identical problem. This is wasted effort; a planner should be able to generate a solution more quickly the second time. A person solving a problem similar to a previous one would probably *recall* the previous solution which would then enable the person to solve the new problem more easily. In DAYDREAMER, the analogical planning mechanism enables recall of a previous real or imagined planning episode which is then adapted to a new goal. Parts of the adapted plan may need to be repaired for the new goal, which is not identical to the previous one. Even so, much of the adapted plan may stand untouched.

Thus, *episodes enable learning by reducing future search in planning.* Episodes are a source of knowledge for choosing among alternative plans for achieving a subgoal—in solving a problem, for each subgoal, the plan that worked best in a similar previously considered situation is chosen. That is, *daydreaming evaluates the consequences of alternative future plans of action, storing the best plans for possible future use.* If the program can perform search in advance through daydreaming, search in the future will be reduced—provided that the program is good at anticipating future problems and in detecting when a problem is similar to a previously examined one. In the terminology of Bundy, Silver, and Plummer (1985), episodes repair *control faults.*

Thus learning in a closed daydreaming system may be defined as follows: *If a program is able to solve a given problem more quickly after daydreaming, the program may be said to have learned from that daydreaming.* Alternatively, we may say that *if, after daydreaming, a program is able to generate a better solution to a problem given a limited amount of time, the program has learned from that daydreaming.*

This form of learning is important because time constraints are imposed in most real-world situations: For example, quick responses are expected in human conversation. Whereas in daydreaming mode the program generates many possible plans, both fanciful and realistic, *in performance mode the program immediately retrieves and employs the best previously daydreamed plan for the current situation.* Thus learning enables the program to generate better solutions when there is no time for daydreaming in performance mode.

Given its initial collection of rules and episodes, DAYDREAMER may not always be able to achieve its personal goals in performance mode; for example, it fails in LOVERS1, its first attempt to achieve a **LOVERS** goal. If the program learns, it may then be able to achieve a personal goal which it was previously unable to achieve. Learning also enables the program to *improve* at achieving its personal goals. Furthermore, learning enables the program to retain the ability to achieve its personal goals in the face of a changing environment.

Thus, *learning is demonstrated when, given one performance mode experience in which the program produces a certain behavior and fails, the program is later given a similar experience in response to which the program produces a different behavior and succeeds.*

Closed-system learning, then, enables the program to survive—achieve its personal goals in the simulated real world. Evolution may thus be considered a form of closed-system learning at the species level. However, unlike the process of evolution, learning in DAYDREAMER is not accomplished through random mutation and natural selection; rather, DAYDREAMER already has heuristics which direct it toward a fruitful evolution. (See, however, Lenat's, 1983, hypothesis that evolution of higher animals and plants is not accomplished through random mutation and natural selection, but through *plausible* mutation and testing—that is, that mutation is directed by heuristic rules contained within DNA which, for example, coordinate simultaneous genetic mutations in an advantageous manner.)

The value of spinning a daydream lies partially in the fact that its value is not always readily apparent. Only by considering many possibly irrelevant, absurd, or unrealistic possibilities—which may take much time—are the benefits of daydreaming achieved. In a real-world situation, it may never occur to a person to try to find an overlooked possibility because it might not be apparent that such a possibility exists, and the person may never know for lack of time to go into a daydream to find out. However, in a particularly difficult situation in which all apparent alternatives have negative consequences, it may in fact pay to step back for a while and daydream.

The following additional benefit is provided by episodes: *Episodes enable reduction of search through the filling in of details (such as persons, animals, locations, and so on) of a daydream.* For example, in generating an imaginary sequence of events taking place in a familiar grocery store, details such as the location of the aisles, vegetable section, and check-out lanes may be filled in based on a memory of that grocery store. In particular, planning rules leave certain entities (such as physical objects or locations) unspecified—this situation occurs when a rule consequent contains variables not present in the antecedent. Normally, general planning is required to find appropriate instances of these entities. However, if this planning has been performed once, it may not have to be performed a second time: Entities may be instantiated based on entities contained in a recalled episode. An example of this occurs in RATIONALIZA-TION1 (although in this case the episode was hand-coded rather than generated through prior daydreaming): The recalled episode of Harrison Ford having a job contains the location of his job (Egypt) which is then employed in the current daydream. Another example is RECOVERY1.

The application of previous episodes in daydreaming is also useful to *further* learning, since past experiences are often a predictor of future experiences. Specifically, (a) *episodes provide initial daydream scenarios* (for example, "what if there is an earthquake?") and (b) *episodes provide continuations of existing daydream scenarios* (for example, "what if there is an earthquake while I am on a date?").

4.2.2   Episodes for Changing Knowledge Accessibility

If the possibilities inherent in a closed system are sufficiently numerous, that system can effectively be considered as a nonclosed system. In fact, if one draws the boundaries big enough, the distinction between closed and non-closed systems breaks down: Everything is a closed system. The important problem then becomes one of *control*: How and when are knowledge elements accessed, out of all those contained in the program? Although a program may *in principle* contain the knowledge necessary to generate a given plan in a given situation, the control algorithm may be such that access to that knowledge is unlikely or even impossible in that situation. For example, as mentioned above, in performance mode only the plan with the highest evaluation at a given point is selected; Other plans are simply not explored. Closed-system learning therefore involves restructuring of knowledge so that access to it is possible in new situations. DAYDREAMER restructures knowledge through (a) the storage of episodes, (b) creation of new rules, and (c) partial processing of a concern. A particular kind of knowledge restructuring which can occur through episode storage is discussed in this section.

Rules in DAYDREAMER are divided into two classes—*generic* and *episodic*. Episodic rules are similar to generic rules, except that they are inaccessible to regular planning. A given episodic rule may only be accessed through analogical application of an episode which contains that rule. That is, *an episodic rule may only be applied if an episode containing it is recalled.*

Episodic planning rules are initially contained within, and can only be applied through retrieval of, hand-coded episodes. However, if a daydream is generated which employs episodic rules borrowed from a hand-coded episode, the storage of this daydream may enable those rules to become accessible in new situations. In short, *generation and storage of a daydream may make a previously inaccessible episodic rule accessible.* In other words, in a certain situation, although the hand-coded episode may not be recalled, the new daydream *is* recalled, and an episodic rule previously inaccessible in this situation is now accessible through this daydream.[4]

---

[4]If an episodic rule is employed often enough, does it become a generic rule? Although

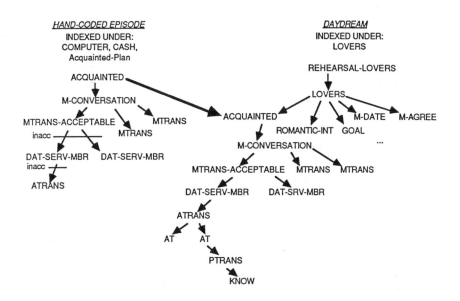

Figure 4.8: Making Inaccessible Rules Accessible

An example of this is provided by COMPUTER-SERENDIPITY (see page 14). As shown in Figure 4.8, a hand-coded episode containing two otherwise inaccessible episodic rules is retrieved through the conjunction of (a) the index of *computer* provided as environmental object input, and (b) serendipity with a active **REHEARSAL** concern for a **LOVERS** goal. A new daydream involving achievement of the **LOVERS** goal is then generated by analogy to this episode. The hand-coded episode is normally not retrievable, since it requires two or more of its indices to be active; thus the episodic rules it contains would normally not be accessed. However, these rules are now contained in the new daydream, which may be retrieved any time a **LOVERS** goal is pursued in the future.

4.2.3   Survival and Growth

Goals in humans are ultimately directed toward survival of the person, growth (Maslow, 1954) of the person (e.g., toward self-actualization), and survival of the species. Goals are thus *tuned* so that if they are continually satisfied the person tends to survive, grow, and reproduce. Thus personal goals—the cognitive representation of needs—should correspond to the true physiological and

---

such a change of status is not performed explicitly in the current version of DAYDREAMER, an episodic rule may be contained in so many diverse episodes that it is effectively accessible in all situations.

psychological needs of the person. In humans, it is clear that oxygen, water, and food are necessary for survival. However, the definition of survival in the case of a computer program such as DAYDREAMER is not as clear.

In computer science terms, survival must refer to the ability of a system to perform its designated tasks. The designated tasks in DAYDREAMER are specified by its personal goals: forming and maintaining interpersonal relationships, keeping a job, and so on. Therefore, in DAYDREAMER, *survival is defined as the ability to achieve, and continue to achieve, personal goals*, while *growth refers to the ability to improve at achieving personal goals over time*, that is, to *learn*.

The performance mode of DAYDREAMER enables us to evaluate the program's success in achieving its personal goals given various external situations as input. Specifically, the success or failure of a personal goal is assessed according to whether, in performance mode, the system carries out certain actions, or receives certain actions or states as input, which are designated to achieve that goal. For example, **ENTERTAINMENT** is satisfied only if the system performs certain actions designated as having entertainment value (such as watching a movie). The **MONEY** goal is satisfied only if money is "given" to the system (via external input). The success or failure of a **LOVERS**, **FRIENDS**, or **EMPLOYMENT** goal is also dependent upon external input, for example, *he dumps me*, *she offers me the job*, and so on. The success of the **SELF-ESTEEM** and **SOCIAL ESTEEM** goals is generally assessed as a function of the success of each of the other goals. Thus there are objective criteria for the success and failure of personal goals.

However, since these criteria refer in part to external input, there is a potential problem: A human could consistently defeat the computer by typing in negative responses. We therefore assume that the responses typed in are those of a "reasonable" person, that the responses present to the system a plausible external world, one which the system might encounter were it a human and not a computer system (since, after all, humans do not usually form true friendships with computers!). The "reasonableness" of the input responses can be varied in order to evaluate how the system is able to cope with different environments.

The criteria for the success and failure of personal goals are contained within DAYDREAMER in the form of inferences and planning rules, some of which were discussed above (and others of which will be presented in the remainder of this chapter). Thus the success or failure of a given goal may be assessed simply by examining the appropriate data structure within DAYDREAMER. This, of course, assumes that the system does not cheat (for example, by altering its **MONEY** data structure to indicate a million dollars). The DAYDREAMER program is constructed in such a way that it does not generally cheat in this manner. However, in certain cases DAYDREAMER does deceive itself: For example, through the **RATIONALIZATION** daydreaming goal, the degree to which a goal is considered to have failed is reduced in order to reduce the negative impact on processing caused by the negative emotional state resulting from that failed goal. Thus **RATIONALIZATION** temporarily tricks the system into thinking its situation is not as bad as it actually is, so that it can get on with the business of actually improving its situation: RATIONALIZATION brings optimism to the system.

If employed to an excessive degree, such strategies can have disastrous effects: By daydreaming away all of its unsatisfied goals, the system would never achieve them in reality. A system which fools itself in this way cannot continue to function for long—although it may lead a blissful existence during its last few moments.

4.3 Daydreaming Goals for Learning

General, abstract planning for future tasks is impossible because of the combinatorial explosion of possible future tasks and world situations. Therefore, daydreaming goals are used to guide the exploration of concrete situations of potential use in the future. In this section, we discuss REVERSAL, RECOVERY, REHEARSAL, and REPERCUSSIONS.

4.3.1 Reversal

The REVERSAL daydreaming goal generates scenarios in which a past real failure is avoided or a future imagined failure is prevented.[5] This goal is activated in response to a negative emotion of sufficient strength resulting from a real or imagined personal goal failure:

IF	NEG-EMOTION resulting from a FAILED-GOAL
THEN	ACTIVE-GOAL for REVERSAL of failure

The REVERSAL daydreaming goal modifies memory in two ways (depending on which of several strategies for REVERSAL is employed): episodes in which the personal goal failure is avoided are stored in episodic memory, and new inference and planning rules are created and stored in semantic memory. Both memory modifications modify future behavior of the program: Episodes are recalled and applied in order to achieve similar future goals, and rules are applied directly.

The experience LOVERS1 (see page 3), in which DAYDREAMER is turned down by Harrison Ford, contains two failures: the failure of a SOCIAL ESTEEM preservation goal and the failure of a LOVERS goal.[6] In the daydream

---

[5] It is also possible to daydream about alternative actions which might have prevented a goal *success* and resulted in a failure. We ignore these daydreams in the present DAYDREAMER for simplicity. Nonetheless, assessing the reasons for success is just as important as assessing the reasons for failure—both enable learning (Heider, 1958, pp. 88-89). Daydreaming of ways success might have been prevented is another way of generating imagined failures for further learning. That is, one may assess how one might have failed in a past experience in order to plan to avoid such possible mistakes in the future.

[6] It also happens to be the case that the SOCIAL ESTEEM goal is a subgoal of the LOVERS goal.

REVERSAL1 (see page 9), a **REVERSAL** of the **SOCIAL ESTEEM** goal leads to success of the **LOVERS** goal: DAYDREAMER imagines having worn nicer clothes. When DAYDREAMER is later confronted with a similar situation in LOVERS2 (see page 9), she puts on nicer clothes.

Here are some other daydreams involving **REVERSAL** of the **LOVERS** goal:

REVERSAL2

I feel upset. What if I had told him I found out his girlfriend was cheating on him? He would have broken up with his girlfriend. I would have asked him out. He would have accepted. I feel regretful.

REVERSAL3

What if I had told him I liked his movies? He would have had a positive attitude toward me. I would have asked him out. He would have accepted. I feel regretful.

The value of the **REVERSAL** daydreaming goal arises from the rule of thumb that *if something has happened once, it may happen again* (or if something is imagined, it is possible that it may happen in reality). This is especially true in a program such as DAYDREAMER which is perpetually concerned with a fixed set of personal goals.

REVERSAL is accomplished in different ways depending upon what caused the personal goal failure: If the personal goal failure was generated through an inference, or chain or inferences, as in REVERSAL1, the **UNDO-CAUSES** strategy is employed. If the personal goal failure resulted from failure of other persons to behave in the manner necessary for achievement of that goal, the strategies of **EXPAND-LEAVES**, used in REVERSAL2, and **EXPAND-ALTERNATIVES**, used in REVERSAL3, are employed. Other possibilities are not handled by **REVERSAL**. We now discuss each strategy in turn.

The **UNDO-CAUSES** strategy applies to the failure of a personal goal which was generated by an inference rule. For example, failure of **SOCIAL ESTEEM** results from a chain of inferences while planning for **LOVERS** in LOVERS1, as shown in Figure 4.9. This strategy may only be applied if one of the *leaves* (ultimate sources), call it N, of the inference chain leading to the goal failure is the negation of some state. (In the two-valued logic of DAYDREAMER contexts—see Chapter 7 for details—the absence of the assertion of a given fact is equivalent to the assertion of the negation of that fact—the *closed-world assumption*.) In this case, there exists such an N:

(**NOT** (**WELL-DRESSED** *Me*))

If the negation of N is deemed a *long-term state*, examples of which are **RPROX** (city of residence) and **INSURED** (having insurance), a new top-level

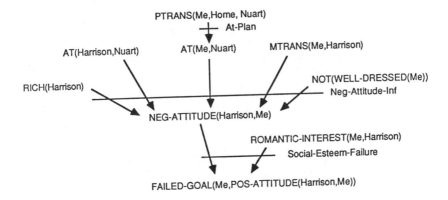

Figure 4.9: Inference Chain for a Social Regard Failure

personal goal whose objective is the negation of N is created and activated. This enables DAYDREAMER to prevent future similar failures once and for all.

Otherwise, if the negation of N is not a long-term state (as is the case for **WELL-DRESSED**), the **UNDO-CAUSES** strategy creates (a) a new inference rule which activates a preservation goal in appropriate circumstances, and (b) a new planning rule for achieving that preservation goal. The new preservation goal should be activated whenever it is possible that the personal goal may fail in the near future (for example, when DAYDREAMER has an active goal to **MTRANS** to a **RICH** person toward whom she has **ROMANTIC-INTEREST**). Planning for this preservation goal should then attempt to prevent N (in our example, not being **WELL-DRESSED**) or to achieve the negation of N (being **WELL-DRESSED**).

Specifically, the antecedent of the new inference rule is given by the following: for each variabilized[7] version VL of all the leaves besides N, either VL is true or a goal whose objective is VL is active. The consequent of the new inference rule is to create a new preservation goal with some unique identifier. The antecedent of the new planning rule is a preservation goal with the same unique identifier as above. The consequent is a variabilized version of the negation of N.

In REVERSAL1, the **UNDO-CAUSES** strategy creates the following two new rules:

---

[7]The operation of *variabilization* converts constants, such as persons and physical objects, into variables which can unify with (match) any instance of the major type from which they were derived.

IF	person is **RICH** or **ACTIVE-GOAL** for person to be **RICH** and person is **AT** location or **ACTIVE-GOAL** for person to be **AT** location and self **PTRANS** to location or **ACTIVE-GOAL** for self to **PTRANS** to location and self **MTRANS** to person or **ACTIVE-GOAL** for self to **MTRANS** to person and **ROMANTIC-INTEREST** toward person or **ACTIVE-GOAL** for **ROMANTIC-INTEREST** toward person
THEN	**ACTIVE-GOAL PRESERVATION UID.1**

IF	**ACTIVE-GOAL PRESERVATION UID.1**
THEN	**ACTIVE-GOAL** for self to be **WELL-DRESSED**

After these rules are added, the past failure episode is then replanned on behalf of the **REVERSAL** daydreaming goal starting from an appropriate context. This results in a **REVERSAL** scenario—REVERSAL1 in this case. These rules are then later employed in the performance mode experience LOVERS2.

Within a concern, planning for a preservation goal has priority over planning for nonpreservation goals. Thus any actions necessary to achieve preservation goals will be taken before further planning for regular goals is performed.

The remaining strategies for **REVERSAL** operate if the personal goal failure was caused by the failure of another person to perform actions necessary to achieve that goal; thus these strategies may be employed only on goal failures which result in performance mode. Performance mode experiences provide important feedback information which would otherwise be unavailable: whether a given plan worked or did not work when applied in the external world. Although one may not know exactly why a given plan failed, one may follow the strategy of continuing to use a plan which seemed to work or, if the plan did not seem to work, attempting to use a different plan. Both **EXPAND-LEAVES** and **EXPAND-ALTERNATIVES** are instances of such a strategy.

The **EXPAND-LEAVES** strategy for **REVERSAL** finds succeeded leaf subgoals whose realism is below a certain threshold and reinitiates planning for those subgoals (on behalf of the current **REVERSAL** daydreaming goal). A leaf subgoal is one for which no further planning was performed because it was considered satisfied at the time of planning. In REVERSAL2, the subgoal that the movie star is **NOT** in a **LOVERS** relationship with another woman is replanned.

The **EXPAND-ALTERNATIVES** strategy for **REVERSAL** finds unexplored branch contexts (if any) generated in planning for the personal goal and reinitiating planning in those contexts (on behalf of the current **REVERSAL** daydreaming goal). REVERSAL3 demonstrates the use of an alternative plan for getting the movie star to have a **POS-ATTITUDE** toward the daydreamer.

Both of these strategies may make use of information provided as input which justifies the action of the person which resulted in the personal goal failure. For example, Harrison Ford may explain that he does cannot go out with DAYDREAMER because he has a girlfriend. Such input may be used to reduce the search: In expanding leaves, only those leaf subgoals stated as reasons for the external action need be considered. In expanding alternatives, only alternative plans for subgoals stated as reasons for the external action need be considered.

What is the emotional impact of the **REVERSAL** daydreaming goal? First we consider daydreams resulting from a real goal failure: An imagined past goal success resulting from an alternative action leads to a negative, rather than positive, emotion: *regret*. This occurs in each of the example daydreams above. Thus successful **REVERSAL** actually has a negative emotional impact— emotional state is sacrificed for the benefit of learning. If, instead, an imagined goal failure results from alternative actions, positive emotions of *relief* result. Thus in a situation in which there simply was no way of avoiding the past failure, **REVERSAL** functions as a form of rationalization—the negative emotional state associated with a past failure is reduced. In the case of imagined goal failures which result in emotions of *worry*, successful **REVERSAL** (generating means by which the imagined failure may be avoided and a success achieved) results in positive emotions, which reduce the original level of worry. However, unsuccessful attempts resulting in further imagined failures produce further *worry* emotions.

4.3.2 Rehearsal and Recovery

The purpose of the **REHEARSAL** and **RECOVERY** daydreaming goals is twofold: to discover problems in a future plan before it is actually carried out in performance mode, and to construct a plan in advance in order to improve efficiency. **REHEARSAL** is activated in response to a halted personal goal concern:

IF	subgoal to **WAIT** and corresponding top-level goal contains no variables and **POS-EMOTION** of sufficient strength connected to top-level goal
THEN	**ACTIVE-GOAL** for **REHEARSAL** of top-level goal

All personal goal concerns are halted just before they are about to perform an action and the program is in daydreaming mode. This insures that the program will get a chance to daydream about future tasks before performing them.

REHEARSAL involves considering possible future completions of a concern and incorporating these plans into episodic memory for future use by that concern. For example, before the date in LOVERS3, DAYDREAMER daydreams about the date in REHEARSAL1.

REHEARSAL may lead to an imagined personal goal failure, which then activates a **REVERSAL** daydreaming goal, which in turn determines how to avoid that failure. The daydream NEWSPAPER (see page 7) is an example.

RECOVERY is activated in response to a negative emotion of sufficient strength resulting from a personal goal failure:

IF	**NEG-EMOTION** resulting from a **FAILED-GOAL**
THEN	**ACTIVE-GOAL** for **RECOVERY** of failure

RECOVERY involves the generation of future scenarios for achieving the same goal which failed.

What is the emotional impact of **REHEARSAL** and **RECOVERY**? An imagined goal success in a future daydream results in the positive emotion of *hope*. Thus **REHEARSAL** and **RECOVERY** can serve to counteract negative emotions resulting from previous imagined failures. By telling oneself that one promises to achieve a goal, or that a goal will eventually be achieved, one tends to feel better (even if that goal is never in fact achieved or planned for—though this may lead to later problems). This is expressed in the proverb "pain is forgotten where gain follows." Rapaport (1951) also noticed the emotional value of future daydreaming: "Daydreams often succeed in allaying, at least temporarily, the fears and urgings with which they deal. Their work of planning—by experimenting with possible solutions—seems to succeed at times in 'binding' mobile cathexes" (p. 463). Of course, an imagined goal failure in a future daydream results in the negative emotion of *worry*, which in turn activates a **REVERSAL** daydreaming goal to cope with that worry and prevent the imagined goal failure.

4.3.3   Repercussions

The **REPERCUSSIONS** daydreaming goal involves exploring the consequences of a hypothetical future situation. In particular, if an important goal failure or success occurs for another person (or animal), a **REPERCUSSIONS** daydreaming goal is activated to consider what it would be like for that failure or success to occur to the daydreamer. If a hypothetical future situation leads to a personal goal failure, the **REVERSAL** daydreaming goal will be activated to find ways to avoid such a future failure.

In the daydream REPERCUSSIONS1 (see page 8), a **REPERCUSSIONS** daydreaming goal is first activated which causes an earthquake in Los Angeles to be hypothesized. Rules fire which infer the collapse of the apartment and thus the failure of goals to stay alive and preserve possessions. These failures then result in the activation of two **REVERSAL** daydreaming goals. The first goal applies the **UNDO-CAUSES** strategy which creates rules causing DAYDREAMER to go underneath a doorway whenever an earthquake begins in the future (in the same city in which she lives). The second **REVERSAL** results in the activation of a new top-level personal goal: to get insurance in order to preserve her possessions. DAYDREAMER then tests the rules created in the first **REVERSAL** by generating another alternative scenario. This time, rules fire which infer a falling plant (hung above the doorway) and the failure of a goal not to be hurt. This in turn activates another **REVERSAL** goal which initiates a new top-level goal to move the plant. The two top-level goals are later pursued in performance mode: DAYDREAMER moves the plant, and goes to the insurance company and purchases the insurance. Although it may be obvious that one can get hurt in an earthquake, it may not be obvious to watch out for a falling plant. Thus only through such a daydream is one able to learn to watch out for a falling plant in case of an earthquake.

4.4   Evaluation and Decision Making

In previous sections we have described how daydreams are produced, stored, and later recalled and applied through analogical planning in a new situation. This section addresses the issues of daydream *evaluation* and *selection*: After generating a multitude of daydream scenarios, how does DAYDREAMER decide which scenario to pursue in the future? How is such a decision fixed in memory? Might the program consider several alternative scenarios and not make a final decision among them? How are choices made among several previously daydreamed courses of action applicable in the current situation? When does DAYDREAMER abandon a daydream or back up and modify it because it is not turning out as desired? What different ways may daydreams be evaluated?

4.4.1   Daydream Evaluation Metrics

Humans often decide to perform a given daydreamed plan based on which "feels best" or which the person is most "comfortable" with. In DAYDREAMER, however, these sensibilities must be made concrete. We therefore present three daydream evaluation metrics: *desirability, realism,* and *similarity.*

Suppose DAYDREAMER generates two alternative future daydreams in which she preserves her possessions in case of an earthquake in Los Angeles: (a) move to New York, or (b) stay in Los Angeles and purchase insurance. The first daydream has the following negative consequence: She will lose her job in Los Angeles. Thus the first daydream has a lower desirability than the second. As a result, DAYDREAMER will select the second daydream rather than the first for execution in performance mode.

While planning for one personal goal, side effects or consequences for other personal goals are detected as follows: Either a personal goal activation, success, or failure happens to be inferred (through application of an inference rule) or such a personal goal happens to result as a subgoal of the personal goal originally being planned. Therefore, *the desirability of a final daydream scenario is calculated according to the sum of the importances of each of the personal goals activated or resolved some time after the activation of the top-level goal of the concern.* Goal successes have positive importance while goal failures and active goals have negative importance. A negative desirability thus indicates an undesirable scenario while a positive value indicates a desirable one. The im-

portance of a goal is initially set according to an *intrinsic importance* assigned
by the inference rule which activated that goal (which in turn is determined by
the programmer). However, this initial importance may be modified through
processes such as rationalization (discussed in Chapter 3.)

Suppose DAYDREAMER instead daydreams that she moves to New York and
becomes a famous artist. This might have a higher desirability than the daydream of purchasing insurance in Los Angeles. However, it has a much lower
realism. Although daydreams of low realism may be useful for further daydreaming (such as in generating rationalizations or in the generation of more
realistic solutions through incremental modification), DAYDREAMER should
not generate external behavior based on an unrealistic daydream—the sequence
of actions taken by the program, when performed in the external world, would
be unlikely to result in the daydreamed outcome. Therefore, DAYDREAMER
decides whether to act on a given daydream based on both the desirability and
realism of that daydream. In this case, staying in Los Angeles receives a higher
overall evaluation based on both desirability and realism, and the program decides to stay in Los Angeles.

On a scale from 0 to 1, *the realism of a scenario is evaluated based on
the plausibilities of the rules employed in generating that scenario.* 1 indicates a completely realistic scenario while 0 indicates a completely unrealistic
one. Methods for calculating scenario realisms are discussed by Shortliffe and
Buchanan (1975), Duda, P. E. Hart, and Nilsson (1976), Charniak (1983b), and
Pearl (1982). *Evaluations of desirability and realism are stored along with daydreams in episodic memory.* Realisms, but not desirabilities, are also associated
with subepisodes (episodes associated with subgoals).

One might object to the absence of such metrics as cost and risk. However,
such metrics are subsumed—albeit in a rudimentary manner—by the above
metrics: The cost of executing a given plan is equivalent to the (negative) impact
of that plan on personal goals: For example, if running to the store before it
closes causes one to become fatigued, this cost is accounted for in a personal goal
not to become fatigued. The risk of a given plan is also determined by negative
impact on personal goals—provided that the scenario has a high realism. That
is, if a plan is risky, it is likely that it will result in negative consequences: For
example, since robbing a bank is risky, a realistic daydream about robbing a
bank would include getting caught. If getting caught were overlooked by the
daydream, it would then have a low realism. Thus cost and risk are taken into
account by the desirability and realism metrics.

Another metric, that of *similarity*, is not applied when storing an episode,
but rather when an episode is retrieved. While desirability and realism evaluate
daydreams in isolation from any particular task, the similarity metric is an evaluation of a daydream in terms of a particular target situation—the similarity of
the source and target goals is evaluated. *The similarity of two goal objectives is
determined through a structural comparison of those two objectives.* Similarity
may be determined according to the distance of the types of two objectives in

the type hierarchy and recursively according to the similarity of the components of the two objectives; differences occurring deeper in the structure have a smaller weighting. For example, in the current program, the similarity of the **REVENGE** goal against Harrison Ford and the **REVENGE** goal against Robert Redford is 2.4. The similarity of the **REVENGE** goal against Harrison Ford and the **REVENGE** goal against the daydreamer's boss is 2.1. These two goals are less similar than the previous two because one of the latter goals involves **REVENGE** for a failed **LOVERS** goal while the other of the latter goals involves **REVENGE** for a failed **EMPLOYMENT** goal. In the former case, both goals involved **LOVERS** goals. The similarity of the **REVENGE** goal against Harrison Ford and a **FOOD** goal is a large negative number.

In the current version of DAYDREAMER, this similarity evaluation is sufficient to select, out of several alternative episodes, the episode most appropriate to the target goal. This metric, for example, will enable the program to select REVENGE3 instead of REVENGE1 the next time an **EMPLOYMENT REVENGE** daydream is generated. However, as the number of indexed daydreams and episodes becomes large, similarity will have to be measured in terms of relevant features of the goal objective and of the enclosing situation. The program will have to determine such features automatically in order for it to learn effectively. For example, DAYDREAMER might build up a large collection of alternative plans for forming **LOVERS** relationships with people depending on the particular personality traits of the person. Given a new person, the best, most similar, plan would have to be selected using those traits as indices.

Daydream evaluation is employed at several points in processing:

- *During production*: As a daydream or external plan is generated, an ongoing assessment of its realism is maintained. A planning branch is abandoned if (a) the current concern is a personal goal or learning daydreaming goal and (b) the scenario realism falls below a certain threshold (higher for personal goals than for learning daydreaming goals). For example, in planning for a future date via **REHEARSAL**, if an earthquake is imagined to occur in the middle of the date, this scenario will be deemed unrealistic and abandoned.

- *After production*: After generation of a daydream or external plan, its desirability and realism are evaluated for later use in deciding whether or not to employ that episode.

- *During application*: When several previous daydreamed or real planning episodes are retrieved as potentially applicable to a given situation, those episodes are evaluated with respect to the current situation. That is, the similarity of the source and target goals is calculated.

episode	simil.	realism	desir.	ordering1	ordering2	ordering3
REVENGE1	2.1	.2	1.8	2.1	.76	0
REVENGE3	2.4	.2	1.8	2.4	.86	0

Table 4.1: Ordering of Retrieved Episodes

4.4.2   Daydream Selection

Several previously daydreamed or real plans for a goal may be retrieved when planning to achieve that goal in a new daydream or performance mode experience. How does DAYDREAMER choose which of these episodes to employ?

In DAYDREAMER, *the ordering of episodes as candidates for potential use is calculated according to the stored desirabilities and realisms of those episodes, and the assessed similarity of those episodes to the current situation.* This calculation is different depending on the type of concern. If the concern is a personal goal or a learning daydreaming goal, then the ordering is based on the product of desirability, realism, and similarity, provided that each is above a certain threshold (higher for personal goals than for learning daydreaming goals). Otherwise, the ordering is based on the similarity, provided it is above a certain threshold.

For example, suppose DAYDREAMER is about to generate another **RE-VENGE** daydream after being fired from her job. Previous **REVENGE** daydreams, REVENGE1 and REVENGE3, are recalled, and an ordering for these episodes is determined as shown in Table 4.1. Since the current concern is a daydreaming goal, the ordering of these daydreams is based only upon the similarity, shown as *ordering1*. Therefore, REVENGE3 is attempted before REVENGE1. (REVENGE3 will succeed and so REVENGE1 will never be attempted.) If the current concern had instead been a learning daydreaming goal, the orderings would have been based upon the product of the three evaluations, shown as *ordering2*. Again, in this case, REVENGE3 would have been attempted before REVENGE1. However, if the current concern had been a personal goal (and thus a performance mode experience), the realism of both of these episodes would have been below the acceptable minimum level of realism for performance mode planning. Thus the *ordering3* would have been 0 for both episodes and neither would have been employed in the generation of external behavior.

If all the retrieved alternative daydreams for achieving a given goal have a low or negative desirability, this is an indication that the goal itself may not be a good idea. Therefore, *whenever all daydreamed attempts to achieve an active goal have negative desirability, that active goal is deactivated.* This enables the program to learn to avoid pursuing certain courses of action. For example, in the case of the movie star, it might be determined that dating movie stars always turns out bad.

4.5 Intention Formation and Application

In daydreaming, one often makes mental notes, commonly called *intentions*, about what one will do in the future: One thinks something along the lines of "next time X happens, I will do Y" or "tomorrow I will do Z." Intentions, once formed, do not remain in consciousness. Rather, an intention is reactivated only at some appropriate time in the future.

We distinguish between *persistent* and *one-shot* intentions. A persistent intention is a generalized plan to be applied whenever a certain situation arises in the future: for example, to run for the nearest doorway in case of an earthquake, to check that the car lights are turned off before leaving the car, to call a store to make sure it is open before driving there, to dress up before going out alone, when meeting a new person to find out the telephone number of that person in order to be able to contact that person in the future, and so on. A one-shot intention, on the other hand, is a more or less specific plan which applies once and is then deactivated: for example, to go grocery shopping on the way home from work, to pick up a copy of a newspaper while shopping, to mail a letter, and so on.

Our distinction between the recall of persistent and one-shot intentions is similar, but not identical to, the distinction of Meacham and Leiman (1982) between habitual prospective remembering and episodic prospective remembering: Habitual remembering is more specific than recall of persistent intentions—it involves the recall of intentions as part of a routine, such as taking vitamins each morning at breakfast. Episodic remembering, however, is basically the same as the recall of a one-shot intention.

It is sometimes difficult to distinguish the two classes: Planning a particular sentence for use in a particular job interview may be viewed as a one-shot intention since it is performed only once; and yet, if the person does not get the job and has to go to more interviews, this same sentence could be employed again and again. For the present purposes, any detailed future plan which has the potential of being applied more than once in the future is considered a persistent intention.

In modeling intentions, we must consider both how intentions are formed and how they are later recalled and acted upon. That is, how is a particular plan generated and then how is a later situation recognized as appropriate for application of this plan? First we consider persistent intentions.

We hypothesize that *persistent intentions correspond to daydreamed episodes having high evaluations.* That is, persistent intentions are formed though the generation, evaluation, and storage of daydreamed scenarios in episodic memory; such episodes are indexed under the goal of the scenario and other indices, such as emotions, persons, and physical objects.[8] Whenever a similar goal is activated in the future (and other indices are active), this episode will be recalled and potentially applied. (If several episodes are recalled it will then be necessary to select from among these episodes.) Proper recall of a persistent intention then depends on proper indexing of the episode.

We further hypothesize that *one-shot intentions correspond to concerns* in DAYDREAMER. In effect, each currently active concern constitutes an intention to complete that concern. A concern (and therefore a one-shot intention) is initiated upon activation of a personal or daydreaming goal. Processing activity on behalf of a given concern is performed when that concern has the highest level of motivation, as provided by emotions. Once initiated, a concern persists until completion. This tendency of humans to complete a task once begun—even if interrupted—has been demonstrated in psychological experiments (Lewin, 1926/1951) and observed by other researchers (Mandler, 1975; Miller, Galanter, & Pribram, 1960). However, a concern may be interrupted in DAYDREAMER if a new concern is initiated with a higher level of motivation or if an existing concern acquires a higher level of motivation. An existing concern can acquire a higher level of motivation if, while performing another concern, something comes up which is relevant to that concern. This is detected through two mechanisms:

- *Fortuitous subgoal success*: While performing a personal goal concern, an active subgoal of another personal goal concern is accidentally achieved. A *surprise* emotion is generated and becomes an additional motivating emotion associated with the second concern.

- *Serendipity*: An input state, retrieved episode, or internally generated state accidentally suggests a solution to another concern. A *surprise* emotion is generated and associated with the second concern. See Chapter 5 for details.

Sometimes an interrupted concern is resumed shortly thereafter—for example, when the concern is interrupted by a short concern such as (in humans) swatting a mosquito—while other times a concern is interrupted by a longer concern. In both cases, there is the potential that yet other concerns will be activated or that other concerns will acquire greater motivation, so that the interrupted concern will not resume execution immediately after completion of the interrupting concern. It is possible for the program to follow a series of interruptions from each concern it is working on, returning to interrupted concerns only when concerns complete or when interrupted concerns acquire

---

[8]However, humans often decide on a *single*, best daydream to perform in the future. This one daydream is *the* intention. In DAYDREAMER, a particular daydream could be designated as such if it had a evaluation sufficiently higher than each of the other candidates. For the time being, however, all daydreams with successful evaluations are stored, and the daydream with the highest evaluation may be considered as the single intention since it is most likely to be employed in future performance mode planning.

higher motivation. There is no special mechanism for resumption of interrupted concerns, since when an interrupting concern completes, it is removed and the program continues processing with the most highly motivated concern at that time (which is often, but not always, the last interrupted concern). Concerns are also interrupted and placed into a *nonrunable* state if the program desires to perform an action and is currently in daydreaming mode or the program is in performance mode and that action contains variables. This distinction in our program between runable and nonrunable concerns accounts for results discussed by Lewin (1926/1951): Resumption of an interrupted activity occurs even without external stimulus related to that activity (as in runable concerns), while unperformed intentions require appropriate external stimulus to activate (as in nonrunable concerns reactivated through fortuitous subgoal success and serendipity).

Humans, of course, may perform more sophisticated planning to interleave several concerns and achieve greater efficiency. For example, if one concern has purchasing a newspaper as a subgoal and another has purchasing groceries as a subgoal, the two subgoals may be accomplished in one, rather than two, trips to the store (which carries newspapers). DAYDREAMER is not generally capable of batching plans in this way, since planning is performed separately for each concern. However, while performing one concern, the program may realize that the current situation is relevant to another concern, thus achieving a similar batching effect in an opportunistic manner: In performing the concern to obtain groceries, the subgoal of being near a newspaper is achieved, thus reactivating the other concern to purchase a newspaper. Once the newspaper is obtained, the concern to obtain groceries may be resumed. Experiments (B. Hayes-Roth & F. Hayes-Roth, 1979) have indicated that human planning is often performed in such a manner, especially when the plans reach a certain level of complexity. In addition, although DAYDREAMER plans for each concern separately, one concern may involve more than one personal goal. In particular, the impact on personal goals other than the top-level goal is taken into account during planning.

There are, of course, still difficulties with this approach: The program will go through the checkout line twice, failing to realize that it can buy the groceries and the newspaper at the same time. Also, once having purchased the newspaper, it may continue with that concern, which could take it out of the store to another location before having purchased the groceries. Avoiding such problems requires more advanced planning techniques (such as those discussed by Wilensky, 1983 and Sacerdoti, 1977) which are beyond the scope of the current work.

Lewin (1926/1951, pp. 97-98) argues that the use of associations cannot account for the recall of an intention on an appropriate occasion since, for example, dropping a letter into a mailbox when it is seen (S. Freud, 1901/1960, p. 152) should *increase the force* of association between the mailbox and the dropping of the letter, and yet once this intention is carried out, seeing a second mailbox results in *no forces* directed toward mailing the letter.[9] Instead, he proposes that intentions result in the formation of a tension state called a "quasi-need" (similar to our emotionally-motivated concerns) to which there corresponds a set of situations and objects (said to have a "valence" for the quasi-need) enabling actions which satisfy the quasi-need. In DAYDREAMER, a "valence" for a concern is determined by whether a situation achieves a subgoal of that concern or suggests a solution to that concern through serendipity (see Chapter 5).

Nonetheless, as J. L. Singer (1975) points out, intentions and unfinished business are often recalled in daydreaming via a chain of associations or remindings. In DAYDREAMER, persistent intentions (stored episodes) can be recalled in this fashion (see Chapter 8) and one-shot intentions (concerns) can be recalled or reactivated through an associational process in the space of rule connections.

Why do we sometimes forget our intentions? A common phenomenon is walking into the kitchen or somewhere else and then wondering what one is doing there. Perhaps such forgetting is caused by daydreaming about things other than the current plan, which pushes that plan out of limited-capacity working memory (as suggested by J. R. Anderson, 1983, p. 130). S. Freud, (1901/1960) discusses the forgetting of intentions caused by obscure motives (see also Lewin, 1926/1951). In DAYDREAMER, a persistent intention might be forgotten if an episode is improperly indexed; a one-shot intention might be forgotten if the serendipity mechanism is unable to find a relationship between the current situation and a concern when it should.

4.6   Related Work in Machine Learning

There has been much work in the past on learning by creating new inference, planning, or production rules and modifying existing ones. Bundy, Silver, and Plummer (1985) and Dietterich and Michalski (1983), for example, review this work. Typical basic strategies for rule creation and modification include: (a) dropping conditions, (b) adding conditions, (c) converting constants into variables (variabilization), (d) converting variables into constants (instantiation), and (e) composing two rules. New rules are created in DAYDREAMER through the **UNDO-CAUSES** strategy for the **REVERSAL** daydreaming goal. (An additional *input induction* strategy, not developed extensively in the present work, is used in DAYDREAMER to create a new planning rule through the variabilization of two consecutive input states in certain specialized situations.) The **UNDO-CAUSES** strategy is used when a personal goal failure results from the absence of some state in some situation. It creates a new rule which initiates plan-

---

[9]Although this may be true in most cases, informal observations suggest that one may sometimes become confused about whether a given intention has been carried out. Perhaps it is possible that daydreamed actions may be confused with real ones or that confusions among similar intentions may occur.

ning to achieve that absent state in future similar situations. Sussman's (1975) HACKER program used a similar strategy for correcting failures or "bugs" in a plan given the reason for the failure.

DAYDREAMER primarily learns through the generation, evaluation, storage, retrieval, and application of daydreamed episodes. Although planning and inference rules are neither created nor modified by this process, new episodes improve the program's ability to select from among alternative rules in any given situation. Rules which were previously inaccessible in a given situation may become accessible. Since daydreamed episodes are applied through a process of analogical planning, the remainder of this section is devoted to a review of previous work in this area.

The analogical planning mechanism of DAYDREAMER is similar to Carbonell's (1983) method for learning by analogy. Both methods employ the following steps to solve a new problem: (a) recall the solution to a previous similar problem, and (b) employ that solution to reduce search in generating a solution to the new problem.

In particular, Carbonell starts with a "means-ends analysis" (Newell & Simon, 1972) framework for problem solving: The task is to find a sequence of operators which transform an initial state in the problem space into a final (goal) state; a difference function is used to compute the differences between one state and another; operators are indexed by the differences they reduce. In general, problem solving involves a repeated process of selecting operators which reduce the differences between the current state and the goal state (and appropriate recursion for subproblems resulting from operators with unsatisfied preconditions).

Problem solving by analogy is then carried out as follows: The initial and goal states and other constraints of the current problem are compared to those of previously solved problems, using the difference function as a similarity metric. In addition, the percentage of preconditions of those operators employed in the previous problem solution which are already satisfied in the new situation is calculated. The previous problem solution most similar to the new problem according to these metrics is then selected (or a number of previous solutions are selected).

The previous solution is then transformed into a solution to the new problem as follows: A process of problem solving is carried out in a higher-level *space of problem solutions* called "T-space" (for transform space), with the previous solution as the initial state and a solution to the new problem as the goal state. This task is performed through a process of means-ends analysis whose available "T-operators" include: inserting an operator into the solution, deleting an operator, reordering operators, substituting operator parameters, and (a) invoking means-ends analysis in the (lower-level) problem space in order to achieve an operator precondition satisfied in the previous situation but not in the new one and then (b) splicing the resulting subsequence into the solution.

DAYDREAMER employs a subgoaling framework rather than a means-ends analysis one. A previous plan, recalled primarily on the basis of the similarity of the previous goal to the new one (but according to other indices as well), functions as a "suggestion" to the process of planning for the new goal. Nonetheless, the operations which DAYDREAMER performs as it generates a new plan are analogous to "T-operators": (a) a subtree for achieving a subgoal satisfied in the original situation but not in the new one may be generated and spliced in, (b) a subtree for a subgoal satisfied in the new situation but not in the original may be deleted, (c) a subtree for achieving a subgoal may be generated and spliced in when the plan for that subgoal was applicable in the original situation but is not applicable in the new one, and (d) parameters may be substituted.

Analogical planning in DAYDREAMER is also related to the "purpose-directed analogy" mechanism of Kedar-Cabelli (1985). Given a goal concept (e.g., a hot cup), purpose of that concept (e.g., to enable one to drink hot liquids), a target example (e.g., a styrofoam cup), and domain knowledge, the mechanism (a) retrieves a known instance of the goal concept (e.g., a ceramic mug), (b) generates a proof that this instance satisfies the purpose of the goal concept (e.g., that a ceramic mug enables one to drink hot liquids), and (c) uses this proof to construct a proof that the target example satisfies the purpose of the goal concept (e.g., that a styrofoam cup also enables one to drink hot liquids). This last step is accomplished in a fashion similar both to analogical planning in DAYDREAMER and to Carbonell's analogical problem solving (i.e., through incremental repair of the proof). A generalization may then be formed from the common features of the two proofs; this generalization provides a definition of the goal concept stated in terms of lower-level structural features (e.g., upward concavity, flat bottom, and so on) instead of higher-level functional ones (e.g., used to drink hot liquids).

Kedar-Cabelli assumes that the source and target concept instances are given, and therefore does not address the problem of finding an analogy in the first place. In contrast, DAYDREAMER selects episodes related to the current situation automatically. Unlike the analogical planning mechanism of DAYDREAMER, neither the scheme of Carbonell nor that of Kedar-Cabelli enables incorrect subgoals in a tentative analogical plan to be themselves repaired through further (recursive) analogical planning. Carbonell, however, has suggested two related ideas: (a) two recalled solutions might be combined into one (p. 144), and (b) future problems in T-space might be solved by analogy to previous T-space problem solutions (p. 149). The mechanisms of Carbonell and Kedar-Cabelli are otherwise quite similar to our analogical planning mechanism. Analogical planning in DAYDREAMER is also similar to the chunking on goal hierarchies of Rosenbloom (1983) and the work on MACROPS in STRIPS (Fikes, Hart, P. E., & Nilsson, 1972).

However, unlike these mechanisms, our mechanism is embedded within a larger framework and set of strategies for learning from *imagined* past and future scenarios: (a) upon a failure in the external world, alternative past scenarios may be generated, evaluated (for example, according to the impact on various personal goals of the program) and stored for possible future application by analogy, and (b) possible future situations and plans may be hypothesized, evaluated, and stored for possible future application by analogy. Analogical planning is thus used as the basic mechanism for applying previous imagined scenarios to the generation of new imagined scenarios, as well as to the generation of external actions in performance mode. In addition, this mechanism enables previous *real* episodes (hand-coded or formed during a performance mode experience) similar to a hypothesized scenario to be employed by analogy in generating possible continuations of that scenario. Furthermore, we consider the possibility and role of *accidentally* becoming reminded of a previous problem directly or indirectly applicable to a current problem (serendipity).

Chapter 5

Everyday Creativity in Daydreaming

Everyday leaps of creativity often occur in the spontaneous and free-flowing internal activity of daydreaming:

- A daydream about an unlikely event or a fanciful solution to a problem may lead, through modification, to a plausible, useful solution. For example, suppose it is raining outside and you would like to get the newspaper without getting wet, but you don't have a raincoat.[1] You happen to gaze at a plant in the room and begin daydreaming that the plant grows at an accelerated rate, reaches the newspaper, coils around it, and then enables you to pull the paper in by pulling on the plant. This fanciful solution may lead you toward a more feasible approach such as fetching the paper with a long stick.

- Potential solutions to problems may be generated through the serendipitous detection of relationships among objects or events during daydreaming. We saw an example of this above: the plant, an accidental input, was related to the current problem of fetching the newspaper. Here is another example: Suppose you want to meet a famous actor. Then suppose you start thinking about a movie the actor starred in. Then you recall that you were in a minor car accident on the way back from that movie. You remember exchanging telephone numbers and other information with the person you rear-ended. You realize that if you know what streets the movie star uses, one way of meeting the star (if it is worth it to you) is to force an accident with the star.

Previous work in the field of artificial intelligence on problem solving (Newell, J. C. Shaw, & Simon, 1957; Fikes & Nilsson, 1971; Sacerdoti, 1977; Wilensky,

---

[1]This example problem (but not the solutions) is borrowed from Wilensky (1983).

1983) has ignored the role of *accident* and the role of *fanciful thinking*—two important aspects of creative problem solving.

In this chapter, we address the problem of constructing a daydreaming computer program—DAYDREAMER—which exhibits these properties: To address the recognition and exploitation of accidental relationships among problems, we propose a *serendipity* mechanism. To address the generation of fanciful possibilities, we propose a mechanism for the arbitrary *mutation* of daydreamed events.

First, we address the problem of how creativity is defined from a computational standpoint. Second, we explore in more detail some of the attributes of creativity as identified by previous investigators. Third and fourth, we describe the mechanisms of serendipity and mutation which address those attributes. Fifth, we relate our mechanisms to the well-known (Wallas, 1926, pp. 79-107) phenomena of incubation and insight. Sixth, we review previous related work in artificial intelligence in detail.

5.1   Definition of Creativity

Creativity is generally defined as the production of something—a scientific theory, work of art, poem, novel, and so on—which is both novel and valuable according to consensual judgment (Rothenberg, 1979, pp. 3-6). DAYDREAMER is concerned with finding creative solutions to problems in the fields of interpersonal relations, employment, and the like.

The value of a solution is measured in DAYDREAMER according to the metrics of *desirability* and *realism: Desirability measures the value of the outcome of a scenario in terms of the degree of satisfaction of each of the personal goals of the system. Realism measures the likelihood that the sequence of actions specified by the solution will result in the outcome.*

A novel solution to a problem is any solution which was previously unknown to the system. However, this definition presents an apparent paradox: It seems that any solution generated by the system without external input cannot be unknown, since if it was able to generate the solution, it must have already known that solution. Nonetheless, novel solutions often arise in closed systems. The answer is as follows: Although every future solution to a problem is indeed inherent in the current state of the program, reaching a future solution may require a significant amount of processing. A given solution should be considered known only if the system can generate that solution in a short amount of time. Any solution which would require an expensive search process to find given the current system state (that is, without the occurrence of accidents sidestepping the search, as discussed below) is an novel solution with respect to that system state. Thus, *a new solution is one which, without the occurrence of an accident, could not have been generated in the past—at all, or without having consumed more than a certain amount of time and space resources.*

5.2 Aspects of Creativity in Daydreaming

In the past, several investigators have related daydreaming and creativity: S. Freud (1908/1962) proposes that, like daydreams, works of fiction result from ungratified wishes, with the difference that in fiction appropriate literary devices are employed to sufficiently soften or disguise those wishes:

> A strong experience in the present awakens in the creative writer a memory of an earlier experience (usually belonging to his childhood) from which there now proceeds a wish which finds its fulfilment in the creative work. The work itself exhibits elements of the recent provoking occasion as well as of the old memory. (p. 151)

Rothenberg (1979) writes that "during the course of [the] process of discovery, the creative person engages in a good deal of fantasy" (p. 129) and "his thoughts may rove freely, but he is constantly alert and prepared to select and relate his thoughts to the creative task he is engaged in" (p. 130). Several studies have investigated the relationship between daydreaming and creativity (J. L. Singer, 1975, pp. 66-67).

In this section, we discuss three attributes of a creative system: *breaking away*, *fanciful possibility generation*, and *finding new connections*. We then propose two mechanisms in DAYDREAMER to address these attributes.

5.2.1 Breaking Away

Several theories of creativity refer to breaking away from established patterns of thought: Wertheimer (1945) proposes that productive thinking involves being sensitive to disturbances, gaps, superficialities, and distortions in the initial structural view of a problem situation, which leads to reorganization, reorientation, and improvement of that view. Guilford (1967) discusses the production of possibilities which diverge from existing sets. DeBono (1970) describes a technique called "lateral thinking" which helps one escape from existing thought patterns. Adams (1974) discusses the importance of overcoming perceptual, emotional, cultural, environmental, and other conceptual blocks.

5.2.2 Fanciful Possibility Generation

The importance of generating possibilities or variations in quantity and suspending judgment have been emphasized by several authors: In 1788, Schiller (quoted by S. Freud, 1900/1965) wrote:

> It seems a bad thing and detrimental to the creative work of the mind if Reason makes too close an examination of the ideas as they come pouring in—at the very gateway, as it were. Looked at in isolation, a thought may seem very trivial or very fantastic; but it

may be made important by another thought that comes after it, and, in conjunction with other thoughts that may seem equally absurd, it may turn out to form a most effective link. (p. 135)

In Osborn's (1953) "brainstorming" technique for group problem solving, participants are encouraged to generate wild ideas and not to criticize those of other participants. The generation of variations on a theme is considered to be the essence of creativity by Hofstadter (1985).

It is useful to generate fanciful possibilities and variations in quantity for two reasons: First, *a fanciful scenario may turn out not to be as unrealistic as initially thought*: Daydreaming about an earthquake might pay off if and when that earthquake actually occurs. Second, *a fanciful scenario may lead through modification to a useful solution to a problem*, as we saw above.

Therefore, a creative system must have the capability of generating many fanciful and realistic possibilities. In addition, it must be able to evaluate and select from among those possibilities, and to modify and repair those possibilities to suit certain purposes.

5.2.3   Finding New Connections

Several theories characterize creativity in terms of recognizing and exploiting new connections among, or combinations of, existing elements.

Mednick (1962) proposes that creative thought consists of the formation of new and useful combinations of elements which are remotely associated with each other—the novelty of the combination being proportional to the associative remoteness of the elements. He proposes that the combination of two remote ideas can be suggested through accidental environmental stimuli, through similarity of the two ideas along certain dimensions, and through a mediating third idea in common with the two ideas.

Koestler (1964) proposes that creative acts result from "bisociation"—the coming together of two previously unrelated associative contexts or frames of reference (such as music and arithmetic in the harmonic laws posited by Pythagoras).

Rothenberg (1979) characterizes creativity in terms of what he calls "janusian" thinking and "homospatial" thinking. He defines janusian thinking as "actively conceiving two or more opposite or antithetical ideas, images, or concepts simultaneously" (p. 55) and homospatial thinking as "actively conceiving two or more discrete entities occupying the same space, a conception leading to the articulation of new identities." (p. 69)

5.2.4   Mechanisms in DAYDREAMER

We propose two mechanisms in DAYDREAMER to (a) break away from possibilities generated by the existing analogical planning mechanism and daydreaming goals, (b) generate fanciful possibilities, and (c) find new connections:

- *Serendipity*: generating a new solution to a problem through the accidental juxtaposition of a recalled experience, input physical object, or input or internally generated state with that problem.

- *Mutation*: generating possibilities through arbitrary modification of existing possibilities, which sometimes leads to new solutions.

5.3 Serendipity in Daydreaming

A *serendipity* typically involves a scenario such as the following: One is having trouble solving a problem. Every possibility that one tries does not seem to work. So, one sets the problem aside for a while. Then at some later time—perhaps while reading a book, watching a movie, or daydreaming—one suddenly realizes that something just experienced or thought of is applicable to the problem. By accident one has stumbled upon a potential solution. One can then go back, verify the solution, and flesh it out. The solution may be discarded if it does not work.

In the performance mode experience RECOVERY3 (see page 13), DAYDREAMER receives an alumni directory from the college she attended which happens to contain the number of Carol Burnett. DAYDREAMER had previously been daydreaming about contacting Harrison Ford in order to ask him out again. Serendipitously, DAYDREAMER realizes that the alumni directory is applicable to the problem of finding out the unlisted telephone number of Harrison Ford. DAYDREAMER could possibly find out Harrison's telephone number by obtaining a copy of the alumni directory from the college Harrison Ford attended, if any.

Serendipity may occur when one notices that something currently under focus suggests a solution for a recent (but not necessarily under current focus) problem. Serendipity may also occur when a new problem is initiated, and one notices that something recently (or currently) under focus suggests a solution for the new problem. It does not matter whether the plan is introduced prior to the goal or the goal is introduced prior to the plan. This is analogous to the commutativity property in the "scientific community metaphor" of Kornfeld and Hewitt (1981).

We distinguish several cases of serendipity in DAYDREAMER:

- *Episode-driven serendipity*: retrieval of an episode or storage of a new episode suggests a previously unknown solution for an active, suspended, or halted concern.

- *Object-driven serendipity*: a special case of the above in which episode retrieval results in response to physical objects provided as input to DAYDREAMER.

- *Rule-induction-driven serendipity:* another special case, employed in RE-COVERY3, in which a new episode is stored which contains a new episodic rule induced from states or actions provided as input to DAYDREAMER.

- *Concern-activation-driven serendipity:* a previously unknown solution to a newly created concern is suggested by a recently retrieved episode.

- *Input-state-driven serendipity:* a state or action provided as input to the program suggests a solution to an active, suspended, or halted concern. This mechanism is used for the recall of previous intentions.

- *Mutation-driven serendipity:* arbitrary mutation of an action suggests a potentially new solution for a particular concern.

In DAYDREAMER, *serendipity enables discovery of a new solution through the use of normally unavailable planning paths or rules.* Specifically, in concern-activation-driven serendipity and episode-driven serendipity, a solution to a problem is generated which could not have been generated previously because the solution employs a rule which was *inaccessible* to regular and analogical planning. An inaccessible rule is an episodic rule[2] which is contained only in inaccessible episodes (e.g., episodes requiring physical objects as retrieval indices). As solution generated by mutation-driven serendipity may not have previously been generated because of "mental blocks" (or failures to consider certain planning paths, to be discussed) inherent in regular planning. A previously blocked solution can be generated by input-state-driven serendipity as a result of particular situations in the external world that had never arisen before.

5.3.1　Serendipity-Based Learning

Serendipity in planning leads naturally to learning: Once a plan is found through serendipity, it is not necessary for another serendipity to occur in order to employ a similar plan in a future similar situation. Instead, the plan is incorporated into episodic memory, indexed under the goal it achieves, so that in the future this plan can be applied directly to a similar goal through analogical planning. *Serendipity enables learning since the result of a serendipity is stored as an episode for possible use in future similar situations.*

For example, suppose DAYDREAMER is again concerned with how to meet Harrison Ford when it happens to have a car accident. As DAYDREAMER is exchanging telephone numbers and other information with the person, it notices that one way of meeting Harrison Ford is to force an accident with him. The next time the program has the goal of meeting someone, the plan of forcing an accident with that person will immediately be retrieved. This solution is one which would have been difficult to generate out of thin air. Thus the program has learned through its previous serendipity experience.

---

[2] Episodic rules, contained initially in hand-coded episodes (and also induced in RECOVERY3), are discussed in greater detail in Chapter 4.

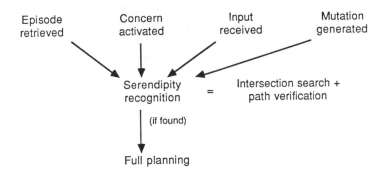

Figure 5.1: Serendipity

5.3.2   Serendipity Recognition

As shown in Figure 5.1, DAYDREAMER attempts to find a serendipity whenever:
(a) an episode is retrieved, (b) a concern is activated, (c) an input state or action
is received, or (d) a mutated possibility is generated. In this section, we address
how DAYDREAMER recognizes and exploits a serendipity in these situations.

In RECOVERY3, the program has an active **RECOVERY** concern to generate a scenario in which she asks Harrison Ford out on a date in the future.
Given the input that Carol Burnett's telephone number is in the UCLA Alumni
Directory, the program recognizes the following connection between these input states and its active **RECOVERY** concern: a subgoal of asking Harrison
out is having a communication path to Harrison; a subgoal of having a communication path to Harrison is calling Harrison on the telephone; a subgoal of
calling Harrison on the telephone is knowing his telephone number; a subgoal
of knowing his telephone number is reading his telephone number in an Alumni
directory; a subgoal of reading his telephone number in an Alumni directory is
the Alumni directory containing his telephone number; a subgoal of the Alumni
directory containing his telephone number is the Alumni directory being that
of the college he attended.

In general, how does the program determine whether a given planning rule
(an inaccessible rule contained in a retrieved episode, an induced rule, or one
corresponding to an input state) is applicable to a given goal? A rule would be
applicable to a goal if that rule could be used to achieve that goal, or if that rule
could be used to achieve some subgoal of that goal, and so on. A rule is thus
applicable to a goal if that rule may be used to achieve an arbitrary subgoal
of that goal. There may be several ways of breaking down each subgoal into
further subgoals. Humans seem to be able to detect such a relationship with a
minimum of effort. What algorithms are available to accomplish this task?

The arbitrary applicability of a planning rule to a goal may be detected by
performing an *intersection search* (Quillian, 1968; Pohl, 1971; Klahr, 1978) in

the *rule connection graph* (Kowalski, 1975) between the given rule and a rule for achieving the given goal: The rule connection graph is composed of *forward connections* and *backward connections* between rules. One rule has a forward connection to another rule for each consequent subgoal of the latter rule which *unifies* (Charniak, Riesbeck, & McDermott, 1980; Robinson, 1965) with the antecedent goal of the former rule. One rule has a backward connection for each consequent subgoal of each other rule having an antecedent goal which unifies with the consequent subgoal of the former rule. Connections refer to the appropriate consequent subgoal. A rule is compiled into the rule connection graph when it is created.

Anderson (1983) has demonstrated experimentally that humans conduct a form of intersection search known as "spreading activation." Since rule intersection graphs are restricted versions of the general graphs constructed and searched in a spreading activation architecture, it is reasonable to hypothesize that *humans construct and search rule intersection graphs* as well. This hypothesis is a topic for future psychological experimentation.

Rule intersection attempts to find paths in the rule connection graph from a given *top rule* to a given *bottom rule*. Quillian (1968) used a breadth-first marking algorithm; Pohl (1971) describes bidirectional search algorithms in more detail. We employ a recursive bidirectional depth-first algorithm: starting from the top rule and bottom rule, the procedure works its way down and up at the same time up to a certain depth limit. The maximum depth cannot be too great, because even with moderate fanout, the search space grows very quickly (i.e., exponentially). So far, this has not proved to be a problem in the current version of DAYDREAMER. Whenever an intersection occurs (with odd or even path length), the complete path is constructed out of the two half-paths and added to a list of found paths. Since there may be cycles in the rule connection graph, the resulting paths may contain cycles. (However, paths are limited to at most two occurrences of a given rule.) Rule intersection search may only use generic planning rules or episodic rules contained in a recent episode. Figure 5.2 shows the path found through rule intersection search in the case of RECOVERY3.

Finding one or more paths through rule intersection search, however, does not mean that a serendipity has occurred. The series of rule applications given by the path might not unify together; that is, a rule might not be applicable to the result produced by the previous rule. A candidate path must be therefore be *verified* for applicability to the specific goal corresponding to the given top rule. As verification is performed, an episode (planning tree) suitable for use in analogical planning is created. Figure 5.3 shows the verification episode which is created in the case of RECOVERY3. The path verification algorithm constructs concrete subgoals starting from the top rule of the path and moving downwards until the bottom rule of the path is reached. The subgoal at each level must unify with the antecedent goal of the rule at that level. Unbound variables are suitably propagated down the tree (i.e., as in *regression* [Nilsson,

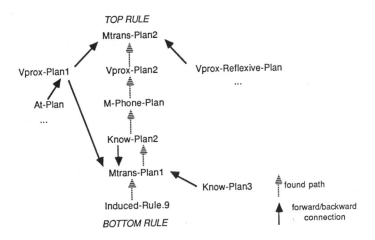

Figure 5.2: Rule Intersection for RECOVERY3

1980]). Episodes associated with rules are also used to supply values for unbound variables. If verification is unsuccessful (because a subgoal fails to unify with a rule), the path and episode are discarded. Otherwise, a serendipity has occurred. In summary, *recognizing a serendipity consists of (a) finding a path through the rule interconnection graph from a rule related to one concern to a rule related to another concern, and then (b) verifying that path.*

5.3.3 Episode-Driven Serendipity

Episodes provide a useful source of knowledge for the generation of creative possibilities. Given an episode and a problem, one may always force a solution to that problem using that episode. *The less an episode is related to a problem, the more bizarre the solution derived from that episode (but the higher the potential for production of a truly new and useful solution).* For example, if one is reminded that the encyclopedia is a source of information, one may generate the idea to look up the movie star's telephone number in the encyclopedia. Although one is not likely to find it there, one may later refine this plan into the more useful one of finding a listing for the movie star in some book that *is* likely to contain his (or his agent's) number, such as a Screen Actor's Guild directory. Seeing a television show about a high-school reunion may give one the idea of calling the movie star's parents and pretending to be an old high-school friend.

An episode which is somehow applicable to a given concern might not be indexed in such a way as to be accessible by that concern. That episode might happen to be accessible only while performing some *other* concern. Therefore, *serendipity recognition is invoked whenever an episode is retrieved* (or derived from input as in RECOVERY3) in order to find inaccessible rules contained

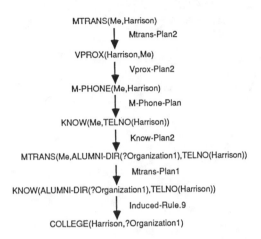

Figure 5.3: Verification Episode for RECOVERY3

in that episode which might be useful in achieving an active concern. Norman (1982, pp. 24-27) reports experiencing such a serendipitous reminding while daydreaming in the shower: A string of associations starting from a recent event led to a forgotten episode containing a plan for purchasing a special kind of photographic slide tray which he was having trouble locating.

5.3.4   Object-Driven Serendipity

The stream of thought does not proceed independently of outside stimuli. As Klinger (1978) has noted (in contrast to earlier conceptualizations of daydreaming as task-irrelevant thought, daydreaming includes thoughts ranging from the most stimulus-dependent to the most stimulus-independent. Physical objects in the environment are especially common triggers and modifiers of stream of consciousness thought (Pope, 1978)—while daydreaming, one will happen to gaze at some object in the room, become reminded of some person, episode, or future plan, and shift to a daydream related to that reminding. You can demonstrate this effect to yourself by spending a minute or so looking around the room at various objects and seeing what thoughts come to you. You will probably experience at least one reminding (followed by a daydream provided you do not cut it short) for each object you look at.

Remindings resulting from objects often depend on, and are useful for, suggesting a solution for an active concern. For example, consider the case in which DAYDREAMER has an active **REHEARSAL** concern for a **LOVERS** goal. The program thus faces the problems of how to initiate a new relationship, how to meet people, and so on. In an empty context, remindings in response to physical objects will most likely be recent episodes involving that object or episodes in which that object is particularly salient. However, *in the context of an active concern, remindings often relate somehow to satisfaction of that concern.*

Remindings result from physical objects in the daydreams COMPUTER-SERENDIPITY and LAMPSHADE-SERENDIPITY (see page 14). One way of generating these remindings might be to store each episode under two indices: a personal goal and a physical object. However, in general an episode might not happen to be indexed under a personal goal to which it could ultimately be applied. For example, the lampshade episode does not involve a **LOVERS** goal; it involves the goal for **ENTERTAINMENT** at a nightclub. Even so, it is indirectly applicable to the **LOVERS** goal through the subgoals of impressing someone and getting someone to be romantically interested. In general, given a physical object, we would like to retrieve episodes in which that object is salient which are related to an arbitrary subgoal of an active concern. Therefore, *(a) episodes are indexed under physical objects salient in those episodes, (b) whenever a physical object is taken as input, serendipity recognition is performed for each episode indexed under that object, and (c) such an episode is only retrieved if it results in a serendipity.* Physical objects are taken as input when DAYDREAMER runs out of things to daydream about; such input may also be forced by interrupting the system.

For example, upon receipt of *Computer* as input, DAYDREAMER finds several episodes indexed under this physical object. It then performs an intersection search from the episodic planning rules in those episodes to the rules associated with active concerns (including **LOVERS**). It finds a path from an episodic planning rule in the computer dating episode (which states that a way of acceptably initiating communication with someone one does not already know is through a dating service) to the rule for achieving **LOVERS**. (The reader may consult Figure 4.8 in Chapter 4 for the specific rules and path.) This path is successfully verified and so a serendipity has occurred and a reminding is produced of the computer dating episode. The episode produced during path verification is added as a new analogical plan for the **LOVERS** goal (actually a subgoal of **REHEARSAL**) and a potential future plan in which DAYDREAMER joins a dating service is produced.

What is the purpose of having episodes be retrieved in response to input physical objects? Why not simply retrieve *all* episodes having inaccessible rules applicable to an active concern? For one thing, people obviously don't perform such a retrieval; this is probably because the number of episodes would be too large to handle. For a similar reason, such a retrieval is not performed in DAYDREAMER. *The use of additional surface-level indices such as physical objects (a) enables appropriate episodes to be retrieved at a more useable rate and (b) introduces a useful random element into the process of problem solving.* Note, however, that recalling episodes at random would in fact accomplish a similar function.

5.3.5   Input-State-Driven Serendipity

A serendipity often occurs when the world state is recognized as applicable in some way to some active concern (F. Hayes-Roth & B. Hayes-Roth, 1979). For example, Shanon (1981) reports the following experience and resulting daydream: On the way back from the supermarket, the subject (S) passes by a public garden; S thinks how it is a shame S cannot pick some flowers for H because it is dark; S then realizes that S can go back to the supermarket to buy some flowers for H. Therefore, in DAYDREAMER, *serendipity recognition is invoked upon receiving an input state or action.* The bottom planning rule for the intersection search is one having a consequent subgoal which unifies with the input state or action.

In LOVERS1 (see page 3), LOVERS2 (see page 9), and LOVERS3 (see page 93), the input state is recognized as applicable to her active concern through the serendipity mechanism. The same mechanism would enable DAYDREAMER to recognize the applicability of the following inputs: someone asking DAYDREAMER out on a date and someone breaking up with his girlfriend.

5.4   Action Mutation

Fanciful possibilities are often generated in daydreaming through arbitrary modification of events or goals. Such variations sometimes lead to accidental discovery of a solution to a problem. For example, in the daydream RECOVERY2 (see page 12), the action Harrison Ford tells me his telephone number is transformed into Harrison Ford tells someone else his telephone number. This person could then tell DAYDREAMER his telephone number. Of course, the movie star has to want to give his number to this person, who also must want to give the number to DAYDREAMER. Suitable plans are generated in order to flesh out this possibility: DAYDREAMER will enlist her friend Karen.

In general, *new possibilities for achieving a goal can be generated by (a) performing a structural transformation on an unachieved action subgoal of that goal, (b) determining whether the resulting action subgoal is applicable to the original goal, and (c) if so, attempting to achieve the new action subgoal.* In this section, we present the mechanism of *action mutation* which enables this process in DAYDREAMER.

5.4.1   The Purpose of Action Mutation

In designing a daydreaming machine there is a conflict between the need to reduce the search space and the need to *increase* the search space: In order to avoid the combinatorial explosion resulting from alternative possibilities, methods must be employed to select from among available alternatives at each point in a daydream. However, if possibilities are cut short, more creative solutions buried within those possibilities may never be found. Therefore, a useful addition to the program is a method for exploring possibilities in a somewhat random fashion and in a quantity which will not overload the program. Thus, *without search reduction, the program would be overloaded with possibilities; with search reduction, the program may miss creative solutions; somewhat random exploration of possibilities enables possible discovery of creative solutions even if search reduction is employed.*

The function of action mutation is to generate and explore possibilities which might not otherwise have been considered by the analogical planning mechanism of DAYDREAMER: Planning in DAYDREAMER consists of repeatedly breaking subgoals down into further subgoals until subgoals that are already satisfied are reached. There are alternative means whereby a given subgoal may be broken down into further subgoals—these means are specified by both generic rules and episodes. Finding a solution to a problem thus consists of a *search* through this space of alternatives. In principle, then, any valid solution to a problem may be found though exhaustive search. However, instead of performing an exhaustive search, the normal planning mechanism of DAYDREAMER employs the following collection of strategies for reducing the potentially large search space:

- *Limitation on the number of attempted rules by evaluation metrics*: Out of all the possible rules for achieving a subgoal, only a certain number of these rules will be attempted. Therefore, some possible plans will be missed.

- *Precedence of satisfied states*: Whenever a subgoal is satisfied by one or more current states, planning rules for achieving those states are not considered.

- *Precedence of existing analogical plans*: If an analogical plan has already been selected for a subgoal and the applicability of this analogical plan is verified, other alternatives for this subgoal will not be generated. (However, if this subgoal fails, the entire analogical planning subtree will fail, and alternative plans for achieving the subgoal at the root of the analogical planning will then be considered).

Plans generated through action mutation thus *are* implicit in existing generic rules and episodes. Nonetheless, such plans still might have not been found by the analogical planning mechanism.

Although the above strategies have the advantage of preventing the program from performing an overly expensive search, they have the disadvantage of sometimes causing the system to miss possible solutions. However, there is considerable psychological evidence that people as well employ strategies for reducing search which prevent them from seeing many possible solutions (J. L. Adams, 1974; Wertheimer, 1945, pp. 169-192; F. C. Bartlett, 1932). The purpose, then, of such strategies as action mutation and serendipity is to counter the "mental blocks" which sometimes result from the use of search reduction strategies.

5.4.2   Strategies for Action Mutation

The following strategies are employed to mutate actions: *permute objects, generalize object*, and *change action type*.

The permute objects strategy juggles around object entities contained in the action, such as persons and physical objects. For example, Harrison Ford tells me his telephone number might be mutated into I tell Harrison Ford my telephone number. If there is only one object in an action, no mutations result from this strategy. If there are two objects, there is one mutation: The objects are switched. If there are three objects, there are five mutations. In general, there will be $n! - 1$ mutations of an action containing n objects.

The generalize object strategy substitutes one of the object entities contained in the action with an abstraction consisting of: (a) a variable whose type is that which would normally fill the slot or slots containing the original element, and (b) the requirement that the variable not be bound to the original element. For example, Harrison Ford tells me his telephone number might be mutated into Someone else tells me Harrison Ford's telephone number or Harrison Ford tells someone else his telephone number (as in RECOVERY2).

The change action type strategy modifies the type of the action and then adjusts the fillers of the slots to be of appropriate type for the new action. In the current version of DAYDREAMER, this mutation is only defined for actions of type **MTRANS**, **ATRANS**, and **PTRANS**. For example, Harrison Ford tells me his telephone number might be mutated into Harrison Ford gives me an object containing his telephone number or Harrison Ford moves an object containing his telephone number toward me. These mutations might lead, respectively, to daydreams in which the movie star gives his card to DAYDREAMER, or in which the movie star accidentally leaves behind his wallet.

Arbitrary mutation of an action can lead to useless or nonsensical possibilities. In order to avoid spending too much time generating and exploring useless possibilities, the serendipity mechanism is employed to evaluate generated mutations. The bottom planning rule for the intersection search is one having a consequent subgoal which unifies with the mutated action. If serendipity is successful, an episode suitable for use in analogical planning is created and the mutated action is fleshed out through analogical application of that episode.

5.5   The Phenomena of Incubation and Insight

From the introspective reports of Poincaré, Helmholtz, and others, Wallas (1926) abstracted four stages of the creative process: One sits down to solve a problem, exerts much effort, but nonetheless reaches a stumbling block (the "preparation" stage). Then one takes a break—either a period of work on a different problem or one of relaxation (the "incubation" stage). At some point, possibly after resuming work on the original problem, a likely solution presents itself—sometimes in a flash of insight (the "illumination" stage). Next one must verify and complete the potential solution (the "verification" stage). One often has the feeling that the break from the problem has allowed things to "settle" in one's mind, making it easier to find a solution the next time the problem is examined. Wallas (p. 81) proposes that the four stages overlap in the daily stream of thought as different problems are considered.

Several theories of this phenomenon have been proposed. Poincaré (1908/1952) proposes that the preparatory work mobilizes a collection of "atoms" out of which a solution might be constructed, although one has not yet been found. These atoms continue moving after one has stopped working on the problem—they enter into various combinations with each other and with other atoms which they happen to collide into. Eventually a good combination occurs, which must always be verified through conscious thought. Hofstadter (1983) implemented a similar molecular analogy in his JUMBO program for finding English anagrams. Wallas (1926) proposes that nonconscious or "fringe-conscious" work may be performed during incubation.[3]

Varendonck (1921) offers an explanation in terms of forgotten daydreaming activity: "Consciousness forgets all the preparatory work, retains only the correct solution, and has rejected the numerous incorrect ones ..." (p. 140). Woodworth and Schlosberg (1954) propose that when the thinker is blocked by a false set of assumptions in the preparatory stage, incubation enables those interferences to drop away, allowing the thinker to find a more correct direction for solution of the problem. Simon (1966) argues that the incubation phenomenon may be explained in terms of familiarization and selective forgetting: In attempting to solve a new problem, working memory may be overloaded with the many alternative, as yet unsuccessful, methods for solving the problem. During this initial effort, however, information toward solution of the problem has been added to long-term memory. When the problem is later reexamined with an uncluttered working memory, solutions may be constructed from scratch with the benefit of the new information in long-term memory. In this case, a solution to the problem is more likely to be constructed right away.

Rothenberg (1979) suggests an explanation in terms of repression:

> The creator is blocked on a problem because some conflict or anxiety is involved in its unconscious meaning or structure. He turns away from the task and, when in a relaxed ego state, he overcomes the anxiety and solves the problem through creative cognitive processes.

---

[3]Theories of nonconscious work are to be distinguished from theories which refer to origins of creativity in the Freudian unconscious or in ungratified infantile wishes (S. Freud, 1908/1962; Kris, 1952; Rothenberg, 1979, p. 41).

Another contributory factor may be that the time lapse and distraction allows for inessential elements of the problem to drop away. (p. 400)

Our explanation of the illumination phenomenon—including the insightful idea and corresponding emotion—is in terms of serendipity recognition during daydreaming or processing of external input: At any time, DAYDREAMER is involved in several active concerns, each motivated by emotions. The concern with the highest motivation is the one that runs, while other suspended concerns will run as soon as the current concern is completed or if their motivation becomes higher than that of the current concern, and still others are tentatively halted. We hypothesize that *human insight corresponds to the detection of a potential solution or partial solution relevant to an active concern through serendipity in response to input environmental objects and events, recalled episodes, and daydreamed possibilities* (such as mutations). The potential solution is then incorporated into the concern. Our view is similar to Ericsson and Simon's (1984, pp. 162-164) hypothesis that insight results from direct recognition of existing structures in long-term memory.

We hypothesize that *the emotion which accompanies insight in humans is necessary in order to divert attention to the new concern*. A similar mechanism is employed in DAYDREAMER: When a serendipity occurs involving a given concern, a *surprise* emotion is attached to that concern as an additional force of motivation, and the concern is made runable if it was not previously runable.

There is a potential problem: People are often unaware of any antecedent thoughts responsible for the insight (Wallas, 1926, p. 95-96) analogous to the real and imaginary objects and events which result in serendipity in DAYDREAMER. Perhaps one tends to focus on the new results and forget how they were attained (James, 1890a, p. 260); perhaps the antecedent thoughts were just below the level of consciousness (Varendonck, 1921, p. 213). In any case, serendipity at least accounts for many instances of insight in which an internal or external antecedent may be identified.

The value of incubation (waiting) and partial preparation may be explained as follows: DAYDREAMER may perform some processing toward a concern and then, before that concern is completed, switch to another concern (this occurs when the emotional motivation of another concern becomes higher than the motivation of the current concern, or the current concern seeks to perform an external action and the system is not currently in performance mode). Since a problem often may not be solved until more information is available, it is useful to stop working on a concern until such a time as new information, stimuli, or concrete objects for accomplishing a plan may be available. DAYDREAMER will then later switch back to the original concern upon recognition that something currently under focus is relevant to that concern (or if, for some other reason, the motivation of that concern becomes greater than the motivation of all other concerns). Therefore, the processing of a concern proceeds in an *incremental* fashion: A concern is performed for a while; then it is set aside in an intermediate state; then that concern is taken up again starting from that intermediate state when something relevant to that concern is discovered; the concern is set aside again; and so on until the concern is completed. Each time the concern is examined, the partial solution is revised—moved closer to a final solution in light of the most recent information.

In addition, another incremental effect is achieved by the fact that previous solutions are stored in episodic memory. When a solution is later attempted for a similar or different problem, previous episodes are available as a starting point.

5.6 Previous Work in Problem Solving

Our view is that creative problem solving is often carried out by humans through daydreaming. (Other ways of problem solving—such as working out a list of actions on paper—are, of course, considered distinct from daydreaming.) In the field of artificial intelligence, there has been much previous work on problem solving (also called planning). In this section, we review some of this work and compare it to DAYDREAMER.

5.6.1 GPS and STRIPS

Early work in artificial intelligence (Feigenbaum & J. Feldman, 1963; Newell, J. C. Shaw, & Simon, 1957) recognized the importance of *search* in solving problems: A problem may be characterized by an initial state, a description of a goal state, and a collection of operators for transforming one state into another. Solving the problem then consists of finding a sequence of operators for transforming the initial state into a goal state. That is, a search is performed through the *problem space*: The set of states reachable (via a sequence of operator applications) from the initial state. The GPS program (Ernst & Newell, 1969; Newell & Simon, 1972) introduced the technique of *means-ends analysis*: Finding differences between the current state and the goal state and selecting operators known to reduce those differences.

The STRIPS program (Fikes & Nilsson, 1971) views problem solving as a process of search through the space of world models. That is, problem solving is regarded as finding a sequence of operators which transform a given initial world model into a world model in which a given goal is true.

DAYDREAMER can be viewed as conducting a search through the space of *scenarios*. Each scenario consists of a sequence of events resulting in some world situation (starting from some initial world situation) and represents a possible solution—partial or complete—to a problem. One set of operators is similar to means-ends analysis: apply a planning rule to achieve (reduce) an active subgoal (difference). The similarity to GPS and STRIPS ends there. Further operators are provided by all the mechanisms of DAYDREAMER from the daydreaming goal strategies to serendipity.

Specifically, DAYDREAMER differs from GPS and STRIPS in several important respects. First, DAYDREAMER addresses the problems of *inaccessibility*, *fanciful solutions*, and *accident*: Consider a domain with a large number of operators, such as the domain of interpersonal relations and common everyday occurrences. GPS and STRIPS, which conduct a search using a set of perfectly accessible operators, fail as models of problem solving in such a domain. All operators cannot be equally accessible at any point: In a computer program, too many operators would overwhelm the system, while in a human, too many operators would exceed the capacity of working memory. Mechanisms must therefore be employed to reduce the number of operators available at a time. But as a result of such mechanisms, potential solutions may be missed. In order to find these missed solutions, additional strategies must then be employed to uncover missed solutions, as well as to recognize when a missed solution has accidentally been stumbled upon. These problems are addressed by DAYDREAMER: (a) available operators are pruned (via episodic rules, for example), (b) existing solutions are perturbed to generate fanciful solutions (via mutation, for example), and (c) accidental solutions are recognized (via serendipity).

Second, DAYDREAMER addresses the problem of *multiple, ongoing goals*: GPS and STRIPS take as input a problem to be solved, solve the problem, and then stop. Intelligent agents must be able to (a) recognize and anticipate problems on their own, (b) handle multiple problems, and (c) decide which problem to concentrate on at any time. These issues are addressed in DAYDREAMER through the use of personal goals, daydreaming goals, and emotion-driven control.

It is true that ongoing planning to achieve and maintain a collection of internal goals may in principle be modeled in terms of the traditional search paradigm for problem solving: For example, one might add formulas or rules which imply certain internal goals in certain world models and a means for directing the search in order to achieve those goals. However, classical work in problem solving has not recognized and elaborated upon this possibility.

Third, DAYDREAMER addresses the role of *experience*: GPS and STRIPS do not (a) use previous solutions to reduce search in generating new solutions, or (b) use previous experiences to generate possible sequences of world events. DAYDREAMER is capable of both through the mechanism of analogical planning: (a) if a creative solution is found through serendipity, that solution is stored so that the program does not have to chance upon the same sequence of accidental events in order to generate that solution—or a similar one—in the future; (b) previous experiences are used to generate continuations of an ongoing daydream (as when Harrison Ford goes to Cairo in RATIONALIZATION1.)

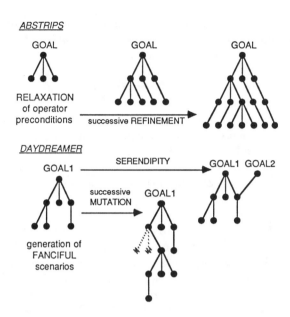

Figure 5.4: ABSTRIPS vs. DAYDREAMER

5.6.2 ABSTRIPS

The ABSTRIPS program (Sacerdoti, 1974), which is based on STRIPS (Fikes & Nilsson, 1971), employs the method of "hierarchical planning" to reduce search. In this approach, a plan for achieving a goal is generated which ignores details (such as how to satisfy operator preconditions); this plan is then used to guide the generation of a plan at a finer level of detail; in turn, the resulting plan is used to guide the generation of a plan at an even finer level of detail; and so on.

In contrast, the various strategies for the generation of fanciful scenarios in DAYDREAMER—such as mutation and daydreaming goal strategies—actually *increase* the amount of search (in order to discover hidden solutions). In DAYDREAMER, plans for achieving a goal are not successively refined into complete plans, but rather are destructively *altered* into complete plans for achieving that goal, or perhaps even some other goal. The difference between the two approaches is illustrated schematically in Figure 5.4.

5.6.3 NOAH

The NOAH planning program (Sacerdoti, 1977) conducts a search through the space of possible plans (or solutions) rather than through the space of world models. Plans are represented as "procedural nets" which specify a partially ordered sequence of actions at multiple levels of detail. "Critics" are responsible for finding and correcting problems or redundancies (by reordering nodes or eliminating actions) in a plan. For example, if the goal is to paint the ceiling and to paint the ladder, NOAH will first expand this goal into two concurrent child nodes: (1) paint the ceiling and (2) paint the ladder. These nodes will be broken down further into two concurrent sequences: (1a) get paint, (1b) get the ladder, (1c) apply paint to the ceiling, and (2a) get paint, (2b) apply paint to the ladder. The "resolve conflicts" critic will then notice that (2b) (applying paint to the ladder) effectively deletes a precondition of (1c) (applying paint to the ceiling)— that of having a ladder. The critic therefore modifies the procedural net so that (1c) is performed before (2b). The "eliminate redundant preconditions" critic combines (1a) and (2a) into a single action. Other critics are called "use existing objects," "resolve double cross," and "optimize disjuncts." Domain-specific critics may be defined as well.

The critics of NOAH enable it to perform some advanced plan optimizations which DAYDREAMER cannot perform. However, NOAH still suffers from the problems described above regarding STRIPS and GPS. Although NOAH is able to refine a slightly incorrect or incomplete solution incrementally, it is unable to generate and exploit highly fanciful, distortive, or seemingly unrelated solutions such as those obtained through mutation and the various daydreaming goal strategies in DAYDREAMER.

5.6.4   PANDORA

One theory of problem solving which does not suffer from all of the shortcomings of traditional views of problem solving is that of Wilensky (1983). His theory, which is partially implemented as the PANDORA computer program (Faletti, 1982), is similar in many respects—but different in others—to DAYDREAMER.

The planner proposed by Wilensky consists of the following components: a *goal detector, plan proposer, projector*, and *executor*. The goal detector activates goals in appropriate situations; this is accomplished using situation-goal pairs similar to the goal activation inference rules of DAYDREAMER. The plan proposer selects plans for achieving a goal. Specific plans have priority over more general ones: If a proposed plan fails, a more general one is then attempted. This is implemented by compiling rules into a hierarchy. The projector tests a plan by simulating its execution in a hypothetical world—that is, by daydreaming. This is accomplished via a mechanism similar to that employed by DAYDREAMER in daydreaming mode: The actions of a plan are asserted into a hypothetical data base and the consequences of those actions, such as preservation goals, are noticed by the goal detector. Presumably, the goal detector may employ a sequence of inferences in noticing a goal, just as in DAYDREAMER. The executor carries out the actions of a plan as specified by a "task network."

Wilensky's emphasis is on general "meta-planning" in which knowledge about the planning process itself is represented in the same form as knowledge of the domain. The same planning mechanism which constructs plans for achieving goals in the domain may thus be used to achieve "meta-goals," or goals of the planning process. Meta-planning is not a focus of the present work; thus meta-planning knowledge is simply hard-coded into the planning mechanism of DAYDREAMER.

The operation of meta-planning in Wilensky's theory of planning is illustrated in the following example: Suppose one has the goal of getting a newspaper from outside while it is raining. The plan proposer first selects the most specific plan, that of going outside and carrying the paper back in. The projector then simulates the execution of this plan (as in daydreaming) and discovers that it results in one getting one's clothes wet. The goal detector thus activates a preservation goal to keep one's clothes dry. Since this preservation goal resulted from planning to achieve another goal (that of getting the newspaper), the **RESOLVE-GOAL-CONFLICT** meta-goal is activated to resolve the conflict between these two goals.

The most specific plan for achieving this meta-goal is **USE-NORMAL-PLAN**, which selects a stored plan for resolving this specific type of goal conflict. Thus a plan for staying dry when going outside while it is raining, if any, is selected. In fact, such a plan exists: putting on a raincoat.

However, if no such plan were found or if such a plan were to fail, **TRY-ALTERNATIVE-PLAN**, the next most specific plan for achieving the **RESOLVE-GOAL-CONFLICT** meta-goal, would be employed. This plan involves selecting an alternative plan for achieving the initial goal which does not contain an action which led (via one or more inferences) to the preservation goal. Since going outside is an action which leads to the dryness preservation goal, a plan not containing this action, if any, is selected. An example of such a plan is getting one's dog to fetch the newspaper.

A problem with this strategy, which Wilensky does not mention, is that a stored plan might not specify all of the actions which will eventually be necessary to realize that plan: In order to carry out an action of the plan, an unsatisfied precondition of that action might first have to be achieved; this might in turn require performance of some other action not specified in the original plan. In general, carrying out a given plan may result in an arbitrarily deep tree of subgoals. For example, in order to read a newspaper, the plan might be to have physical control over the newspaper; in order to have physical control over the newspaper, the plan might be to grab the newspaper; in order to grab the newspaper, the plan might be to be near the newspaper; in order to be near the newspaper, the plan might be to go to the location of the newspaper. Thus the plan for reading the newspaper does not directly mention that one must go to the location of the newspaper, since that is only one out of many possibilities: Another plan for being near the newspaper, for example, might be to ask someone to bring it to you. Thus the method for alternative plan selection suggested by Wilensky does not work unless plans are stored in fully expanded form. However, the number of possible expanded plans for each type of goal is likely to be quite high: A combinatorial explosion results whenever there is more than one plan for achieving each subgoal.

If **TRY-ALTERNATIVE-PLAN** is unsuccessful, a final strategy called **CHANGE-CIRCUMSTANCE** is employed: This plan involves finding states of the world which lead to the preservation goal and generating a plan to alter those states. For example, if there is nothing between the rain and an object, that object will get wet; a plan for avoiding getting one's clothes wet is therefore to place some appropriate object, such as a newspaper, in between the rain and one's clothes. The **CHANGE-CIRCUMSTANCE** strategy is similar to our **UNDO-CAUSES** strategy for **REVERSAL**. Wilensky notes that **CHANGE-CIRCUMSTANCE** is a method whereby stored plans for resolving specific goal conflicts (such as putting on a raincoat to keep from getting wet) may be derived. However, unlike DAYDREAMER, plans derived via this strategy are not stored for use in planning future similar situations—these plans must be rederived each time. Thus Wilensky does not exploit the potential for ongoing *revision* of plans (which we argued in Chapter 1 is important for creativity).

Unlike traditional problem-solving systems, but like DAYDREAMER, Wilensky's planner infers its own goals. However, unlike DAYDREAMER, Wilensky's planner lacks a control mechanism for: (a) selecting which of many active goals to form and execute plans for at any given time, and (b) interleaving the formation and execution of plans. Wilensky's planner copes with multiple goals in part through meta-planning: For example, meta-themes such as **DON'T WASTE RESOURCES** result in efficient plans for achieving several goals. However, although meta-planning can schedule the execution of a few goals, it cannot in general schedule the execution of all future goals. Furthermore, although time constraints on the execution of plans can in principle be handled through meta-planning, some goals may be so urgent that there simply is no time to perform such meta-planning. Consequently, a planner must have an appropriate mechanism for interleaving the formation and execution of plans for various goals and for directing attention to those current and future goals that are deemed most important at any given time. Emotion-driven control is the mechanism employed in DAYDREAMER.

Wilensky's planner creates a task network through planning whose actions are later executed. In DAYDREAMER, future plans or *intentions* are formed and carried out via two mechanisms: (a) the generation, storage in memory, and application, via analogical planning, of daydreamed plans, and (b) partial completion of a concern, suspension of that concern, and resumption of that concern. The stored plans and suspended concerns of DAYDREAMER are similar to Wilensky's task networks, with the following significant difference: Wilensky's planner must construct a plan from scratch each time a problem is encountered, even if that problem is similar to some previously solved problem, while in DAYDREAMER, once a plan is formed, that plan may be applied to any future similar problem.

The rules of DAYDREAMER provide a more economical representation of planning knowledge that those of PANDORA. While in DAYDREAMER a single rule is used to represent the fact that performing a certain action results in a certain goal state, in PANDORA two rules must be employed: one for representing the fact that that action may be employed as a plan for achieving a goal whose objective is that state, and another for representing the fact that that action results in that state. For example, the two PANDORA rules (Wilensky, 1983, p. 139):

```
(PlanFor
  (Goal
    (Goal
    (Planner ?X) (Objective (At (Actor ?X) (Location ?L)))))
  (Action
    (Drive-vehicle
      (Driver ?X) (To ?L) (Vehicle ?V)))))
(Result
  (Action
    (Drive-vehicle
      (Driver ?X) (To ?L) (From ?FL) (Vehicle ?V)))
  (State
    (At (Actor ?X) (Location ?L))))
```

may be expressed as the single DAYDREAMER rule:

```
(RULE subgoal (DRIVE-VEHICLE actor ?Person1
                             from ?Location1
                             to ?Location2
                             vehicle ?Vehicle1)
      goal (AT actor ?Person1
               obj ?Location2))
```

Furthermore, action preconditions, causal inferences, and the specification of how a complex action—such as driving a car—is broken down into more simple actions, are represented in their own unique way in PANDORA, while in DAYDREAMER each of these forms of knowledge is represented in the same way—as a rule (with the exceptions that rules whose goal is an action have slightly different semantics, and that rules may be restricted to application only as a plan or as an inference). This uniformity of representation in DAYDREAMER simplifies the operation of the analogical planning mechanism employed in learning.

Both Wilensky's planner and DAYDREAMER evaluate a scenario based on the sum of the values of the goals in that scenario as well as according to crude judgements of the plausibility of the scenario.

Wilensky presents advanced meta-planning techniques for handling forms of goal competition, goal subsumption, goal concord, and goal overlap which are not dealt with in the current DAYDREAMER. In addition to the issue of plan generation, Wilensky is concerned with the issue of plan *understanding*—that is, how knowledge of planning may be used to explain the behavior of characters in stories.

Wilensky does not describe his planner at a level of detail sufficient to enable one to implement the mechanism. It is not always clear which aspects of his theory of planning have been implemented. In effect, DAYDREAMER, as described in the present work, provides an explicit implementation of many of the mechanisms only hinted at by Wilensky. Naturally, we are indebted to Wilensky's previous work.

In our view, daydreaming and problem solving are closely related phenomena; humans perform problem solving through daydreaming. Although Wilensky (1983) is "interested in how people work out a plan in mundane situations ..." (p. 7), he does not address many of the important aspects of human planning which are dealt with in DAYDREAMER: the role of emotions and motivation, the role of episodic memory, the role of serendipity and mutation, and the relationship between planning and learning.

5.6.5   Opportunistic Planning

An idea related to problem solving through serendipity is the idea of opportunistic planning (B. Hayes-Roth & F. Hayes-Roth, 1979). In opportunistic planning, the planner takes advantage of world states which happen to be true, thus (a) saving it the effort of achieving those states, or (b) enabling it to achieve a goal that could not be achieved previously because those states could not be attained at all. A similar exploitation of fortuitous states was accomplished in PLANEX (Fikes, Hart, P. E., & Nilsson, 1972).

As illustrated in Figure 5.5, opportunistic planning is a special case of serendipity. Whereas in opportunistic planning the world state must be identical to the subgoal objective, in serendipity the world state may be related to the subgoal via (a) parameter substitution and (b) a chain of planning rules. Opportunistic planning is the same as what we have been calling fortuitous subgoal success.

5.7   Previous Work in Creativity

This section reviews previous theories of creativity implemented as computer programs.

5.7.1   TALE-SPIN

The TALE-SPIN program (Meehan, 1976) generates simple stories in English through the simulation of the goal-directed behavior of multiple characters. The stories generated by TALE-SPIN are determined by various parameters supplied as input: the characters of the story—such as fox, crow, ant, and

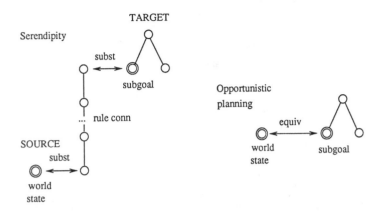

Figure 5.5: Opportunistic Planning vs. Serendipity

canary—one of which is designated as the main character, a problem of the main character—such as hunger or thirst—which becomes the motivating topic of the story, physical objects such as breadcrumbs, cheese, berries, and water, personality traits and relationships of the characters, beliefs about the traits of other characters, and the initial knowledge states of the characters.

The functions of daydreaming and writing stories are quite different: Whereas daydreaming is a private activity concerned mostly with personal learning and emotion regulation, an author writes a story to entertain, inform, or otherwise communicate with the reader. A daydream is a production of the moment; it is generated on the fly; it wanders freely from topic to topic and need not be comprehensible to others. A story, on the other hand, is a deliberate, finished product intended for others; it is determined in part by the literary or artistic goals of the writer.

Nonetheless, daydreams are often story-like (consider revenge daydreams, for example), just as stories are often daydream-like (consider the later works of James Joyce, for example). The similarities in content between stories and daydreams arise because daydreaming is one source of ideas for stories, or, for that matter, any creative output.

What similarities and differences are there between DAYDREAMER, a program intended to address stream of consciousness thought, and TALE-SPIN, a program intended to address story generation? The basic generation mechanisms of the two programs are rather similar: Both employ planning, forward inferencing, and planning by one character to modify the mental states (and thus behavior) of other characters. Facts in TALE-SPIN's global, objective data base represent physical states of the world and beliefs of each of the characters. In DAYDREAMER, all facts in a context are beliefs of the daydreamer; some of those are beliefs about the physical world, while others are beliefs about the beliefs of other persons. If one views the global data base of TALE-SPIN as the beliefs of the storyteller, the distinction almost dissolves. The one difference is that the "storyteller" in DAYDREAMER—the daydreamer—is permitted to have goals and carry out actions in the world. In TALE-SPIN, planning and inference rules are represented as Lisp procedures rather than in a declarative form such as that employed by DAYDREAMER.

In DAYDREAMER, only the daydreamer is able to simulate hypothetical future events and make decisions based on those simulations. In TALE-SPIN, each character is able to perform limited hypothetical simulations: Specifically, the consequences of performing an action may be simulated in order to decide whether or not to perform that action. This is accomplished by running forward inferences of an action in a hypothetical data base from the point of view of a given character. However, this mechanism does not simulate the further planning behavior of other characters, as in the forward other planning of DAYDREAMER. Furthermore, since this mechanism is invoked to evaluate a single action, it does not enable debugging of the larger plans in which those actions are embedded. Thus characters in TALE-SPIN do not perform detailed rehearsal in order to plan for future events analogous to that of DAYDREAMER.

Since the output of the global simulation performed by TALE-SPIN is intended to be a story rather than a daydream, this simulation is never rolled back in order to consider alternative possibilities. In order to generate different stories, one simply runs the program again. The course of each story in TALE-SPIN is determined exclusively by the initial goals, traits, and relationships of the characters. That is, there is nothing in TALE-SPIN analogous to daydreaming goals and emotions for directing the course of the simulation according to some higher-level purpose, such as a point which the author wishes to convey to the reader in the case of story generation. Even if TALE-SPIN were able to roll back in order to produce alternative stories it would have no criteria by which to evaluate those alternatives. It takes a reader to appreciate any serendipitous sequences of events which may be generated by TALE-SPIN; these go unnoticed by the program, which merely simulates. Thus TALE-SPIN is quite limited as a model of creativity.

TALE-SPIN does not address the following other issues dealt with in DAYDREAMER: learning from experience, episodic memory storage and application (which is critical to the *revision* of stories), future plan formation and use, regulation of emotional state, and motivation by emotions.

5.7.2 AM and EURISKO

Lenat's (1976, 1983) AM program explores new mathematical concepts under the direction of a collection of heuristic rules. Rather than generating proofs, AM produces mathematical conjectures which it tests empirically. The control mechanism for AM is based on an ordered list of tasks, called an "agenda." Each task contains a list of reasons for performing that task (such as "there is only one known generalization of primes so far") and associated numerical values. Although the reasons are called "symbolic," the only semantic information derived from a reason is whether it is identical or different to another reason—this enables the system to assign low value to duplicate reasons. The priority of a task is calculated as a function of these values, as well as values associated with the task's action, slot, and concept. The control mechanism consists of repeatedly picking the highest priority task off the agenda and executing it, subject to time and space resource limitations.

Although in DAYDREAMER the motivating emotions (corresponding to resolved goals) of daydreaming goal concerns might be considered as "reasons," the assignment of priorities is not in general based on reasons. Rather, priority is initially determined by goal activation rules containing intrinsic importances, and later modified as a result of serendipity, rationalization, and side-effect personal goal successes and failures which occur in the concern. A similar effect to that of AM is achieved in that concerns of greater "interest" are concentrated upon.

The exploratory activity of AM involves the fortuitous discovery of new concepts and relationships inherent in the initial set of concepts and heuristics. For example, the concepts of addition and multiplication, which arise via different routes, are later noticed to be related (specifically, AM notices that $n + n = 2n$). Since everything in AM can be considered a serendipity, the notion of serendipity becomes diluted. Unlike DAYDREAMER, AM cannot trigger a serendipity in response to external input. AM has no fixed goal other than to explore concepts and relationships which it finds interesting. Serendipity in DAYDREAMER, on the other hand, seeks to model serendipity in human daydreaming, and is thus concerned with finding relationships which have an impact on active personal and daydreaming goals chosen from a fixed set (however, other mechanisms and strategies—such as fanciful planning, other planning, daydreaming goals, and mutation—enable the program to generate a variety of fanciful daydream scenarios which may have no direct application to any particular personal goal).

The major differences between AM and DAYDREAMER result from their dissimilar domains and tasks: AM seeks to discover new concepts in the domain of mathematics, while DAYDREAMER seeks to discover new plans in the domain of events in the world involving people and their relationships. Like DAYDREAMER, AM has a "dynamic memory" (Schank, 1982)—it modifies its memory in a way that affects future behavior. However, the memory of AM contains *concepts* (consisting of definitions, examples, generalizations, operations, and so on), while the memory of DAYDREAMER contains *episodes* or sequences of previous real or imaginary world actions and states.

Much of the success of AM in finding new mathematical concepts was due to the fact that the characteristic functions of concepts were represented as Lisp code and that syntactic mutations on this code—such as elimination of AND clauses, changing an AND to an OR, and so on—turned out to yield mostly useful new concepts (Lenat & J. S. Brown, 1984): the Lisp representation was ideally *matched* to the mathematical domain. The interpersonal domain of DAY-DREAMER presents a more difficult problem in representation and discovery using those representations than the mathematical domain of AM—reasoning about world events is more complex than reasoning about mathematical objects and operations: In the mathematical domain, an experiment yields definite results—for example, if a concept's definition predicate returns true when applied to an object produced by an experiment, it is known with complete certainty that the object is an instance of that concept. However, the results of an experiment conducted in the interpersonal domain of human daydreaming are not definite—asking a person out on a date, for example, can have any number of responses depending on whether the person is interested, is currently involved with someone, is free that night, and so on. Furthermore, even an experiment carried out in the external world is not assured of definite results: If one is turned down after asking someone out on a date, one might not know *why* one was turned down.

In general, the interpersonal domain is far less formal and certain, and more complicated as a result of its concern with the mental states (beliefs, attitudes, goals, and emotions) of others (and other events in the external world). Finding a correct and complete set of rules for this domain would be tantamount to a complete unraveling of human cognition. Given our current state of knowledge on this topic, we have to settle for part of the story in DAYDREAMER.

Once AM had built up many higher-level concepts, the heuristics it had were no longer very useful in the resulting domain. That is, it had no common sense about number theory and tried, for example, to discover numbers that were both odd and even. EURISKO (Lenat, 1983) is a descendant of AM which is able to discover new heuristics. Initial heuristics given to the program are in fact heuristics for discovering heuristics. However, some heuristics for mathematical functions turned out to be applicable to heuristics as well. For example:

> IF the results of performing f are only occasionally useful
> THEN consider creating new specializations of f

Discovery of heuristics in EURISKO would correspond in DAYDREAMER to the discovery of new strategies for daydreaming (such as daydreaming goals and meta-plans), which is left as a topic for future research.

5.7.3 BACON

BACON (Langley, Bradshaw, & Simon, 1983) is a program which discovers empirical laws. The strong point of BACON is its ability to induce compact formulas which characterize a set of numerical data using domain-independent heuristic rules. However, BACON is not particularly successful as a model of human creative processes. For example, although BACON discovers Archimedes law of displacement, it arrives at this formula via a path quite unlike that which Archimedes is supposed to have taken. As the legend goes, Archimedes was entering the tub to take a bath when it dawned on him that the volume of the displaced water is equal to the volume of his submerged body (Langley, Bradshaw, & Simon, 1983, p. 315). This somewhat accidental discovery is similar to the object-driven serendipities modeled by DAYDREAMER. In contrast, BACON models the discovery simply by taking as input a collection of numerical data and inducing a formula which fits those data. BACON has no representations for water, objects, containers, or displacement.

Chapter 6

Daydreaming in the Interpersonal Domain

In order to daydream in a certain content area, DAYDREAMER must have an in-depth representation of that area. People daydream about a variety of things from everyday concerns to technical matters to unlikely science-fiction-like adventures. Some of the more common topics of daydreaming are human relationships, sexual activities, good fortune, career achievement, heroic feats, hostility, and guilt (J. L. Singer & Antrobus, 1972). We have singled out the natural and frequent daydream topics of interpersonal relations (J. L. Singer & McCraven, 1961)[1] and employment for detailed representation in DAYDREAMER.

Dyer (1983a) and Schank and Abelson (1977) have previously addressed the representation of interpersonal relationships (such as friends, lovers, and so on) for use in story understanding. Although the representations for understanding and planning overlap to some degree (Wilensky, 1983), the emphasis is different: Previous research in story understanding has been concerned with tracking existing relationships between story characters in order to generate expectations for connecting up the events of a story. In contrast, we address how an individual takes actions in order to initiate, maintain, and terminate relationships with others. We also provide an implementation of Heider's (1958) analysis of attitudes—positive or negative evaluative judgments toward a person, object, or idea—which are important in interpersonal relations.

In this chapter, we first review the representation of simple states, mental states, and actions. Second, we discuss extensions to the basic planner required for planning in the interpersonal domain. Third, we discuss the representation

---

[1]Although the domain of interpersonal relations might be taken to include erotic fantasies, which are a common aspect of daydreaming (Stoller, 1979; Hariton & J. L. Singer, 1974; Friday, 1973; J. L. Singer & Antrobus, 1972), this level of representation is beyond the scope of the present work.

of attitudes. Fourth, we discuss the representation of interpersonal relationships. Fifth, we present some strategies for generating fanciful scenarios in the interpersonal domain.

It should be pointed out that the representation elaborated here presents a somewhat simplified and idealized view; this is also true of other current artificial intelligence representations of natural phenomena. Some readers may find our representations too stereotypical to capture the richness of what is probably the most complicated, varied, pleasurable, and difficult area of human experience: human relationships.[2] However, the purpose of constructing representations is in fact to capture certain generalizations—stereotypes—which humans have, both in order to construct theories of human cognition and in order to construct intelligent computer programs. The representations provided here are a first approximation to these stereotypes.

6.1 Basic States and Actions

The basic representation in DAYDREAMER for states and actions in the world is based on Schank's (1975; Schank & Abelson, 1977) *conceptual dependency* (CD) representation. In this section, we review CD and present extensions to CD employed in DAYDREAMER.

6.1.1 Conceptual Dependency Representation

CD was originally developed to enable computer programs to represent the meanings of English sentences (Schank, 1975). There are two types of CD conceptualizations (often referred to simply as CDs): *actions* and *states*. An action CD consists of a *primitive action*, an *actor* (the animal performing the action), an *object* (the animal or physical object on which the action is being performed), a *direction* of the action (including origin and destination), and an optional *instrument* for the action. A state CD consists of a *state*, a current *value* of that state, and an *object* (which is in the given state with the given value). CD conceptualizations are connected via causal links (dependencies). Some primitive actions of CD (Schank & Abelson, 1977) are:

ATRANS The abstract transfer of possession, by an animal, of a physical object from one animal to another; an abstraction of giving, taking, buying, selling, and so on.

MTRANS The mental transfer of information, by an animal, from an animal or physical object to an animal; an abstraction of talking, reading, gesturing, and so on.

---

[2]Quinn (1981) presents some alternatives to the goal subsumption model of relationships (Schank & Abelson, 1977) adopted here.

PTRANS The physical transfer from one location to another, by an animal or physical object, of an animal or physical object; an abstraction of walking, driving a car, riding a bicycle, and so on.

6.1.2 Conceptual Dependency in DAYDREAMER

DAYDREAMER employs a modified conceptual dependency representation for states and actions. The functional behavior of representations in the program is specified by planning and inference rules such as the following:

IF	**ACTIVE-GOAL** for person to be **AT** location
THEN	**ACTIVE-GOAL** for person to **PTRANS** to location

This rule states that in order to achieve the state of being **AT** a particular location, one employs the **PTRANS** action to go to that location; conversely, if one **PTRANS**es from one location to another, one is then **AT** the new location and no longer **AT** the original location. **PTRANS** is further broken down as follows:

IF	**ACTIVE-GOAL** for person to **PTRANS** to location
THEN	**ACTIVE-GOAL** for person to **KNOW** location

That is, in order to **PTRANS** to a particular location, one must **KNOW** where that location is.

In DAYDREAMER, mental states of persons are called *beliefs*. A belief consists of an *actor*—the person whose mental state is described, and an *object*—the mental state.[3] The mental state may itself be the mental state of another person—yet another belief. One typical use of beliefs is the representation of mental states of others which refer to the mental states of the self—such as Harrison Ford believing that DAYDREAMER has the goal to go out with him. Most often, beliefs are used to represent the attitudes and goals of others.

6.2 Other Planning

Daydreaming in the interpersonal domain often involves *other planning*—monitoring and attempting to modify the mental states (attitudes, goals, and so on) of other persons. We distinguish between *forward* and *backward* other planning.

Forward other planning involves running the planning mechanism from the viewpoint of another person. This enables the program to determine in a given situation what inferences that person might make, what beliefs that person might form, what goals that person might initiate, what actions that person

---

[3]A belief of another person is actually a belief of DAYDREAMER about the belief of another person. However, since *all* representations in the program are assumed to be beliefs of the daydreamer, this extra level is ignored in the discussion for simplicity.

might take on behalf of those goals, and so on. That is, forward other planning provides a simulation of the behavior of another person. For example, when DAYDREAMER imagines she is going out with Harrison Ford in RATIONAL-IZATION1, DAYDREAMER predicts that the movie star will have the goal to work and thus will travel to Egypt to shoot a film.

Backward other planning involves attempting to achieve some mental state in another person. For example, in LOVERS1, DAYDREAMER has a goal to go out with the movie star; in order to achieve this goal she must plan for the star to have the goal to go out with her.

In order to perform (both forward and backward) other planning, DAYDREAMER must have a model of the inference rules, planning rules, and initial mental states of other persons. The general assumption is made that the behavior of others is similar to the behavior of DAYDREAMER. This enables an economical representation of rules for both personal and other planning. However, rules may be added and deleted from this basic set for particular persons or classes of persons. Thus, for example, movie stars may have a different set of goals than DAYDREAMER.

The mental states of other persons are maintained in the same planning contexts as those containing self mental states; any mental state of another person is marked as a belief of that person. In order to perform forward other planning for a person, the planning algorithm described in Chapter 7 is employed. However, this algorithm (a) only retrieves beliefs of the other person and (b) marks as beliefs of the person any newly asserted subgoals or facts. The algorithm is sufficiently general to handle the simulation by another person of another person (including the self), the simulation by another person of the simulation by yet another person of yet another person, and so on. However, such nested planning is avoided in DAYDREAMER since it rapidly leads to a deep regress.

Backward other planning is employed in the following situation: Suppose a goal exists whose objective is some mental state of another person. The program has two ways of achieving this goal: Either it can employ a rule stated in terms of another person having the given mental state, or it can employ a rule stated in terms of the *self* having the given mental state. The former is attempted first, and then the latter is attempted. In the latter case—backward other planning—the rule is applied as if the other person were the self.

6.3   Attitudes

An *attitude* is an evaluation or generalized judgment made by a person of a physical object, another person, or an idea. Attitudes are important in the interpersonal domain, since the behavior of people is guided by their attitudes. If, for example, a person has a negative attitude toward another person, then the person is less likely to assist the other person in achieving that person's goals. It is insufficient, however, merely to model the impact of attitudes on the behavior of a person. In addition, it is necessary to model how a person models the attitudes of others, and how this modeling in turn affects that person's behavior: We are often concerned with the attitudes of other persons toward us and frequently behave in order to manipulate those attitudes (Goffman, 1959).

In this section, we address: (a) the representation of attitudes, (b) the formation of attitudes, (c) the monitoring of the attitudes of others, and (d) the modification of the attitudes of others.

6.3.1 Basic Attitude Representation

The representation of attitudes in DAYDREAMER is based on previous work by Heider (1958) (who calls them "sentiments"). While emotions (such as anger directed toward an object) have a short duration, attitudes are stable dispositions toward objects. Attitudes may be classified into *positive* and *negative* attitudes. A **POS-ATTITUDE** toward something means liking that something; a **NEG-ATTITUDE** means disliking it. There is a special positive attitude called **ROMANTIC-INTEREST**[4] as well.

A taxonomy of attitudes has been constructed by Schank, Wilensky, Carbonell, Kolodner, and Hendler (1978). However, their emphasis is on the inferences which can be made on the basis of attitudes in the comprehension of stories, rather than on the relationship of attitudes to behavior. As a result, the set of attitudes which they propose is too large for our purposes—a sufficiently difficult problem is presented by only positive, negative, and romantic attitudes.

Attitudes are mental states. Because all mental states are those of the daydreamer, an attitude by itself represents a self mental state. An attitude of another person is represented as a belief containing an attitude. However, for simplicity, we use "person has a **POS-ATTITUDE** toward an object" to mean "a **BELIEVE** of person that **POS-ATTITUDE** toward object."

6.3.2 Assessing, Forming, and Modifying Attitudes

In social psychology, there are a number of similar *cognitive balance theories* which contend that the attitudes which a person holds tend to proceed toward balance: Festinger's (1957) theory of "cognitive dissonance," Heider's (1958) "preference for balanced states," the "principle of congruity" of Osgood and Tannenbaum (1955), as well as other related theories. Heider (1958, pp. 174-217) discusses various cases of this principle, some of which are encoded as rules in DAYDREAMER. Each rule has several functions: The rule may be employed (a) as an inference rule in the formation of the attitudes of DAYDREAMER, (b) as a planning rule for DAYDREAMER to achieve certain self attitudes, (c)

---

[4]This attitude might have been called the **LOVE** attitude, but since DAYDREAMER concentrates on modeling the initial phase of a relationship, the term more appropriate to this stage is employed.

as an inference rule for the monitoring of the attitudes of others, and (d) as a planning rule for the modification of the attitudes of others.

According to Heider (1958), a person will tend to like another person having similar beliefs and attitudes. Thus in DAYDREAMER we have the rule that people tend to like those who like the same people and things they do:

IF	**ACTIVE-GOAL** to have **POS-ATTITUDE** toward person
THEN	**ACTIVE-GOAL** for person to have **POS-ATTITUDE** toward something that self has **POS-ATTITUDE** toward

Thus, for example, if one likes Indian food, one tends to like other people who also like Indian food. Since people tend to like themselves and aspects of themselves, the above rule also implies that one person will tend to like a second person who likes the first person or an aspect of the first person. Another consequence of the above rule is that if one person likes a second person, the first person tends to believe that the second person likes the first person.

Another principle is that a person will tend to like a similar person and dislike a dissimilar person (Heider, 1958): "Birds of a feather flock together." That is, one will form a positive attitude toward others with similar personality traits (e.g., social class, appearance) and a negative attitude toward those with dissimilar personality traits. An instance of this principle is provided by the following DAYDREAMER rule which states that well-dressed rich people do not like others who are not well-dressed:

IF	person is **RICH** and self is **AT** same location as person and self not **WELL-DRESSED**
THEN	person has **NEG-ATTITUDE** toward self

How is the attitude of **ROMANTIC-INTEREST** generated and modified? In DAYDREAMER we will simply assume that a certain level of this attitude may be generated if one has a positive attitude toward the other person, and one finds the other person attractive in appearance; we assume the **ROMANTIC-INTEREST** attitude may be strengthened if the constituent positive attitude and attractiveness are strengthened, as well as through continued success of a **LOVERS** relationship. We have the following rule:

IF	**ACTIVE-GOAL** to have **ROMANTIC-INTEREST** in person
THEN	**ACTIVE-GOAL** to have **POS-ATTITUDE** toward person and person to be **ATTRACTIVE**

In order to get another person to have a particular personal goal, one must achieve the antecedents of the rule which activates that goal. For example, the following rule states that a goal to be **LOVERS** with someone is initiated if one has **ROMANTIC-INTEREST** in that person and one is not currently **LOVERS** with someone else:

> | IF | **ACTIVE-GOAL** for self to have **ACTIVE-GOAL** of **LOVERS** with person |
> |---|---|
> | *THEN* | **ACTIVE-GOAL** for **ROMANTIC-INTEREST** in person and |
> | | **ACTIVE-GOAL** for not **LOVERS** with anyone |

Although this rule is stated in terms of the self rather than another person, it can still be applied to another person through backward other planning.

6.3.3 Attitudes and Meeting People

In order for DAYDREAMER to form a **FRIENDS** or **LOVERS** relationship, she must meet people. In this section, we investigate how one person approaches another when the two are not already acquainted. That is, we explore the situations in which is it acceptable for one person to initiate communication with another, and how acceptable "opening lines" are produced.

The **ACQUAINTED** relationship in DAYDREAMER models the fact that two people know each other. This relationship has the inference that whenever the two persons meet, each will recognize the other and say hello to the other. Being acquainted with someone is a prerequisite to forming a more important **FRIENDS** or **LOVERS** relationship with the person.

In DAYDREAMER, two become acquainted when they have participated in an **M-CONVERSATION**:

> | IF | **ACTIVE-GOAL** to be **ACQUAINTED** with person |
> |---|---|
> | *THEN* | **ACTIVE-GOAL** for **M-CONVERSATION** with person |

The **M-** prefix is from Schank (1982) and refers to a complex action made up of more primitive actions. Thus, an **M-CONVERSATION** consists of a series of exchanges, where each exchange is one or more **MTRANS**es by the first person followed by one or more **MTRANS**es by the second person.

If two persons are not already **ACQUAINTED**, there are certain constraints on the first **MTRANS** which initiates the conversation—the **MTRANS** from the first person to the second person must be acceptable to the second person, which means:

- the second person does not form a negative attitude toward the first person

- the second person does not believe that others (or society) would form a negative attitude toward the second person if the second person failed to behave as if the second person had a negative attitude toward the first person.

That is, it is acceptable to initiate communication with someone if neither oneself nor the other person is embarrassed as a result.

The acceptability criteria are encapsulated under a state called **MTRANS-ACCEPTABLE** which may or may not exist between two individuals in a given situation. A conversation is then achieved if **MTRANS-ACCEPTABLE** is true of the two persons and if each person introduces oneself to the other:[5]

IF	**ACTIVE-GOAL** for **M-CONVERSATION** between person1 and person2
THEN	**ACTIVE-GOAL** for **MTRANS-ACCEPTABLE** between person1 and person2 and
	ACTIVE-GOAL for person1 to **MTRANS INTRODUCTION** to person2 and
	person2 to **MTRANS INTRODUCTION** to person1

However, if one person **MTRANS**es to another without the state **MTRANS-ACCEPTABLE**, the other will form a negative attitude toward the first:

IF	not **MTRANS-ACCEPTABLE** between person1 and person2 and
	person1 **MTRANS** anything to person2
THEN	person2 has **NEG-ATTITUDE** toward person1

What, then, are the criteria for **MTRANS-ACCEPTABLE**? That is, what determines whether the second person forms a negative attitude (or believes that such an attitude *should* be formed) toward the first person? Such attitude formation is sensitive to the particular goal context in which the communication occurs, as well as the particular existing attitudes and traits of the persons involved. In general, a negative attitude will not be formed (and the communication will be acceptable) if the communication is performed as a subgoal of some goal situation which the two people have (or are contrived to have) in common.

DAYDREAMER contains various planning and inferences rules for **MTRANS-ACCEPTABLE**. For example, a request for a small favor, such as the time, is an acceptable initial communication:

IF	**ACTIVE-GOAL** for **MTRANS-ACCEPTABLE** between self and person
THEN	**ACTIVE-GOAL** for self to **MTRANS** to person that self has **ACTIVE-GOAL** to **KNOW** the time

In DAYDREAMER, communication is also acceptable with another person if (a) one is already acquainted with that person, (b) that person has already spoken to the self, (c) that person already has the goal to be acquainted with the self (as is inferred in the situation of a party), and (d) if the other person smiles at the self.

---

[5]The subgoal for **MTRANS-ACCEPTABLE** is not strictly (causally) necessary for achievement of the goal. It is introduced to prevent certain negative consequences. That is, although a conversation really requires only the exchanges themselves, if an initial communication by one person is deemed unacceptable by another, that party will form a negative attitude toward the first person. Therefore an additional subgoal is added to the rule to prevent formation of such a negative attitude. When actions to prevent negative consequences are *not* already compiled into a rule, the **REVERSAL** daydreaming goal may be used to create new rules in which such actions *are* compiled in; see Chapter 4.

6.4 Interpersonal Relationships

There are three interpersonal relationships currently represented in DAY-DREAMER: **FRIENDS**, **LOVERS**, and **EMPLOYMENT**. Two persons may go through a sequence of relationships over time: For example, they may start out as **FRIENDS**, which is terminated by the two becoming **LOVERS**, which then is terminated by the two becoming **FRIENDS**, and so on. In general, there may be more than one relationship between two persons—for example, the same two persons could participate in both a **FRIENDS** and an **EMPLOYMENT** relationship. However, not all combinations are possible. In our scheme, **LOVERS** subsumes **FRIENDS** so both relationships should not be present between two people at a given time.

6.4.1 Lovers

The **LOVERS** relationship consists of the following phases: initiation, mainte-nance, and termination.

Once a personal goal to form a **LOVERS** relationship with someone is acti-vated, this goal may be achieved by: (a) being acquainted with a person, (b) being romantically interested in the person, (c) the person having the goal to be in the relationship, (d) going on a date with the person, and (e) agreeing to initiate the relationship. These steps are expressed as the following planning rule in DAYDREAMER:

IF	**ACTIVE-GOAL** for **LOVERS** with person
THEN	**ACTIVE-GOAL** for **ACQUAINTED** with person and **ACTIVE-GOAL** for **ROMANTIC-INTEREST** in person and **ACTIVE-GOAL** for person to have **ACTIVE-GOAL** of **LOVERS** with self and **ACTIVE-GOAL** for **M-DATE** with self and person and **ACTIVE-GOAL** for self and person to **M-AGREE** to **LOVERS**

It would be possible to introduce a loop into the planning: the **M-DATE**s would continue until the two parties agreed to form the more permanent **LOVERS** relationship (or decided they were not interested).   Since DAY-DREAMER does not represent a great variety or degree of detail of the activities performed on a date, a loop would just add more repetition to the traces. There-fore, only one **M-DATE** is performed before initiating a **LOVERS** relationship in the current system.

We have discussed planning for **ACQUAINTED** and **ROMANTIC-INTEREST** in the previous section. Another rule states that a person will have the goal to initiate a relationship with another person if the first person is romantically interested in the second person and the first person is not currently in a relationship:

> **IF** **ACTIVE-GOAL** for self to have **ACTIVE-GOAL**
> of **LOVERS** with person
> **THEN** **ACTIVE-GOAL** for **ROMANTIC-INTEREST** in person
> and
> **ACTIVE-GOAL** for not **LOVERS** with anyone

This rule, although stated in terms of the self, is employed through the backward other planning mechanism to plan for a person other than DAYDREAMER.

An **M-DATE** in DAYDREAMER consists of: (a) agreeing to perform an activity toward which both parties have a positive attitude (such as **M-RESTAURANT** or **M-MOVIE**), (b) enabling future **MTRANS**es (in order to arrange meeting time and location), (c) at a later time going to the other person's home, (e) performing the activity (including going to an appropriate location to perform the activity), (f) going back to the other person's home, and (g) kissing each other. This is expressed as the following rule:

> **IF** **ACTIVE-GOAL** for **M-DATE** with self and person
> **THEN** **ACTIVE-GOAL** for self and person to **M-AGREE**
> to **M-RESTAURANT** with self and person and
> **ACTIVE-GOAL** for self and person to **ENABLE-FUTURE-VPROX**
> and
> **ACTIVE-GOAL** for it to be **FRIDAY-NIGHT** and
> **ACTIVE-GOAL** for self to be **AT** location of
> person and
> **ACTIVE-GOAL** for **M-RESTAURANT** with self and
> person and
> **ACTIVE-GOAL** for self and person to be **AT**
> initial location of person and
> **ACTIVE-GOAL** for self and person to **M-KISS**
> and
> **ACTIVE-GOAL** for self to be **AT** initial location
> of self

An agreement to perform an activity or initiate a state for the mutual benefit of the parties results when the parties **MTRANS** the fact that they have the same goal to each other:

> **IF** **ACTIVE-GOAL** for self and person to **M-AGREE**
> to thing
> **THEN** **ACTIVE-GOAL** for self to **MTRANS** to person
> that self has **ACTIVE-GOAL** for that something
> and
> **ACTIVE-GOAL** for person to **MTRANS** to self
> that person have **ACTIVE-GOAL** for thing

The **LOVERS** relationship consists of the following conditions and inferences: (a) each party supports various goals of the other, (b) each party maintains a **ROMANTIC-INTEREST** attitude toward the other, and (c) each party is not in a **LOVERS** relationship with another (and especially must not engage in **M-SEX** with another).

Violations of requirements activate *preservation goals* on the relationship. For example, the following rule is employed in RATIONALIZATION1 to generate a preservation goal on the relationship when Harrison Ford leaves Los Angeles.

IF	**LOVERS** with person who is **RPROX** city and self is not **RPROX** city
THEN	**ACTIVE-P-GOAL** of **LOVERS** with person

If the requirements of the relationship are violated sufficiently, one or both parties may decide to terminate the relationship:

IF	**ACTIVE-GOAL** for **FAILED-GOAL** of **POS-RELATIONSHIP** with person
THEN	**ACTIVE-GOAL** for person to **MTRANS** to self that person has **ACTIVE-GOAL** not to be in **POS-RELATIONSHIP** with self

This rule applies to any positive relationship (i.e., **FRIENDS, LOVERS,** and **EMPLOYMENT**).

6.4.2 Friends

A **FRIENDS** relationship is achieved whenever two people are acquainted and they have positive attitudes toward each other:

IF	**ACTIVE-GOAL** for **FRIENDS** with person
THEN	**ACTIVE-GOAL** for self and person to be **ACQUAINTED** and **ACTIVE-GOAL** to have **POS-ATTITUDE** toward person and **ACTIVE-GOAL** for person to have **POS-ATTITUDE** toward self

6.4.3 Employment

The **EMPLOYMENT** relationship is initiated in the following manner: (a) an employer has an opening, (b) an employer wants a particular employee for the job, (c) the employee wants the job, and (d) the employer and employee agree to the job. These steps are expressed in terms of the following rule:

IF	**ACTIVE-GOAL** for **EMPLOYMENT** with person
THEN	**ACTIVE-GOAL** for person to have **ACTIVE-GOAL** of employing someone and **ACTIVE-GOAL** for person to have **ACTIVE-GOAL** of employing self and **ACTIVE-GOAL** for self and person to **M-AGREE** to **EMPLOYMENT** of self with person

One can quit and one can be fired; this is accomplished via the rules given above for termination of **LOVERS**. In addition, one's job fails if one is not living in the same region as the city in which one is employed (this rule is employed in RATIONALIZATION1 to generate failure of employment upon following Harrison Ford to Cairo):

IF	have **EMPLOYMENT** with organization which is **RPROX** city and self is not **RPROX** city
THEN	**FAILED-GOAL** of **EMPLOYMENT** with organization

6.5   Strategies for Fanciful Planning

Consider the following hypothetical (but typical) daydream, given by S. Freud (1908/1962), of a poor orphan boy on his way to see an employer where he might find a job:

> He [daydreams that he] is given a job, finds favour with his new employer, makes himself indispensable in the business, is taken into his employer's family, marries the charming young daughter of the house, and then himself becomes a director of the business, first as his employer's partner and then as his successor. (p. 148)

In this daydream, the goals most important to the orphan—to be employed, to have a family, to be married, to be the director of a business—are achieved without considering all of the necessary planning steps and obstacles. He does not imagine knocking on the employer's door, the opening of the door, talking with the employer about a job, being offered (or refused) a job, accepting the job, performing the activities of the job each day for many days, and so on. Furthermore, he ignores the likely behavior of the employer, family, and daughter as well as his own probable limitations.

Planning in daydreaming *relaxes constraints*: A constraint relaxed in the above daydream is the probable *behavior of other people*. Another constraint often relaxed in daydreaming is the *identity or attributes of the self*. One might imagine being a famous movie star (as in REVENGE1), an Olympic athlete, or even a physical object or animal. Often *physical constraints* are relaxed in daydreaming: One might imagine being able to fly, being invisible, being able to slide underneath doors, and so on. Finally, *social and cultural constraints* may be relaxed as when one imagines yelling "fire!" in a crowded movie theater.

In one daydream, one might imagine walking around in Paris without considering the plane trip required to get there. But after seeing a news report on terrorism, one might daydream about the plane trip itself. How does DAYDREAMER determine when to relax a constraint and when to omit details?

Three strategies are employed in DAYDREAMER for the generation of fanciful daydream scenarios in the interpersonal domain: *plausible planning, subgoal relaxation,* and *inference inhibition.* We consider each of these methods in turn.

6.5.1 Plausible Planning

Daydream scenarios which are not completely realistic may be generated through the use of *plausible* planning and inference rules (Shortliffe & Buchanan, 1975). Associated with each rule is a number from 0 to 1 which roughly specifies how plausible that inference or planning rule is. *A somewhat fanciful daydream scenario results whenever an inference or planning rule is employed whose plausibility is less than 1.* For example, DAYDREAMER employs the following plausible planning rule in generating REVENGE1:

IF	**ACTIVE-GOAL** for self to be **STAR**
THEN	**ACTIVE-GOAL** for self to **M-STUDY** to be an **ACTOR**

This rule, which states that one may become a star by studying to be an actor, is assigned a plausibility of 0.4. The *realism* of a scenario, also measured on a scale from 0 to 1, is calculated based upon the plausibilities of the planning and inference rules employed in generating that scenario. The *particular* values are not important. What *is* important is that (a) rules not guaranteed to succeed may be employed in planning, (b) some rules are more likely to succeed than others, and (c) numbers can be used to provide a rough measure of scenario realism.

For any given subgoal, there may be several applicable plausible planning rules. How does DAYDREAMER decide which rules to employ and in what order to attempt each of these rules? When planning for personal goals and learning daydreaming goals—**REVERSAL**, **REHEARSAL**, **RECOVERY**, and **REPERCUSSIONS**—rules are attempted in order of decreasing plausibility. When planning for emotional daydreaming goals—**RATIONALIZATION**, **REVENGE**, and **ROVING**—rules are attempted in random order. Furthermore, when planning for personal goals and learning daydreaming goals, a rule is only selected if its application would result in a scenario whose realism is above a certain level. This level is higher for personal goals than for learning daydreaming goals. Such a restriction is not imposed on emotional daydreaming goals. Thus personal goals result in the most realistic scenarios, learning daydreaming goals result in moderately realistic scenarios, while emotional daydreaming goals may result in fairly unrealistic scenarios.

Assessment of the realism of a daydream scenario is important since daydreams are potential plans for future action. If a particular daydream contains unrealistic world events in response to actions of the daydreamer, it should not be employed in the world; in performance mode it does not matter what the imagined positive consequences of the daydream are if those consequences are derived from an unrealistic scenario. If DAYDREAMER failed to make realism assessments, it would act on its unrealistic daydreams and be disappointed with the results—that is, its personal goals would fail.

6.5.2 Subgoal Relaxation

In addition, *fanciful daydream scenarios may be generated by hypothesizing the success of an unsatisfied subgoal without generating a plan—even a plausible plan—to achieve that subgoal.* Such subgoal relaxation is performed in the following situations:

- All plans to achieve a subgoal have failed and the objective of the subgoal is either: a mental state (e.g., goal, attitude, belief) of another or a self attribute (e.g., appearance, social class). Examples occur in LOVERS1, in which DAYDREAMER assumes Harrison Ford is interested in her and thinks she is attractive.

- A daydreaming goal strategy (implemented as rules) explicitly specifies that the success of a given subgoal is to be hypothesized. Here the *purpose* of a daydream, as specified by the current daydreaming goal, determines which subgoal success to hypothesize. For example, one plan for the **RATIONALIZATION** of a past goal failure involves hypothesizing the success of that same goal, without worrying about how that goal might have been achieved.

Subgoal relaxation, by itself, is similar to the relaxation of operator preconditions employed in ABSTRIPS (Sacerdoti, 1974). See Chapter 5 for a review of ABSTRIPS in the full context of DAYDREAMER.

6.5.3 Inference Inhibition

In addition, *fanciful daydream scenarios may be generated by failing to infer negative consequences of a situation.* In particular, social or cultural consequences (such as being embarrassed or being thrown in jail) are ignored whenever a special daydreaming goal[6] is invoked to explore a socially or culturally unacceptable goal (such as causing chaos or stealing). For example:

> I'm sitting on the deck of Pacific Princess with lots of people swimming and sun-bathing not far away from me. A waiter comes around passing out desserts. Then someone gives a signal and everybody starts throwing food all over the place. (McNeil, 1981)

> I imagine myself as the notorious bank robber who is wanted by the law everywhere. I'm entering a majestic bank full of security guards. I still carry out my plan and achieve my goal despite . . . the security network. (McNeil, 1981)

Daydreams produced using inference inhibition have low realism values, so that they are not acted upon in performance mode. Alternatively, inferences

---

[6]This daydreaming goal is not yet implemented in the current version of DAYDREAMER.

can be turned on during generation of socially unacceptable actions; although the realism of such a scenario may then be high, the *desirability* evaluation of the scenario will be low because of its negative consequences. Thus such a daydream will also not be acted upon in performance mode.

Inference inhibition probably occurs in humans only after a person has learned which actions are socially unacceptable through daydreaming *without* inference inhibition. Thereafter, the generation of such daydreams simply serves as a form of self-entertainment (J. L. Singer, 1975). Generation of negative consequences would reduce the entertainment value of such daydreams.

Chapter 7

Implementation of DAYDREAMER

This chapter describes the implementation of DAYDREAMER in detail: First, we present statistics on the program. Second, we present the GATE representation language. Third, we discuss the representation of DAYDREAMER planning and inference rules in this language. Fourth, we discuss the basic procedures for daydreaming. Fifth, we revise these basic procedures to incorporate analogical planning and episodic memory. Sixth, we discuss the modifications necessary to support forward and backward planning from the viewpoint of others. Seventh, we discuss the procedures for serendipity and mutation.

7.1 Summary Statistics

DAYDREAMER consists of 12,739 lines of code, broken down by major sections as follows:

Component	Lines
Control	717
Planner	2186
Episodic memory	915
Reversal	553
Serendipity	762
Mutation	447
Misc	1315
Generator	2122
Representations	3722
Total	12,739

A large percentage of the code is devoted to the planner and the representations and planning knowledge for the interpersonal domain. The rest of the components are relatively compact.

DAYDREAMER consumes 2,521,088 bytes of memory (interpreted) in T version 3.0 on an Apollo DN460. This figure does not include the GATE language, which consumes 346,112 bytes of memory (compiled).

A number of statistics on the program were obtained from the traces of a complete run; annotated trace excerpts are provided in Appendix A.

The following statistics were obtained for the planner:

Event	Count
Rule firings	322
Context sprouts	186
Planning rule firings	186
Inference rule firings	136
Subgoal relaxations	15
Input states/actions	6
Output actions	20
Side-effect personal goals	8

The planning rule firings broke down as follows:

Type	Count	Percentage
Regular planning rules	130	69.9 %
Analogical planning rules	44	23.6 %
Coded (Lisp) planning rules	9	4.8 %
Backward other planning rules	3	1.67 %

Thus almost one quarter of the planning rule applications were carried out through analogical application of a planning episode. There were four serendipities, and one mutation.

The following statistics were obtained for concerns:

Event	Count
Initiation	22
Success	15
Failure	4

The program switched from daydreaming mode to performance mode four times, and switched from performance mode to daydreaming mode three times.

The following statistics were obtained for episodic memory:

Event	Count
Stored episodes	12
Retrieved episodes	8
Activated indices	71

Sixteen hand-coded episodes were initially provided to the program. After adding 12 more episodes, the program contained 28 episodes.

Five new planning rules were added to the program's collection of rules. The program ended up using two of them later on.

The distribution of rule utilization was as follows:

Times Used	Number of Rules	Sample Rules
22	1	*Believe-plan1*
20	2	*Mtrans-plan2 At-plan2*
12	1	*Ptrans-plan*
10	1	*Enable-future-vprox-plan1*
9	1	*Vprox-plan2*
7	3	*Vprox-plan3 Vprox-plan1 Mtrans-plan1*
5	6	*Rprox-plan M-phone-plan1 M-agree-plan*
4	10	*Romantic-interest-plan Reversal-theme*
3	8	*Know-plan2 Employment-theme-plan*
2	31	*Under-doorway-plan Reversal-plan*
1	52	*Well-dressed-plan2 Star-plan*

A total of 116 rules were used; 114 of those rules were initially provided to the program. Since the program was provided with 135 rules initially, 21 rules ended up never being used.

7.2 The GATE Representation Language

DAYDREAMER is implemented using GATE (E. T. Mueller, 1987a; E. T. Mueller & Zernik, 1984), a graphical artificial intelligence program development tool for the T language (Rees, N. I. Adams, & Meehan, 1984), a version of the Scheme dialect of Lisp (Winston & Horn, 1981; McCarthy, Abrahams, Edwards, Hart, & Levin, 1965), running on Apollo Domain and Hewlett-Packard Bobcat workstations.

GATE consists of *slot-filler objects*, a *textual representation* for these objects, operations on these objects such as *unification, instantiation*, and *variabilization*, and a *context* mechanism. GATE also includes a graphical interface which provides visual animation of a running program as it manipulates slot-filler objects, and enables one to create, view, and modify objects graphically.

7.2.1 Slot-Filler Objects

Slot-filler objects are similar to the slot-filler objects of Schank and Riesbeck (1981), frames (Minsky, 1975), Lisp a-lists (McCarthy et al., 1965), and the structures or records of traditional programming languages such as Pascal (Wirth, 1971).

Slot-filler objects consist of: (a) one or more *object names*, (b) a *type*, and (c) zero or more pairs, where each pair consists of a *slot name* and a *slot value*. Object names and slot names are Lisp atoms. Types are organized into a hierarchy; types are themselves slot-filler objects. A slot value is either a slot-filler object or some other Lisp object (such as a character string or a procedure). Several pairs with the same slot name are permitted.

Basic operations which can be performed on slot-filler object include: *create*, *add-name*, *get-names*, *set-type*, *get-type*, *add-slot-value*, *remove-slot-value*, *get-slot-values*, *copy*, and so on.

A textual representation for slot-filler objects enables *printing* to and *reading* from a terminal, window, file, string, or Lisp list. For example:

<pre>
(PTRANS actor <i>John1</i>
         from (RESIDENCE obj <i>John1</i>)
         to <i>Store1</i>
         obj <i>John1 Mary1</i>)
</pre>

The slot-filler object represented above is of type **PTRANS** and consists of five pairs. One pair of the object consists of the slot name actor and slot value *John1*. This slot value refers to another slot-filler object whose name is *John1*. Whenever a slot value is a slot-filler object having a nonautomatically generated name, only the name is printed. The slot value of the pair whose slot name is from is another slot-filler object which does not have a name. In this case, the full textual representation of this slot-filler object is recursively printed. There are two pairs of the top-level slot-filler object whose slot name is obj. The slot value of one pair is *John1* while the slot value of the other is *Mary1*.

7.2.2   Unification

Unification (see, for example, Charniak, Riesbeck, & McDermott, 1980) is a pattern-matching operation performed on two slot-filler objects. The objects may contain a special kind of slot-filler object called a *variable*. Two slot-filler objects unify if values for variables can be found such that substituting the values for those variables in the objects would produce equivalent structures.[1] For example, if unification is invoked on

<pre>
(PTRANS actor <i>?Person</i>
         to <i>?Location</i>)
</pre>

and

<pre>
(PTRANS actor <i>John1</i>
         to <i>Store1</i>)
</pre>

the result is the following list of variable *bindings*:

<pre>
    ?Person = John1
    ?Location = Store1
</pre>

---

[1] Actually, the structures resulting from substitution do not have to be equivalent. Rather, one structure must be a substructure of the other: one slot-filler object unifies with another if each pair in the first object unifies with a unique pair in the second object; however, each pair in the second object need not have been accounted for. Thus unification in GATE is asymmetrical.

Unification takes two objects and an initial list of bindings, normally empty, and returns an augmented list of bindings if the two objects unify. Three cases involving variables in unification must be considered: First, an unbound variable unifies with a slot-filler object if the object is an instance of the *type* of the variable. A variable of type **TYPE** may be represented as either of the following:

>*?name:***TYPE**
>*?Type...*

When an unbound variable successfully unifies with an object, that object becomes that variable's value in the list of bindings. Second, a bound variable unifies with an object if the value of that variable unifies with the object. Third, two variables unify if the type of one variable is an improper supertype of the type of the other.

GATE provides an extended syntax and semantics for unification. In particular, it supports several *special* slot-filler objects of type **UAND**, **UOR**, **UNOT**, and **UPROC**:

(**UAND** obj *obj1* obj *obj2* ...)

A **UAND** object unifies with another object if all of *obj1 obj2* ... unify with that object. Bindings are augmented in a cumulative manner from each unification.

(**UOR** obj *obj1* obj *obj2* ...)

A **UOR** object unifies with another object if any of *obj1 obj2* ... unifies with that object. Bindings are augmented by the first successful unification.

(**UNOT** obj *obj*)

A **UNOT** object unifies with another object if *obj* does not unify with that object. Bindings are not augmented.

(**UPROC** proc *proc*)

A **UPROC** object unifies with another object if *proc* (a Lisp lambda expression) applied to that object returns a non-**NIL** value. Bindings are not augmented.

These features were inspired by the pattern matcher of the DIRECTOR language (Kahn, 1978). We have extended the constructs to full unification— that is, the constructs may now be used in both arguments to the matcher, with a well-defined semantics. We have also added appropriate extensions for typed variables and slot-filler objects, multiple slot values per slot name, and cyclic data structures. E. T. Mueller (1987a) discusses the GATE unification language in greater detail.

7.2.3 Instantiation

Instantiation (see, for example, Charniak, Riesbeck, & McDermott, 1980) takes a slot-filler object and binding list, and returns a copy of the object in which any variables have been replaced by their values in the binding list. For example, if the slot-filler object:

(**PTRANS** actor *?Person*
        to *?Location*)

is instantiated using the binding list:

    ?Person = John1
    ?Location = Store1

the returned slot-filler object is:

(**PTRANS** actor *John1*
        to *Store1*)

Any unbound variables remain as variables in the copy. The slot-filler object may contain other slot-filler objects—the complete structure with the given object as root is copied, with any cycles preserved in the copy.

7.2.4 Variabilization

Variabilization (see, for example, Charniak & McDermott, 1985) takes a slot-filler object, a predicate, and a procedure, and returns a complete copy of the slot-filler object (with cycles preserved) in which any enclosed objects answering true to the predicate have been replaced by unique variables of a type determined by applying the procedure to the object. Multiple occurrences of the same object will become the same variable. For example, given:

(**PTRANS** actor *John1*
        to *Store1*
        obj *John1 Mary1*)

and an appropriate predicate and procedure, variabilization returns:

(**PTRANS** actor *?Person1*
        to *?Location1*
        obj *?Person1 ?Person2*)

7.2.5 Contexts

A context is a collection of slot-filler objects—called *facts*—which specify the state of a possible world. GATE contexts are similar to other context mechanisms such as OMEGA viewpoints (Barber, 1983) and AP3 contexts (Goldman, 1982), all of which derive from the original contexts of QA4 (Rulifson, Derksen, & Waldinger, 1972).

Several operations may be performed on contexts: The *create* operation returns a new context containing no facts. The *assert* operation adds a fact to a context. The *retract* operation removes a fact from a context. The *retrieve* operation returns: (a) a list of facts in the context unifying with a given retrieval pattern,[2] and (b) the corresponding variable bindings. The *sprout* operation creates a new context which is a copy of an existing context. Initially, the new context consists of the same collection of facts as the old context. Thereafter, if the new context is modified, the old context remains unaffected.[3] The new context is called a *child* of the old context; the old context is called the *parent* of the new context; we also speak of *ancestor* and *descendant* contexts.

For example, suppose that the following two objects are asserted into a newly created context:

(**PTRANS** actor *John1*
     to *Store1*)

(**PTRANS** actor *Mary1*
     to *Store1*)

A retrieval from this context using the pattern:

(**PTRANS** actor *?Person*
     to *Store1*)

will return the first two slot-filler objects, and the following two binding lists which correspond to those objects:

    ?Person = John1

and

    ?Person = Mary1

7.3 Planning and Inference Rules

DAYDREAMER planning and inference rules are represented in the GATE language as follows:

(**RULE** subgoal *subgoal-pattern*
    goal *goal-pattern*
    delete *delete-pattern1 delete-pattern2* ...
    emotion *emotion1 emotion2* ...
    is 'INFERENCE-ONLY | 'PLAN-ONLY | 'ACTION-PLAN | 'GLOBAL-INFERENCE
    plausibility *number*)

---

[2]Hashing on the types of facts is employed in the current version of GATE to improve the efficiency of this operation.

[3]We do not specify whether modifications to the old context affect the new context, since the procedures which we will present here do not modify the old context once a new context has been created.

The delete, emotion, is, and plausibility slots are optional. By default, a rule may be employed both as a planning rule and as an inference rule. However, if the is slot of the rule is 'Inference-only, the rule may be employed only as an inference; if the is slot of the rule is 'Plan-only, the rule may be employed only as a plan.

7.3.1   Planning Rules

When a rule functions as a planning rule, the **goal** slot specifies the antecedent and the **subgoal** slot specifies the consequent. That is, if the objective of an active goal unifies with *goal-pattern*, the active goal may be split into the subgoals whose objectives are specified by *subgoal-pattern*, instantiated with the bindings resulting from the unification with *goal-pattern*. Specifically, if *subgoal-pattern* is of the form:

(**RSEQ** obj *subgoal-pattern1 subgoal-pattern2* ...)

then *subgoal-pattern1*, *subgoal-pattern2*, and so on, suitably instantiated, are the subgoals. Otherwise, *subgoal-pattern*, suitably instantiated, is the subgoal. Here is a sample planning rule employed in DAYDREAMER:

Rule *Lovers-plan*

(**RULE** subgoal (**RSEQ** obj (**ACQUAINTED** actor *?Self ?Other*)
                        (**ROMANTIC-INTEREST** obj *?Other*)
                        (**BELIEVE** actor *?Other*
                                obj (**GOAL** obj (**LOVERS** actor *?Self ?Other*)))
                        (**M-DATE** actor *?Self ?Other*)
                        (**M-AGREE** actor *?Self ?Other*
                                obj (**LOVERS** actor *?Self ?Other*)))
      goal (**LOVERS** actor *?Self ?Other*)
      is 'PLAN-ONLY
      plausibility 0.95)

IF	**ACTIVE-GOAL** for **LOVERS** with person
THEN	**ACTIVE-GOAL** for **ACQUAINTED** with person and **ACTIVE-GOAL** for **ROMANTIC-INTEREST** in person and **ACTIVE-GOAL** for person to have **ACTIVE-GOAL** of **LOVERS** with self and **ACTIVE-GOAL** for **M-DATE** with self and person and **ACTIVE-GOAL** for self and person to **M-AGREE** to **LOVERS**

7.3.2   Inference Rules

When a rule functions as an inference rule, the **subgoal** slot specifies the antecedent and the **goal** slot specifies the consequent. That is, whenever *subgoal-pattern* is shown to be true in a given context for some collection of bindings, the instantiated *goal-pattern* is asserted in that context and the instantiated *delete-pattern1*, *delete-pattern2*, and so on, if any, are retracted from the context. If *subgoal-pattern* is of the form:

(**RAND** obj *subgoal-pattern1 subgoal-pattern2* ...)

or

(**RSEQ** obj *subgoal-pattern1 subgoal-pattern2* ...)

then all of *subgoal-pattern1*, *subgoal-pattern2*, and so on, must be shown to be true in the given context. If *subgoal-pattern* is of the form:

(**ROR** obj *subgoal-pattern1 subgoal-pattern2* ...)

then one of *subgoal-pattern1*, *subgoal-pattern2*, and so on, must be shown to be true in the given context. Finally, if *subgoal-pattern* is of the form:

(**RNOT** obj *subgoal-pattern1*)

then *subgoal-pattern1* must *not* be shown to be true in the given context. **RAND**, **ROR**, and **RNOT** are to be distinguished from their counterparts **UAND**, **UOR**, and **UNOT**. A **RAND**, for example, is shown to be true if each of its elements matches some, possible different, fact in the context; a **UAND** in the same position would require each of its elements to match the *same* fact in the context. Here is a sample inference rule employed in DAYDREAMER:

Rule *Neg-attitude-inf*

```
(RULE subgoal (RAND obj (RICH actor ?Other)
                       (AT actor ?Other
                           obj ?Location)
                       (AT actor ?Self
                           obj ?Location)
                       (MTRANS actor ?Self
                               from ?Self
                               to ?Other
                               obj ?Anything:NOTYPE)
                       (RNOT obj (WELL-DRESSED actor ?Self)))
          goal (BELIEVE actor ?Other
                        obj (NEG-ATTITUDE obj ?Self))
          is 'INFERENCE-ONLY
          plausibility 0.9)
```

IF	person is **RICH** and
	self is **AT** same location as person and
	self not **WELL-DRESSED**
THEN	person has **NEG-ATTITUDE** toward self

7.3.3 Rules for Actions

Inference rules are commonly used to specify the effects of actions. For example, DAYDREAMER employs the following inference rule for **PTRANS**:

Rule *At-plan*

(**RULE** subgoal (**PTRANS** actor *?Person*
                              from *?Location1*
                              to *?Location2*
                              obj *?Person*)
        goal (**AT** actor *?Person*
                   obj *?Location2*)
        delete (**AT** actor *?Person*
                    obj *?Location1*)
        initial (**AT** actor *?Person*
                     obj *?Location1*)
        plausibility 1.0)

IF	person **PTRANS** from location1 to location2
THEN	person **AT** location2 and
	delete person **AT** location1

IF	**ACTIVE-GOAL** for person to be **AT** location
THEN	**ACTIVE-GOAL** for person to **PTRANS** to location

This rule specifies that when a person performs a **PTRANS**, it is necessary to retract the old location (**AT**) of the person and assert the new one.

The above rule may also function in reverse as a planning rule: When an **AT** goal is present, a **PTRANS** subgoal may be asserted. Such action subgoals are treated similarly to other subgoals—the *preconditions* of the action become further subgoals to be achieved. For example, DAYDREAMER employs the following rule for **PTRANS**:

Rule *Ptrans-plan*

(**RULE** subgoal (**KNOW** actor *?Person*
                             obj *?Location2*)
        goal (**PTRANS** actor *?Person*
                      from *?Location1*
                      to *?Location2*
                      obj *?Free-obj*)
        is 'ACTION-PLAN
        plausibility 1.0)

IF	**ACTIVE-GOAL** for person to **PTRANS** to location
THEN	**ACTIVE-GOAL** for person to **KNOW** location

Once the **KNOW** subgoal succeeds—that is, once the person knows the location being **PTRANS**ed to—the action is *performed* (asserted as an action into the appropriate context). Since it is an undesirable behavior for an action to be performed whenever its preconditions happen to be satisfied, rules whose goal is an action are not run as forward inferences.

Some rule must correspond to each possible action goal. If there are no preconditions for a given action goal, the corresponding subgoal should be:

(**RTRUE**)

Only one rule should apply to each action goal and each rule should contain all the necessary preconditions for any action goal to which that rule applies.

7.3.4 Rules for Concern Initiation

A new concern is initiated by an inference rule (called a *theme*, after Schank & Abelson, 1977) of the following form:

```
(RULE subgoal activation-condition
      goal (ACTIVE-GOAL obj goal-objective)
      emotion (POS-EMOTION strength emotion-strength)
            ...
      is 'INFERENCE-ONLY)
```

When *activation-condition* is shown in the context, a new concern is initiated whose top-level goal is specified by the **goal** slot. The **emotion** slot specifies motivating emotions to be instantiated and associated with the new concern.

Here is a sample rule employed in DAYDREAMER for initiating an **EM-PLOYMENT** personal goal concern:

Rule Employment-theme

```
(RULE subgoal (RNOT obj (EMPLOYMENT actor ?Self))
      goal (ACTIVE-GOAL obj (EMPLOYMENT actor ?Self ?Other
                                        organization ?Organization))
      emotion (POS-EMOTION strength 0.85)
      is 'INFERENCE-ONLY
      plausibility 1.0)
```

IF	self not have **EMPLOYMENT** with anyone and satisfaction level of **MONEY** or **POSSESSIONS** need below threshold
THEN	**ACTIVE-GOAL** for **EMPLOYMENT** with person

Here is a sample rule employed in DAYDREAMER for initiating a **RATIO-NALIZATION** daydreaming goal concern:

Rule Rationalization-theme

```
(RULE subgoal (DEPENDENCY linked-from ?Failed-goal
                          linked-to (UAND obj ?Neg-emotion
                                          (UPROC proc '>THRESH?)))
      goal (ACTIVE-GOAL obj (RATIONALIZATION obj ?Failed-goal))
      emotion ?Neg-emotion
              (NEG-EMOTION strength 0.05)
      is 'INFERENCE-ONLY
      plausibility 1.0)
```

IF	**NEG-EMOTION** of sufficient strength resulting from a **FAILED-GOAL**
THEN	**ACTIVE-GOAL** for **RATIONALIZATION** of failure

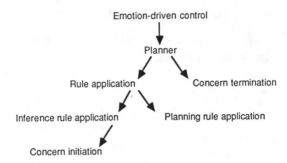

Figure 7.1: Basic Procedure Dependencies

7.4   Basic Procedures for Daydreaming

DAYDREAMER, in its basic form, is built out of seven procedures, whose dependencies are shown in Figure 7.1. The top-level *emotion-driven control* procedure repeatedly selects the most highly motivated concern, and invokes the *planner* on this concern. The planner invokes *rule application* to carry out one step of planning for the concern. If the top-level goal of the concern succeeds or fails, the planner invokes *concern termination*. Rule application invokes *inference rule application* and *planning rule application*. Inference rule application invokes *concern initiation* if a new top-level goal is inferred.

7.4.1   Emotion-Driven Control

The emotion-driven control procedure is as follows:

1. To get off the ground, invoke inference rule application.

2. Select concern with highest emotional motivation.

3. Invoke planner on that concern.

4. Go to step 2.

In performance mode, only personal goal concerns may be selected, while in daydreaming mode, both personal goal concerns and daydreaming goal concerns may be selected. The program switches modes if no concerns are eligible in the current mode. Optional physical objects are accepted as input upon switching modes from daydreaming to performance. If input is provided, the switch is not carried out.

On each cycle of the control loop, the strengths of needs and nonmotivating emotions are subject to decay. Emotions which decay below a certain threshold are removed.

7.4.2   Planner

A planning rule specifies how to split a goal into subgoals: The antecedent of the rule specifies the goal and the consequent specifies the subgoals. There may be several planning rules whose antecedent matches a given goal. That is, for any particular goal, there may be more than one way of decomposing that goal into subgoals. Each such method specifies an *alternative continuation* of the current hypothetical world state.

DAYDREAMER keeps track of alternative worlds using GATE contexts—snapshots of a hypothetical or real world at some moment in time. When several planning rules apply to a given subgoal, each alternative set of subgoals may be activated in a separate context. This enables DAYDREAMER to follow one line of planning, and then, should that line of planning prove unsatisfactory, back up and attempt an alternative plan already activated in its own isolated context. This is similar to the notion of nested transactions (Reed, 1978; Moss, 1981; E. T. Mueller, J. D. Moore, & Popek, 1983) in distributed operating systems and recovery blocks (Randell, 1975) in programming languages: If one means for accomplishing a subtask is unsuccessful, any changes already performed are undone and an alternative method may be invoked.

A planning rule is applied to a given *active* goal in a given context as follows: A new context is sprouted from the given context, and the subgoals specified by the consequent are asserted as new active subgoals in the sprouted context. Thus each alternative set of subgoals is maintained in a separate context. Subgoals in each such context are in turn split via planning rules into further subgoals, sprouting further contexts. In general, a tree of contexts results as shown in Figure 7.2. Eventually, subgoals are reached whose objectives are already true (retrieved in the context): a subgoal *succeeds* in a context when its objective is retrieved in the context or when all of its subgoals have succeeded. The objectives of some subgoals are *actions*; when such a subgoal succeeds, the action is *performed*. A concern terminates successfully when its top-level goal succeeds in a context. If all possibilities are exhausted without success, the concern terminates unsuccessfully and the top-level goal *fails*.

The planner conducts a *depth-first search* (Nilsson, 1980) through the space of contexts. That is, when several contexts are sprouted from the current context (corresponding to several alternative planning rule applications), one of those contexts is selected as the new current context. Contexts are in turn sprouted from this new current context. One of those contexts is again selected, and this process continues. If a dead end is reached—because no planning rules are applicable to the remaining subgoals—*backtracking* is performed. It should be noted that *the creative problem-solving aspect of* DAYDREAMER *arises, not from the depth-first search algorithm, but from the analogical planning, mutation, and serendipity mechanisms which are superimposed upon this basic algorithm.*

Specifically, the planner procedure for a given concern is as follows:

1. Invoke rule application in the current context of concern.

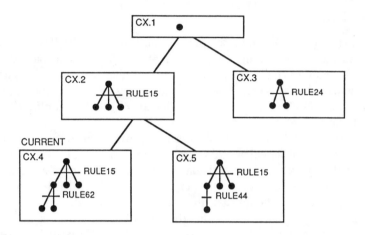

Figure 7.2: Tree of Planning Contexts

2. Mark current context as applied.

3. If top-level goal succeeded, invoke concern termination with "success."

4. Otherwise, if no contexts were sprouted:

   (a) *(Attempt to backtrack)* Until there are unapplied sprouted contexts of current context or until current context is equal to the root context of the concern,[4] set current context to the parent of the current context.

   (b) *(If backtrack unsuccessful, terminate concern)* If current context is root, invoke concern termination with "failure."

   (c) *(Else if backtrack successful, set current context)* Otherwise, set current context to an unapplied sprouted context of current context.[5]

5. Otherwise, set current context to one of the sprouted contexts.[6]

Traces of the execution of the planning algorithm are given in Appendix A.

DAYDREAMER explores hypothetical past, present, and future worlds. However, the program must also have a notion of the "real" world—both past and present. At any time, one context is designated as the *reality context*. This context contains the current state of the world.

---

[4]The root context of a concern is the real or imagined world state in which that concern was first initiated.

[5]The "best" context is chosen according to various heuristics: For nonanalogical planning, the *plausibility* of the planning rule which resulted in the context is used. For analogical planning, the metrics of episode *similarity*, *desirability*, and *realism* (see Chapter 4) are used.

[6]Again, the "best" context is chosen according to various heuristics.

Hewitt (1975) has pointed out that using contexts it is difficult to reason about situations in multiple contexts. He gives the example of two assertions, *Neil Armstrong is standing on the earth* and *Neil Armstrong is standing on the moon*, which are asserted in different contexts. The assertion, however, that *Neil Armstrong's weight in the first context is greater than his weight in the second context* cannot be made in either of these contexts alone. Instead of a context mechanism, Hewitt proposes the explicit use of situational tags (McCarthy & Hayes, 1969) within assertions (no longer confined to any particular context). In DAYDREAMER, references to contexts may be incorporated into facts if desired; however, since all facts must be asserted into some context, a global context may have to be created for this purpose.

Even so, as Sacerdoti (1977) points out, it is difficult to represent and use knowledge about planning in a context scheme. For example, critics (Sacerdoti, 1977) and meta-goals (Wilensky, 1983) are not easily implemented in such a scheme. Since DAYDREAMER is not concerned with daydreaming or learning about the planning process itself (although a complete model of daydreaming would certainly have to account for such metacognitions [Flavell, 1979]), meta-planning strategies are hard-coded in DAYDREAMER.

Contexts are employed in DAYDREAMER for two reasons: to enable backtracking to a previous choice point in daydreaming, and to maintain a trace of each explored alternative. A stack-based undo mechanism (such as that employed in Prolog [Clocksin & Mellish, 1981]) enables backtracking, but does not enable retention of alternatives once abandoned. It is important in DAYDREAMER to retain alternative traces because each may be indexed in episodic memory for possible future use; furthermore, the order of backtracking need not conform to a stack discipline.

It would not be particularly convenient to implement DAYDREAMER in Prolog (Clocksin & Mellish, 1981) because of the specialized control structure which daydreaming requires. The control structure of Prolog—planning rule application (backward chaining) to achieve a single goal—does not facilitate implementation of multiple active goals, interleaved application of inference rules (forward chaining), and the storage and later application of planning episodes.

7.4.3   Concern Termination

The concern termination procedure given a termination status is as follows:[7]

1. Destroy otherwise unconnected emotions associated with concern.

2. Create a new positive or negative emotion, whose strength is equal to the dynamic importance of the top-level goal associated with the concern.[8]

---

[7] Recall that concerns are initiated by inference rules. The procedure for concern initiation will be presented once we have presented the procedure for inference rule application.

[8] Response emotions are only created here if the top-level goal is a personal goal or the **REVENGE** daydreaming goal.

3. If termination status is "success," assert objective of top-level goal in the reality context.

4. If termination status is "failure," change top-level goal from active to failed.

5. Destroy concern.[9]

Before the emotion-driven control procedure is first invoked, the program must first be *initialized*. This consists of:

- initializing various global variables

- creating an initial reality context and loading initial facts into this context

- invoking the inference rule application procedure in the initial reality context, to start up the very first concerns of the program

7.4.4 Rule Application

The rule application procedure in a given context is as follows:

1. For each active subgoal in context whose subgoals have all succeeded or whose objective is retrieved in the context:[10] [11]

    (a) Change subgoal from active to succeeded.[12]

    (b) If objective was not retrieved in context, assert objective in context.

    (c) *(Detect possible fortuitous subgoal success)* If the concern of subgoal is not the current concern, create a *surprise* emotion and associate it with the concern of the subgoal.

2. If any objective was asserted in Step 1, invoke inference rule application in context.

3. If any subgoals succeeded in Step 1, go to Step 1.

---

[9]Any planning structure associated with the concern in the reality context is also garbage collected.

[10]If the subgoal contains variables, (a) a context must be sprouted for each set of bindings resulting from retrieval, (b) the body of this loop must be performed for each such sprouted context, and (c) the body of the loop must also instantiate all the subgoals of the current concern with the bindings, so that any instances of a given variable anywhere in the plan will take on the value specified in the bindings.

[11]It is also necessary to handle *protection violations*: If a succeeded subgoal is no longer retrieved in the context, it must be replanned. Script-like plans (for **M-DATE, M-RESTAURANT**, and so on) are not subject to this check.

[12]If subgoal happens to be an instance of a personal goal, an emotional response must be generated and associated with the current concern. The rule for emotional responses is discussed in Chapter 3.

4. Invoke planning rule application in context.

While planning for one concern, a subgoal of another concern will sometimes succeed. This is the case of a *fortuitous subgoal success*. In such a situation, a new *surprise* emotion is created and added as an additional force of motivation to the other concern. If, as a result, the other concern has a higher motivation than the current concern, processing will switch to that other concern on the next cycle.

The objectives of goals may be states or actions. If the objective of a goal is an action, a planning rule for that goal specifies preconditions for that action as subgoals. Once all the subgoals have succeeded, the action is performed by asserting it as an action into the context in Step 1b above.[13]

In performance mode, when the actor of an action goal is another person and all precondition subgoals of that goal have succeeded, the program stops and waits for input. The parsed input is compared (through unification) to the expected action. If the input action is not the same as the action goal objective, the goal fails (although some other subgoal or concern may still succeed as an indirect result, through subgoal success detection and serendipity). For example, in LOVERS1, when DAYDREAMER waits for input which is expected to be Harrison Ford agreeing to have dinner together, and this input is not received, this subgoal fails, in turn causing failure of the top-level goal.

In addition, after an action of the daydreamer is performed in performance mode, the program stops and waits for optional English input of an action or state in the external world. This input concept is then asserted into the context. The applicability of the concept to the current top-level goals and subgoals of the program, if any, is determined through (a) serendipity, to be described, and (b) detection of fortuitous subgoal success in the above procedure.

7.4.5 Planning Rule Application

The planning rule application procedure in a given context is as follows:

1. For each active subgoal in context whose concern is the current concern and having no subgoals;[14] for each planning rule whose antecedent unifies

---

[13]However, actions may be performed only in performance mode. If an action is about to be performed and the program is in daydreaming mode, the concern is placed into a *waiting* state. Only concerns in the *runable* state are eligible for selection in the emotion-driven control procedure. When the program switches from daydreaming mode to performance mode, all waiting concerns are changed to runable concerns. Furthermore, actions with variables in them may never be performed. If such an action is about to performed, the concern is placed into a *halted* state. A halted concern is only made runable when a serendipity or fortuitous subgoal success occurs for that concern.

[14]Subgoals of a given goal are expanded in a left-to-right sequence, rather than in all possible orders. For example, if a given goal has two subgoals, the second subgoal is not selected for expansion until the first subgoal has succeeded.

with the objective of the active subgoal:[15] [16]

(a) Sprout a new context from context.

(b) For each subgoal S in the consequent of planning rule, create a new subgoal of the active subgoal in the sprouted context whose objective is the instantiation of S with the bindings from the above unification.[17]

7.4.6   Inference Rule Application

Unlike planning rules, when the antecedent pattern of an inference rule is retrieved in a context, a fact specified by the consequent pattern is asserted in the *same* context. Thus whereas planning rules sprout contexts, inference rules do not (since inference rules specify inevitable results rather than alternative possibilities). Inferences rules also have *deletion consequents* (specified by the **delete** slot) which specify facts to be retracted from the context. In English descriptions of rules, any consequent preceded by "retract" is a deletion consequent.

The inference rule application procedure in a given context (which also may initiate new concerns), is as follows:

1. For each inference rule R whose antecedents are all retrieved in context;[18] for each set of retrieval bindings:[19] [20]

   (a) Instantiate each consequent and deletion consequent of R with the above bindings.

   (b) If the (single) instantiated consequent of R specifies a new top-level goal T, and (a) T is a daydreaming goal or (b) the current concern is a personal goal concern, invoke concern initiation on T, R, and the above bindings.[21]

---

[15]These two loops are nested.

[16]Finding planning rules whose antecedent unifies with a subgoal is optimized through use of the rule connection graph: only planning rules connected to the planning rule from which the subgoal was derived are considered. The rule connection graph is created when rules are loaded during program initialization (and updated if new rules are added).

[17]Instead of specifying subgoals, the consequent may specify code to be executed. It is the responsibility of this code to create appropriate subgoals. This feature is used in planning for daydreaming goals such as **RATIONALIZATION** and **REVERSAL**.

[18]This is optimized using the rule connection graph: An inference rule is only considered if it relates to a "touched" fact in the context (i.e., a fact asserted or modified since the last application of inference rules).

[19]These two loops are nested.

[20]Multiple applications of a given inference rule to the same antecedent facts with the same set of bindings must be inhibited.

[21]Concern initiation is not invoked if an identical top-level goal already exists.

(c) Otherwise, assert each instantiated consequent in the context,[22] and retract each instantiated deletion consequent from the context.[23]

2. If an assert or retract was made in Step 1, go to Step 1.[24]

The action taken in the above procedure when a new top-level goal is activated depends on whether the current concern is a personal goal concern or a daydreaming goal concern: When the current concern is a personal goal concern, any new top-level goals—both personal and daydreaming goals—result in the creation of a new concern. When the current concern is a daydreaming goal concern, only daydreaming goals result in the creation of a new concern. Personal goals activated in a daydreaming goal concern do not create new concerns, but rather are performed under the existing current concern.

Why should some top-level goals be performed under a separate concern, and others not? Each concern is intended to correspond to a real top-level goal of the system. Top-level goals which are activated in daydreaming goal concerns are not real top-level goals. Rather, they are top-level goals which *would be* activated in a particular hypothetical situation. If such top-level goals should motivate activity, such as planning to avoid an anticipated goal failure, then a daydreaming goal to fulfill this function, rather than a personal goal, is activated. Daydreaming goals activated in a daydreaming goal concern *do* create new concerns.

Although a personal goal concern modifies the reality context through planning, the only modifications which a daydreaming goal concern can perform to the reality context are: (a) assertion of new top-level goals, and (b) assertion of the eventual outcome—success or failure—of the daydreaming goal.[25]

Planning for daydreaming goal concerns takes place in imaginary contexts which are sprouted off of past or current reality contexts. Any imagined changes to the world state do not affect the reality context but rather are performed in their own isolated contexts. Personal goal concerns are carried out in the external world (e.g., LOVERS1 and FOOD1) and thus repeatedly sprout a context from the old realityindexcontexts;reality context which becomes the new reality context. As shown in Figure 7.3, personal goal concerns may be thought of as

---

[22] If the instantiated consequent can already be retrieved in the context, (a) the *strength* of the retrieved fact is incremented by the strength of the new instantiated consequent, and (b) no assertion is performed.

[23] If an instantiated consequent is a personal goal outcome, an emotional response must be generated and associated with the current concern. The rule for emotional responses is discussed in Chapter 3.

[24] One may insert at this point a limit on the number of times inferences are applied. There is no need for this in the current version of DAYDREAMER since the particular set of rules does not result in infinite inferencing loops. Johnson-Laird (1983, pp. 34-39) discusses how humans might limit inferencing.

[25] Daydreaming goal concerns have other side effects, including: (a) modification of emotional state, and (b) storage of generated planning sequences in episodic memory for future use.

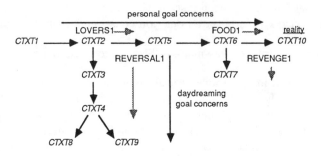

Figure 7.3: Personal and Daydreaming Goal Contexts

proceeding horizontally to the right with reality being the rightmost context, and daydreaming goal concerns (e.g., REVERSAL1 and REVENGE1) may be thought of as proceeding vertically, sprouting off of present or past reality contexts.

7.4.7   Concern Initiation

The concern initiation procedure for a given top-level goal, inference rule, and bindings is as follows:

1. Create a new concern.

2. For each emotion E specified by inference rule:

   (a) *(Daydreaming goal concern activated and motivated by existing emotion.)* If E is an existing emotion, associate E with the concern.

   (b) *(Concern motivated by new emotion.)* Otherwise, create a new emotion by instantiating E with bindings and associate the new emotion with the concern.[26]

3. If top-level goal is a personal goal, set the current context of the concern to the reality context and assert the top-level goal in this context.

4. Otherwise, set the current context of the concern to a new context sprouted from the reality context and assert the top-level goal in this new sprouted context.

7.5   Episodic Memory and Analogical Planning

In this section, we revise and augment the above procedures for analogical planning. Figure 7.4 shows the revised procedure dependencies.

---

[26]If the top-level goal is a personal goal, the strength of E specifies its intrinsic importance. If the top-level goal is a daydreaming goal, E specifies an additional motivating emotion.

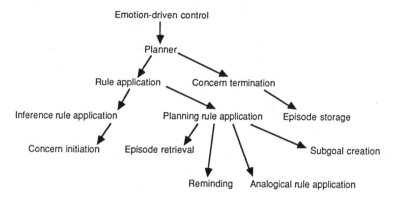

Figure 7.4: Revised Procedure Dependencies

7.5.1 Episode Storage

The *episode storage* procedure given an episode (a planning tree) and an index (either a concept not containing variables or a planning rule) is as follows:

1. *(Intern the index)* Set I to the index, if any, in a global list of unique indices which is equal to, or unifies with, the given index. If there is no such index, add the given index to the global list of unique indices and set I to the given index.

2. Create a link from I to episode.

Episode storage is invoked in two situations:

- *Top-level goal success*: The planning tree for the top-level goal and the subtrees for each of its descendant subgoals are stored as episodes. Each goal is indexed under the planning rule which was used to break that goal down into subgoals. The top-level goal is also stored under additional indices (for persons, physical objects, locations, organizations, and the like) which are derived automatically from the planning context. The *desirability* and *realism* values (see Chapter 4) are also evaluated and stored along with the episodes.

- *Program initialization*: Hand-coded episodes are stored. The programmer may specify various indices in addition to the planning rules corresponding to each subgoal.

7.5.2 Episode Retrieval

The *episode retrieval* procedure given a list of indices is as follows:

1. For each index I in indices:

(a) *(Intern the index)* Set I to the index, if any, in the global list of unique indices which is equal to, or unifies with, I.

(b) If there is such an index, for each episode E linked from I:

    i. Increment the number of marks associated with E.

    ii. Add E to a list of marked episodes.

    iii. If the number of marks associated with E is equal to the number of marks required for retrieval of that episode,[27] add E to a list of retrieved episodes.

2. For each episode E in the list of marked episodes, reset the number of marks associated with E to zero.

3. Return the list of retrieved episodes.

7.5.3   Reminding

A global *list of recent indices* is maintained. This is a short list (currently of length at most 6); addition of a new index (when the list is full) causes the oldest index to be dropped. Whenever new indices are added to the global list of recent indices, the *reminding* procedure is invoked for each episode retrieved using that list. The current overall emotional state (a positive or negative emotion) is included in the global list of recent indices.

The *reminding* procedure for a given episode is as follows:

1. If episode is already a member of a global *list of recent episodes* (currently of length at most 4), exit this procedure.

2. Add episode to the global list of recent episodes; drop the oldest episode from the list (if the list is full).

3. Reactivate emotions associated with episode.

4. For each index I of episode, if I is not already contained in the global list of recent indices, add I to that list and invoke episode retrieval on I and reminding procedure recursively on each retrieved episode.

5. (Unless this procedure was itself invoked as a result of an object-driven serendipity), invoke the serendipity recognition procedure for all concerns and episode.

Step 4 above is what enables the generation of associative streams of remindings: when an episode is retrieved by one index, its other indices are activated; these other indices potentially result in the retrieval of other episodes, whose other indices are in turn activated, retrieving further episodes; and so on.

---

[27] Two different retrieval thresholds are employed: the number required for retrieval during analogical planning (generally one index, the rule, is sufficient), and the number required in other situations (generally set as some percentage of the total number of indices). These thresholds may be set by the programmer for hand-coded episodes.

7.5.4 Revised Planning Rule Application

The planning rule application procedure, described above, is revised for analogical planning. This procedure, given a context, is now as follows:

1. For each active subgoal S in context whose concern is the current concern and having no subgoals:

   (a) *(Continue with existing analogical plan if still applicable in the target domain)* If an analogical episode E is associated with S and the objective of S unifies with the antecedent of the planning rule R associated with E, invoke the analogical planning procedure on context, S, E, R, and the bindings from the unification.

   (b) Otherwise, for each planning rule R whose antecedent unifies with the objective of S:[28]

      i. Set F to the value returned by invoking the episode retrieval procedure on a list containing R.

      ii. *(If analogical episodes retrieved, begin new analogical plans)* If F is not empty, for each episode E in F, (a) invoke the reminding procedure on E, (b) invoke the analogical rule application procedure on context, S, E, R, and the bindings from the unification.

      iii. *(Else, perform regular planning)* Otherwise, invoke the subgoal creation procedure for context, S, NIL, R, and the bindings from the unification.

7.5.5 Analogical Rule Application

The *analogical rule application* procedure for a given context, subgoal, episode, planning rule, and bindings is as follows:

1. *(Use episode to instantiate unbound variables in current plan)* Augment bindings with bindings resulting from the unification of (a) the antecedent and consequents of planning rule with (b) the concrete goal and subgoals of episode.[29]

2. Invoke the subgoal creation procedure on context, subgoal, episode, planning rule, and bindings.

7.5.6 Subgoal Creation

The *subgoal creation* procedure for a given context, subgoal, episode, planning rule, and bindings is as follows:

---

[28]Episodic planning rules are not eligible here. They may only be applied when they are contained within analogical episodes (for example, as a result of serendipity).

[29]Whenever a variable in the given subgoal is provided with a value, instances of that variable in *other* subgoals of the current plan must take on the same value.

1. Set C to a new context sprouted from context.

2. *(Break subgoal into further subgoals)* For each subgoal S in the consequent of planning rule:

   (a) Create a new subgoal of subgoal in C whose objective is the instantiation of S with bindings.

   (b) *(Carry along episode for analogical planning of subgoal)* Associate the appropriate subepisode (if any) of episode (if not NIL) with the new subgoal.

7.6   Other Planning

This section describes the modifications to the above procedures necessary to implement *forward other planning* and *backward other planning*.

7.6.1   Forward Other Planning

Mental states or *beliefs* of the daydreamer and of others coexist in each context. Every fact in a given context is assumed to be a belief of the daydreamer. Some of these beliefs are beliefs about the beliefs of others. While a belief of the daydreamer that it is raining in a certain location is represented as:

(**RAINING** obj *Location1*)

the belief that *Other* believes that it is raining in that location is represented as:

(**BELIEVE** actor *Other*
            obj (**RAINING** obj *Location1*))

"Objective" beliefs of others cannot be represented in this scheme. All facts are assumed to be beliefs of the daydreamer; thus one may only represent beliefs of the daydreamer about the beliefs of others. A belief of a belief of the daydreamer (who is represented as the slot-filler object *Me*), such as:

(**BELIEVE** actor *Me*
            obj (**RAINING** obj *Location1*))

is not currently permitted. Such a structure would be required to model metacognition (Flavell, 1979), which is beyond the scope of the present work.

The planning procedures described in the previous sections are modified to make use of a list called the *belief path*. This list determines the planning viewpoint. Normally, the belief path is set to (*Me*). However, in order to simulate the behavior of *Harrison*, the belief path would be set to (*Harrison Me*). This path means "Harrison as seen through my eyes." In order to simulate the simulation of the behavior of another person *Debra* by *Harrison*, the belief path is set to (*Debra Harrison Me*). This path means "Debra as seen through Harrison's eyes as seen through my eyes." The last element of a belief path must always be (*Me*).

DAYDREAMER rules are expressed relative to the person who is planning, specified by the *?Self* variable. When the daydreamer is planning, *?Self* is bound to *Me*. If planning for another person is being performed, *?Self* is bound to that person. In general, *?Self* is bound to the first element of the belief path.

All assertions, retractions, and retrievals in the planning procedures are performed relative to the current belief path. For example, if the belief path is (*Harrison Me*), then before performing an assertion, retraction, or retrieval on a given fact, it is enclosed in the following way:

```
(BELIEVE actor Harrison
         obj Fact)
```

If the belief path is (*Debra Harrison Me*), then the fact is enclosed as follows:

```
(BELIEVE actor Harrison
         obj (BELIEVE actor Debra
                      obj Fact))
```

The rule application procedure is revised to: (a) perform rule application with a belief path of (*Me*), and (b) perform rule application with a belief path of (*Other Me*) for each other person of current interest, indicated by the presence of an assertion of the form:

```
(OTHER actor Other)
```

in the current planning context. (This assertion is added, for example, when DAYDREAMER is in a **LOVERS** relationship with another person.)

7.6.2 Backward Other Planning

Instead of finding a rule whose antecedent goal unifies with:

```
(BELIEVE actor Other
         obj ?Mental-state)
```

one is found whose antecedent goal unifies with *?Mental-State*. In this unification, *?Self* is bound to *Other* rather than to *Me*. Then the subgoals are instantiated and enclosed in a **BELIEVE** before being asserted:

```
(BELIEVE actor Other
         obj ?Subgoal)
```

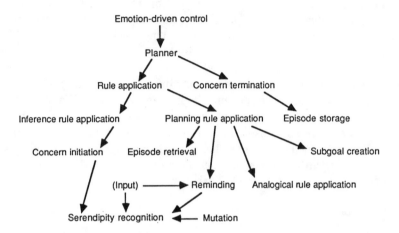

Figure 7.5: Further Revised Procedure Dependencies

7.7   Serendipity

In this section, we further augment the above procedures for serendipity. Figure 7.5 shows the revised procedure dependencies.

7.7.1   Serendipity Recognition

The *serendipity recognition* procedure given a concern and a concept or episode is as follows:

1. Find a top rule T whose antecedent unifies with an appropriate goal G of the concern.[30]

2. If given an episode, find an episodic rule, the bottom rule B, contained in the episode; otherwise, find a rule B having a consequent subgoal which unifies with the given concept.

3. Perform an intersection search from T to B in rule connection graph.[31] [32]

---

[30] If the concern is a personal goal, the personal goal is used. If the concern is a learning daydreaming goal, the associated personal goal is used. If the concern is an emotional daydreaming goal, the daydreaming goal is used.

[31] This search finds up to a certain number of paths (currently 3), including those with cycles, up to some maximum length (currently 8). A recursive procedure is employed which works its way down from T and up from B simultaneously; whenever an intersection is detected, the two partial paths are appended together and added to a list of found paths. To reduce search somewhat, a given rule may occur at most twice in a path.

[32] Only (a) nonepisodic planning rules, or (b) episodic planning rules contained in the given episode or the global list of recent episodes, are eligible in the search.

4. If a path is found:

   (a) Verify the path by (1) unifying the objective of G with the antecedent of the first rule of the path, and then (2) propagating bindings through progressive unification from the beginning to the end of the path while simultaneously constructing an episode (planning tree).[33]

   (b) If verification succeeds:

      i. If concern is a daydreaming goal concern, set C to a new context sprouted from the context in which G was first activated, and set the current context of concern to C.

      ii. Otherwise, set C to a new context sprouted from the reality context (which becomes the new reality context).[34]

      iii. Add the constructed episode as a new analogical plan for G in C.

      iv. Generate a *surprise* emotion and associate it with the concern.

The serendipity recognition procedure is invoked in several situations:

- *Input-state-driven serendipity:* When a (state or action) concept is received as input, the serendipity recognition procedure is invoked for all concerns and the given concept.

- *Object-driven serendipity:* When a physical object is received as input, the serendipity recognition procedure is invoked for all concerns, and for all episodes retrieved using the indices[35] of: (a) the given physical object, and (b) indices in the global list of recent indices. The reminding procedure is invoked for any retrieved episode which results in a serendipity.

- *Concern-activation-driven serendipity:* When a new concern is activated, the serendipity recognition procedure is invoked on that concern for all episodes in the global list of recent episodes.

- *Episode-driven serendipity:* When the reminding procedure is invoked for an episode (unless that procedure has already been invoked from object-driven serendipity), the serendipity recognition procedure is invoked for all concerns and that episode.

- *Mutation-driven serendipity:* When an mutated action is generated, the serendipity recognition procedure is invoked on the current concern and

---

[33]This procedure is discussed in greater detail and an example is given in Chapter 5.

[34]Any existing planning structure for G in C is also garbage-collected here.

[35]In this case, an episode is retrieved whenever the number of marks is one less than the number normally required: the serendipity, if any, serves to provide the remaining "index."

the mutated action.[36] If a serendipity occurs, then the mutation is deemed successful.

7.7.2 Action Mutation

The *action mutation* procedure for a given concern is as follows:

1. For each of the descendant leaf contexts C of the root context of concern; for each active goal in C whose objective A is an action; for each mutation[37] M of A:[38]

   (a) Invoke serendipity recognition on concern and M.

Action mutation is currently invoked in DAYDREAMER when all plans for achieving a given daydreaming goal have failed.

---

[36]In this case, the top rule for the intersection search is derived from the parent goal of the action subgoal being mutated.

[37]The strategies for mutating an action are described in Chapter 5.

[38]These three loops are nested.

Chapter 8

Comparison of Episodic Memory Schemes

In the human stream of thought, thinking about one past experience or episode naturally leads one to think about another. In DAYDREAMER, associated with each episode is a collection of *indices*. An episode is retrieved when a sufficient number of its indices are currently *active*. Whenever an episode is retrieved, its *remaining* indices are made active. Therefore, retrieval of one episode may activate new indices which result in the retrieval of a second episode, which in turn may activate new indices resulting in the retrieval of a third episode, and so on, producing a stream of recalled episodes. Retrieval of a given episode may in fact require indices resulting from more than one previously retrieved episode.

How are episodes stored? Given a collection of active indices, how are episodes which are stored under those indices retrieved? In this chapter, we examine three approaches which have been proposed in the past: *discrimination nets*, *spreading activation*, and *connectionist nets*. We then propose an alternative mechanism called *intersection indexing*.

8.1 Discrimination Nets

A discrimination net (Feigenbaum, 1963; Charniak, Riesbeck, & McDermott, 1980) is a tree structure whose branching nodes correspond to decisions, and whose leaves correspond to the final results of a series of such decisions. Discrimination nets were first used in the EPAM program (Feigenbaum, 1963; Feigenbaum & Simon, 1984) which models human learning of nonsense syllables, and have since been employed in a number of other artificial intelligence programs (Goldman, 1975; Kolodner, 1984).

Discrimination nets may be used as a mechanism for episode retrieval in the following manner: Each branching node consists of the possible values of a given

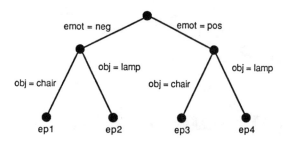

Figure 8.1: Discrimination Net

index, and each leaf designates a particular episode. As an example—which we will use to compare alternative indexing mechanisms—suppose that episodes are indexed under emotions and objects, and that each index takes on two values: negative and positive, chair and lamp, respectively. Further suppose that there are four episodes, each indexed under one of the four possible combinations of indices and index values. A discrimination net which organizes these episodes for retrieval is shown in Figure 8.1. If one wishes to retrieve an episode given the indices and index values of positive emotion and chair, one starts at the root of the tree, then traverses the positive emotion branch, traverses the chair branch, and finally arrives at the node for ep_3.

Such a discrimination net, however, does not facilitate retrieval when indices are presented in the wrong order (if, say, *chair* is presented before *neg*), when one or more indices are unavailable (if, say, only the index of *neg* is given), or both (if, say, only the index of *chair* is given). In such cases, traversal of the net cannot proceed beyond a certain point because an index is not available, or because the next available index does not appear on any of the branches at the current point. Humans, on the other hand, are able to retrieve episodes given partial sets of indices presented in various orders.

One solution to this problem is to employ a *redundant discrimination net* in which episodes are stored under several different index orders. A *fully redundant* discrimination net—an example of which is shown in Figure 8.2—enables a set of indices to be supplied in *any* possible order. If a partial set of indices is supplied, the traversal process will terminate at a nonleaf node. At this point, it is possible simply to retrieve all of the leaf episodes below this node.

However, protocols (Kolodner, 1984; Reiser, 1983) and other evidence (M. D. Williams & Hollan, 1981) suggest that humans must still generate the remaining indices before recall of an episode can occur. Kolodner gives the following example of this process of *elaboration*: Suppose former Secretary of State Cyrus Vance is asked to recall any diplomatic meetings about the Camp David accords with Menachim Begin. Further suppose there are several such episodes in his memory indexed under different places—Israel, Belgium, and the United States.

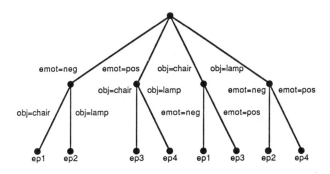

Figure 8.2: Fully Redundant Discrimination Net

At first, without a place index no episodes might be recalled. However, he may infer the place in which an event might have taken place from the nationalities or countries of residence of the participants. In this case, Israel and the United States may be inferred as places, enabling the remaining branches to be traversed and the episodes recalled. (But note that the episode which took place in Belgium will not be recalled unless suitable indices for *that* episode are somehow generated.) Kolodner provides a collection of general and domain-specific strategies for elaboration in memory retrieval. These strategies are implemented in the CYRUS computer program which stores events in the life of former Secretaries of State Cyrus Vance and Edmund Muskie, and answers English questions about those events.

Kolodner employs a redundant discrimination net in which each node is called an E-MOP and viewed as a *generalization* of all the leaf episodes below that node. Thus each E-MOP indexes individual episodes (leaf nodes connected directly to the E-MOP) and other more specific E-MOPs (connected nonleaf nodes). These episodes and E-MOPs are indexed by their differences from each other (in terms of different possible indices and values). For example, in Figure 8.2, the E-MOP reached by traversing the *emot = neg* link from the root generalizes those episodes involving negative emotions—ep_1 and ep_2—which are distinguished from each other by whether they involve a chair or a lamp. In CYRUS, E-MOPs are more than nodes: They are actually structures which contain information useful in elaboration (such as default index values, causal relationships between E-MOPs, and other inferences). In addition, E-MOPs provide generalizations useful for disambiguation and filling in details in text understanding (Lebowitz, 1980).

While the number of nodes in a nonredundant discrimination net grows linearly with the number of stored episodes,[1] a fully redundant discrimination

---

[1]Specifically, a nonredundant discrimination net requires $(eb-1)/(b-1)$ nodes, where e is the number of episodes and b is the number of branches from each node (assumed to be the same for each node).

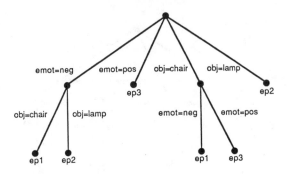

Figure 8.3: Redundant Discrimination Net with Delayed Expansion

net exhibits combinatorial growth.   Kolodner copes with this potential problem
in two ways: First, strategies are used to constrain the possible indices for
indexing episodes and E-MOPs under a given E-MOP—the net is thus only
partially redundant. Second, E-MOPs are created only when a new episode is
to be added into the net at a point where an episode is already present. When
a new E-MOP is created in this way, the two episodes are then indexed by their
differences. Thus expansion of the net is delayed. For example, Figure 8.3 shows
what the net looks like in this scheme before ep_4 has been added.

Because of the unrealistic space requirements of redundant discrimination
nets and further because we are not concerned with elaboration processes in
the current work, such nets are not employed in DAYDREAMER for episode
indexing and retrieval.

8.2   Spreading Activation Models

In general, spreading activation models of cognition (Quillian, 1968; Collins
& Loftus, 1975; J. R. Anderson, 1983) consist of a *processor* and a *long-term
memory* composed of a collection of *nodes*, each of which is connected to other
nodes by bidirectional *links*.[2] Associated with each node is an analog level of
activation, which may be viewed as modeling human short-term memory. The
processor may access any node whose activation is above a certain threshold;
such nodes are therefore said to be a part of *working memory*. The more active
the node, the faster that node may be accessed (or the higher priority that
node has to be processed). The activation of a node may thus be regarded as a
heuristic for the relevance of that node to current processing (J. R. Anderson,
1983). A node is said to be *retrieved* from long-term memory when it is brought

---

[2]The various spreading activation models differ in certain details from the one presented
here, which is a combination of the model of J. R. Anderson (1983) and that of Langley and
Neches (1981).

into working memory—when its activation rises above a certain level.

Activation of a node is continually spread to linked nodes via the process of *spreading activation*. That is, every node with a nonzero activation continuously leaks some of its activation to connected nodes. The purpose of spreading activation is to increase the activation of nodes related to those currently active; presumably, if active nodes are relevant to processing, so are those nodes associated with active nodes. The activation spread from a node to a linked node is proportional to the numeric *trace strength* of the linked node relative to the sum of the trace strengths of all the linked nodes. In other words, activation spread to connected nodes is divided up among those nodes in proportion to their relative strengths. As a result, the more links a node has to other nodes, the less activation each of those linked nodes will receive from the node. This is called the *fan effect* (J. R. Anderson, 1983).

Since activation is continually spread from one node to another, *decay* must be built into the mechanism in order to prevent the activation levels of nodes from forever rising. Decay will cause a node to leave working memory after a certain length of time (unless the activation of the node is replenished)— thus decay models the time-limited aspect of short-term memory. Continuous spreading of activation eventually leads to an even distribution of activation throughout the network; with the addition of decay, that activation tends toward zero. However, certain nodes, called *source nodes*, enable the introduction of new activation into the network. Source nodes may be created and added into working memory by the processor or through perception of the external environment.

Miller (1956) found that the "channel capacity" of short-term memory is limited to seven—plus or minus two—meaningful chunks (Simon, 1974) of information. This limited capacity is modeled in spreading activation models by imposing a limit on the total amount of activation permitted in the system; this imposes a (variable) limit on the number of nodes which can be in working memory at any given time.

How does working memory relate to subjective experience and consciousness? J. R. Anderson (1983) avoids any mention of consciousness, except in a note where he writes: "I do not intend to correlate active memory with what one is currently conscious of" (p. 309). Ericsson and Simon (1984, pp. 221-223) assume that thoughts which may be verbalized (after suitable recoding) correspond to the contents of working memory. They thus imply that (objective) data structures in working memory are identical with the subject's (subjective) focus of attention. However, they identify neither attention nor working memory with consciousness, since, as James (1890a, p. 285) has pointed out, attention picks out only some of the sensations which are a part of consciousness. Ericsson and Simon (1984, p. 222) admit that they do not choose to identify working memory with consciousness because their concern is with information relevant in task-oriented situations. However, as argued in Chapter 10, the entire gamut of conscious thoughts (including, for example, feelings) is subject to verbalization. Although Ericsson and Simon (1984) claim that "day-dreams and similar thoughts contain imagery that never is heeded directly" (p. 223), people are in fact able to describe such imagery to others (Varendonck, 1921; Klinger, 1971; Pope, 1978). Therefore, we can only ask the question: If these other components of consciousness are not a part of working memory, what are they a part of?

Spreading activation and network models of memory have been used by several researchers (Clark & Isen, 1982; Bower & Cohen, 1982) to explain emotion-dependent memory recall effects: Emotional states are linked in memory to the experiences involving those states. When one is in a certain emotional state, spreading activation primes experiences connected to that emotional state, making it more likely for those experiences to be recalled.

The basic architecture for connectionist models (Rumelhart, McClelland, & PDP Research Group, 1986; Waltz & Pollack, 1985) is similar to that for spreading activation as described here, with the addition of *inhibitory links* for spreading negative activation to connected nodes. Thus a node with a high activation level tends to reduce, rather than increase, the activation level of nodes connected by inhibitory links. An inhibitory link from one node to another indicates that the activation of one node is evidence that the other node should not be activated. Such links force the system to select one of the two nodes, but not both. Rumelhart, Hinton, and McClelland (1986) present a generalized model for parallel distributed processing which includes many spreading activation and connectionist models as special cases.

Continual flow of activation between connected nodes is computationally expensive in a serial system: The activation for every node in the network must be calculated based on the activation of each of its neighboring nodes on every clock tick. Because of this problem, spreading activation programming tools such as the PRISM language (Langley & Neches, 1981) require the programmer to invoke spreading activation explicitly and to specify the starting nodes for the process. In PRISM, activation does not spread back along paths from which it originated and spreading terminates when a node is reached which has already had activation spread to it from another path (however, although spreading does not continue beyond such a node, the activation from multiple paths is in fact summed at such intersection nodes). If the "spreading to a depth" strategy is selected, the process also terminates when a certain depth is reached. If the "spreading to a limit" strategy is selected, spreading terminates when the activation level decays below a certain threshold.

How may spreading activation be employed in retrieving episodes from memory? Figure 8.4 shows a network (loosely based on an example of J. R. Anderson, 1983, pp. 91-92) representing four actions (which are more specific versions of the example episodes we are using to compare various indexing mechanisms): failing at sitting down in a chair, failing to purchase a lamp, successfully purchasing a chair, and successfully turning on a lamp. For simplicity in this example, we assume that an episode consists of a single action performed on

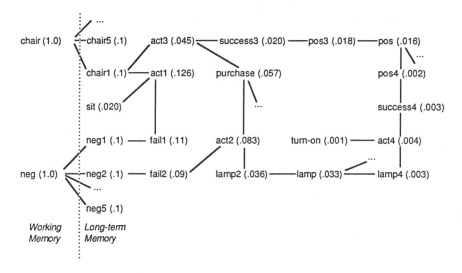

Figure 8.4: Network for Spreading Activation

behalf of a goal. In order to retrieve, say, episodes involving negative emotions and chairs, the nodes *neg* and *chair* are first activated, and then spreading activation is performed starting from those nodes. After this process completes,[3] $action_1$ ends up with an activation of 0.126, $action_2$ with 0.0830, $action_3$ with 0.0450, and $action_4$ with 0.00428. A relatively high activation level is attained by $action_1$ since activation is spread to it from two different paths. That is, $action_1$ is a node at which the paths emanating from *neg* and *chair* intersect. The node we desire to retrieve is in fact $action_1$ since it is an episode involving negative emotions and chairs. However, another such intersection node is $action_2$, because this node involves negative emotions and purchasing, which in turn involves chairs via $action_3$. Thus spreading activation may find episodes that are only indirectly related to the given retrieval indices. The threshold for adding items into working memory is set depending on how close of a match is desired for retrieval: if the threshold is set to 0.1, $action_1$ is added to working memory; if the threshold is 0.08, both $action_1$ and $action_2$ are added.

Thus spreading activation provides a promising mechanism for retrieval: For example, if it is desired to retrieve episodes similar to an episode which is currently active, one may simply spread activation from the current episode to

---

[3]The sample values shown in Figure 8.4 were calculated according to the "spreading to a limit" strategy of PRISM (Langley & Neches, 1981) with all trace strengths equal to 1.0, spread decay equal to 0.9, and nodes other than *neg* and *chair* having an initial activation of 0.0. It was further assumed that relation or class nodes (such as *neg*, *chair*, *purchase*, and so on) were connected to a total of 10 nodes (not shown).

related nodes. This works without regard to how these episodes are represented since a link between two nodes implies some relationship between those nodes in any representation scheme. The more relationships the current episode has in common with another episode, the more activation that other episode will receive. In effect, spreading activation is able to discover its own indices—it would appear that higher-level patterns can emerge out of the relationships already present in the network. In addition, the fan effect automatically assigns higher priority to more unique indices, since nodes which are less unique have more links to other nodes and thus are less likely to contribute to retrieval.

Unfortunately, there are several problems with spreading activation: The first problem is that the activation levels of nodes along the path to a desired episode—and even those not along such a path—are often higher than the activation of the episode itself. In fact, unless intersection effects are significant, they *must* be higher. As a result, all of these nodes are retrieved in addition to the desired episode nodes. These extraneous nodes are difficult for most programs to handle. For example, suppose we would like to retrieve $action_2$; we therefore set the working memory activation threshold to 0.08. In this case, not only are $action_1$ and $action_2$ added to working memory, but so are neg_5, $chair_5$, $chair_1$, neg_1, $fail_1$, neg_2, and $fail_2$. Many of these nodes are irrelevant; we only wanted to know about actions directly or indirectly involving negative emotions and chairs. One possible solution (although it is somewhat against the spirit of spreading activation) is to limit retrieval to certain classes of nodes, such as actions.

A second problem is that as more nodes and links are added to the network, or as desired episodes becomes more distant from their indices, the activation level of a desired episode becomes lower and lower. This is because the fan effect divides the leaked activation of each node among all its connected nodes. The result is that the activation levels of retrieval indices are very large in comparison with the activation levels of retrieved episodes. A possible solution (which again seems against the spirit of the mechanism) is to bring the activation level of an episode up to a certain high level once it is retrieved.

A third problem is that the difference in activation level—in a particular situation—between a desired episode and an undesired episode may be very slight. To an extent this effect occurs in our example: although $action_1$ is much more related to negative emotions and chairs than is $action_2$, their activation levels differ by a relatively small amount. Even so, after carefully setting the working memory threshold in order to retrieve the desired episode in one situation, the programmer may discover that this threshold does not retrieve desired episodes in other situations: the activation levels of episodes are in a completely different range.

Finally, spreading activation mechanisms often present a temptation to try to tune the system manually by modifying trace strengths or the network itself to get a more intuitive result: Why, for example, does $action_2$ end up with a higher level of activation than $action_3$? After all, $action_3$ is closer to *chair* than $action_2$ is to neg. It is not because $action_2$ is involved in an intersection. Rather, it is because of the somewhat irrelevant fact that $chair1$ is involved in two actions (and thus its activation is divided in two) while neg_2 only results from $fail_2$ which is only associated with $action_2$ (and thus in this case the activation is largely preserved). At this point, one may attempt to tune the network. However, this often degenerates into a never-ending process of tuning and retuning. Something is wrong if the proper operation of an artificial intelligence program depends on a very fine tuning of numbers by the programmer. Of course, if the system is able to tune itself, that is another matter entirely. Learning in connectionist models is in fact accomplished via automatic tuning of connection strengths (McClelland, Rumelhart, & Hinton, 1986). We consider these models in the next section.

The rule intersection search employed for serendipity detection in DAYDREAMER does not suffer from the same problems as general spreading activation, since activation spreads only through rule interconnections, rather than through every concept link or relationship. Rule intersection is therefore a more "content-based" scheme (Reiser, 1983) for spreading activation.

A redundant discrimination net trades space for fast retrieval time, while spreading activation trades time for a compact memory representation: Spreading activation is expensive in terms of processing time because the activation level of every node connected to retrieval indices up to a certain depth must be calculated. In contrast, with redundant discrimination nets it is merely necessary to perform a linear traversal of a number of levels of the tree corresponding to the number of indices (nonetheless, Kolodner's [1984] retrieval algorithm exhibits a combinatorial increase with the number of indices because it considers all possible paths "in parallel" until an episode is retrieved). Because of the problems with general spreading activation discussed above, this mechanism is not employed in DAYDREAMER for episode retrieval.

8.3 Connectionism

McClelland and Rumelhart (1986) present a connectionist model of memory storage and retrieval. The essential features of this model are as follows: Given a collection of episodes encoded in terms of a set of microfeatures, the mechanism automatically finds salient features of those episodes. That is, it classifies the episodes into one or more *schemata*—central tendencies of similar sets of episodes. Although single episodes cannot be retrieved from the network, similar episodes which have been repeated several times can be retrieved. Thus episodes and schemata are not distinguished in the network—all episodes are effectively stored as schemata.

Schemata are represented as a stable pattern of activation in the network. Learning of schemata is accomplished in an incremental fashion through alteration of the strengths of links when presented with a new episode. Retrieval is viewed as reinstating the activation pattern of a prior schema given part of that schema.

A subset of the nodes of the network (those that are not hidden) are used for the input and output of schema microfeatures. These external nodes are used to provide indices for retrieval (input) and to provide information retrieved about a schema (output). In order to perform a retrieval, appropriate index nodes are activated—any of the external nodes may be used. Retrieval then consists of reading the activation levels of the remaining external nodes. Thus retrieval consists of filling in the remaining portion of a schema given a part of that schema.

How might we use such a mechanism in DAYDREAMER? One task is for one or more active or recent episodes to produce a reminding of another episode which is related to those episodes. Since a connectionist network can only represent one episode at a time (McClelland & Rumelhart, 1986, p. 176), there must be multiple copies of the same network; the number of copies will be determined by the maximum number of episodes we wish to allow to be active at the same time. The external nodes of each of these networks are then connected together in order to enable episodes active in source networks to produce a reminding of an episode having similar external microfeatures in a target network. The connections must be unidirectional—that is, the external nodes of the source networks must charge the external nodes of the target network without the external nodes of the target network affecting the external nodes of the source networks. Without unidirectional connection of external nodes, all networks would settle into the same episode. Furthermore, this directionality will have to change depending on which networks are currently source networks and which is the target.

Each network will contain a collection of hidden nodes for representing episodes in a distributed fashion, and a set of external nodes for input and output. The external nodes will employ a local representation of the microfeatures of interest external to the module.[4] In particular, we will use the microfeatures of the example we have been using to compare various retrieval mechanisms. As shown in Figure 8.5, there will be one external node for each of the retrieval indices (negative emotion, positive emotion, chair, and lamp) and nodes for the information which we desire to retrieve, namely, the action performed in the episode. We then incorporate the relevant episodes into the network through use of a learning algorithm (see, for example, McClelland & Rumelhart, 1986, pp. 179-181). At this point, the connection strengths among the hidden nodes

---

[4]Local representations of external microfeatures—already determined to be significant by the programmer—are used in most of the example connectionist networks discussed by Rumelhart et al. (1986). One example (Rumelhart, Smolensky, McClelland, & Hinton, 1986, pp. 22-38) uses objects in a room as external microfeatures, another (Smolensky, 1986a, pp. 240-250) uses the resistance, voltage, and current in parts of an electronic circuit. In fact, these examples even involve local representations of the network itself, constructed manually by the programmer. An exception is provided by the learning mechanism of Rumelhart and Zipser (1986), which discovers salient features starting from an arbitrary initial set of microfeatures.

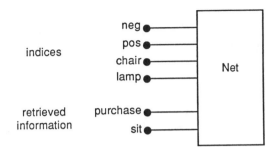

Figure 8.5: Connectionist Net for Episodes: First Attempt

encode the relevant episodes in a distributed fashion. If we now active various indices, the network will respond with the prototypical actions for those indices. Furthermore, we can activate actions and see what indices respond, or any other combination.

However, there are several problems with the above: First, for actions we need to be able to represent the type of action (such as *purchase* and *sit*), and the slots—sometimes called roles and cases (such as the agent, object, destination, instrument, and so on)—of the action. How do we represent actions in terms of the activation levels of external nodes? How, in particular, do we implement slots? For example, if an object is active, how do we know which slot it fills of the active action? Slots may be implemented by using clusters of mutually inhibitory nodes (Rumelhart, Smolensky, McClelland, & Hinton, 1986, pp. 33-34). (This assumes whoever is reading the external nodes is able to interpret this output network properly.) Thus we might use a network such as that shown in Figure 8.6. The drawback of this, however, is that each entity (such as *person*1) must be replicated for each possible slot which that entity can fill. If there is only one action per episode, this may not be so bad. However, if there are several actions in each episode, then each action type and slot must be repeated for each of the actions in the episode, as shown in Figure 8.7. All of these nodes, of course, have to be allocated in advance.

Second, observe what is happening here: Instead of the representation emerging automatically within the network itself, the representation is being constructed manually by the programmer at the external nodes of the network, as shown in Figure 8.8. A major claim of connectionist models is that they can be used to find a set of salient features in input patterns already encoded in terms of some set of microfeatures (McClelland, Rumelhart, & Hinton, 1986; Rumelhart & Zipser, 1986). Unfortunately, it is difficult to find an appropriate set of microfeatures when the system has to deal with complex concepts such as emotions, goals, actions, beliefs, persons, objects, and causal relationships. For example, what set of microfeatures should be used to represent actions? How, for example, will the system distinguish between a sequence of two sim-

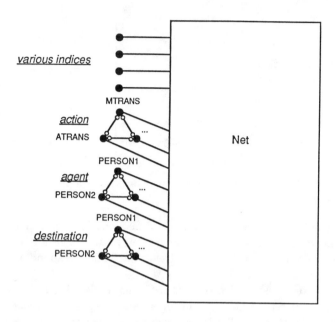

Figure 8.6: Connectionist Net for Episodes: Second Attempt

ilar actions performed in one episode, and two similar actions which occur in different episodes? In the first case, one would like to retain two actions. In the second case, one would like to form a generalization. The connectionist model of memory presented by McClelland and Rumelhart (1986) would generalize in both cases. In order to avoid difficulties such as this, the connectionist designer is forced to examine such questions as: What is an episode? How are episodes represented? Should an episode be represented as a sequence of actions? Should the sequence of actions be ordered? Do episodes involve causal relations among those actions? What kinds of causal relations are there? These are exactly the kinds of problems which knowledge structures such as scripts, plans, and goals (Schank & Abelson, 1977) are intended to address.

Although connectionist models do not enable one to bypass the difficult problems of representation, they do provide a degree of automatic generalization once appropriate representations have been found. Nonetheless, it is important to recognize the limitations of this generalization mechanism: First, unlike mechanisms for explanation-based generalization (DeJong, 1981), connectionist learning procedures cannot distinguish relevant features from irrelevant features in a small number of trials. Consider a story about kidnapping which describes the kidnap victim as a woman whose hair is dark brown and whose father is rich. A connectionist model will treat each of these features equally. A human, on the other hand, knows that certain features are more relevant to the goals of

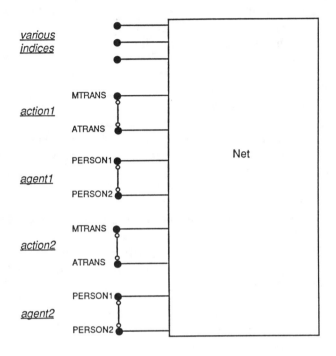

Figure 8.7: Connectionist Net for Episodes: Third Attempt

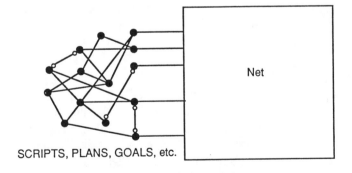

Figure 8.8: Connectionist Net for Episodes: Fourth Attempt

the kidnapper and hence are more useful as components of a generalization.

Second, since generalizations are emergent properties of the network rather than representational structures, connectionist mechanisms have difficulty in using those generalizations as a basis for the representation of further concepts (McClelland & Rumelhart, 1986, pp. 209-214). For example, one task in DAY-DREAMER is as follows: When the system is reminded of an episode which is related to the current daydream, components of that episode may be incorporated into the daydream. A connectionist scheme has one method of combining episodes—this method might be described as *interpolation*. Daydreaming requires more interesting ways of composing episodes: For example, employing an episode as a subepisode of the current daydream, employing only parts of the episode in the current daydream, modifying parts of the episode before employing them in the current daydream, and so on.[5] Viewing episodes and schemata as emergent properties of a network is an intriguing idea, but—in its current formulation—of little use in the generation of complex behaviors such as daydreaming.

At the very least, connectionist models provide a mechanism for episode retrieval similar to that provided by spreading activation. How do connectionist models differ from spreading activation models? Distributed connectionist models create new concepts by tuning the weights of the links in the network to create a new stable pattern of activation. However, only one concept can be represented at a time in such a scheme. This presents a serious difficulty since it is often necessary to relate two concepts and it is often necessary to have multiple instances of a given concept. One solution is to use a separate network for each concept instance (Hinton, McClelland, & Rumelhart, 1986; Hinton, 1981). McClelland (1986) presents a mechanism called *connection information distribution* which replicates modular networks on a demand basis. In local spreading activation schemes, a new concept is created simply by adding a node to the network and connecting it via appropriate links to the nodes to which it is related. A local connectionist scheme which allows creation of new nodes, however, is quite similar to spreading activation networks.

8.4   Intersection Indexing

Instead of employing a general spreading activation mechanism for indexing, we propose the use of what we call *intersection indexing*. This scheme employs a network consisting of a node for each index and a node for each episode. Each index node has a connection to those episode nodes indexed under that index. An intersection net for our example is shown in Figure 8.9.

A marking algorithm (Quillian, 1968; McCarthy et al., 1965, pp. 42-43) is employed for retrieval. Given a collection of retrieval indices, the connections from each index to each episode are traversed and a mark count associated

---

[5]The application of episodes in daydreaming is discussed in Chapter 4.

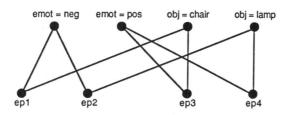

Figure 8.9: Intersection Indexing Net

with the episode (initially zero) is incremented. Whenever the mark count of a given episode reaches a certain threshold specified by that episode, that episode is retrieved. The number of indices required for retrieval is determined on an episode by episode basis. If all indices are required for retrieval of a given episode, the threshold specified with that episode should be equal to the number of connections to that episode. Often, however, it is desired to retrieve an episode when some percentage of its indices are active. In DAYDREAMER, episodes are indexed in a redundant fashion so that retrieval is possible even if only some of a given episode's indices are active.

Index intersection is in the spirit of spreading activation, but without all its disadvantages. It is similar to the nets used in PANDEMONIUM (Selfridge, 1959). Because only one level of links is traversed, the retrieval algorithm is not expensive. The number of links traversed is equal to the total number of episodes connected to the given indices. Nor does intersection indexing have the prohibitive space requirements of a fully redundant discrimination net.

At any time, DAYDREAMER maintains a short list of *active indices*. *Whenever an episode is retrieved (using the currently active indices), the remaining indices of that episode are activated.* Active indices and intersection indexing are illustrated in the daydream ROVING1 (see page 5). An episode at Gulliver's Restaurant (recalled directly by the **ROVING** daydreaming goal) activates the indices of Marina del Rey (the location of the restaurant) and Steve (a character in the episode). The index of Marina del Rey in turn leads to retrieval of an episode (whose reminding threshold is one index) involving a job located there. Retrieval of this episode in turn activates the index of Venice (a town along the bus route to the job). Now the combined active indices of Steve and Venice, resulting from two different episodes, lead to the retrieval of a third episode (whose reminding threshold is two indices) involving Steve at Venice Beach.

Chapter 9

Review of the Literature on Daydreaming

Although the present work is, to our knowledge, the first theory of daydreaming implemented as a running computer program, over the years many investigators have discussed, researched, and theorized about daydreaming and the related phenomenon of night dreaming. Our theory has been enormously influenced by—and is indebted to—much of this previous work, which is reviewed in this chapter.

9.1   Previous Work in Daydreaming

This section reviews past work in daydreaming, from the associationist theories of long ago to modern psychological research into the human stream of thought.

9.1.1   Associationist Theories of Thought

The earliest theories of daydreaming are based on the principle of *association* first formulated by Aristotle (Beare, 1931)—that ideas experienced together or in close succession tend to become linked, so that when one of those ideas is later thought of, the others are likely also to follow. Aristotle was concerned with the role of association in *memory*. He proposed that when one wishes to recall a given experience, one first selects an experience associated with that experience (via contiguity in space or time, similarity, or contrast) and then the desired experience follows automatically (H. C. Warren, 1921). In *Leviathan*, Hobbes (1651/1968) proposed that *thinking* is partially determined by the principle of association:

211

> All Fancies are Motions within us, reliques of those made in the
> Sense:[1] And those motions that immediately succeeded one another
> in the sense, continue also together after Sense. ...But because in
> sense, to one and the same thing perceived, sometimes one thing,
> sometimes another succeedeth, it comes to passe in time, that in the
> Imagining of any thing, there is no certainty what we shall Imagine
> next; Onely this is certain, it shall be something that succeeded the
> same before, at one time or another. (ch. 3, pp. 8-9)

Hobbes distinguished two kinds of thinking governed by this principle—*unguided*
and *regulated*:

> This Trayne of Thoughts, or Mentall Discourse, is of two sorts. The
> first is *Unguided, without Designe*, and inconstant ...In which case
> the thoughts are said to wander, and seem impertinent one to an-
> other, as in a Dream. Such are Commonly the thoughts of men,
> that are not onely without company, but also without care of any
> thing; though even then their Thoughts are as busie as at other times
> ...And yet in this wild ranging of the mind, a man may oft-times
> perceive the way of it, and the dependance of one thought upon an-
> other. For in a Discourse of our present civill warre, what could seem
> more impertinent, than to ask (as one did) what was the value of a
> Roman Penny? Yet the Cohærence to me was manifest enough. For
> the Thought of the warre, introduced the Thought of the delivering
> up the King to his Enemies; The Thought of that, brought in the
> Thought of the delivering up of Christ; and that again the Thought
> of the 30 pence, which was the price of that treason: and thence
> easily followed that malicious question; and all this in a moment of
> time; for Thought is quick.

> The second is more constant; as being *regulated* by some desire,
> and designe. ...From Desire, ariseth the Thought of some means
> we have seen produce the like of that which we ayme at; and from
> the thought of that, the thought of means to that mean; and so
> continually, till we come to some beginning within our won power.
> (ch. 3, p. 9)

Thus Hobbes recognized a classic form of daydreaming—undirected thought,
carried out when one is otherwise unoccupied—and the fact that this thought
follows a certain progression. He also recognized the form of thought which is
directed by desires (goals) and consists of what we would now call planning.

After Hobbes, other British associationists—John Locke, David Hume,
David Hartley, James Mill, John Stuart Mill, Alexander Bain, among others—
continued to develop and explore the principle of association; however, the basic

---

[1]By "Sense," Hobbes means sensory experience.

notions remained the same. H. C. Warren (1921) and Rapaport (1974) provide detailed reviews, not repeated here, of the work of these and other philosophers and psychologists.

There are several problems with strictly associationist theories of daydreaming: First, associationist theories do not account for the fact that, out of the multitude of ideas associated with a given idea, only a small number of those ideas actually come to mind. After all, when one sees a cat, one is not overwhelmed with every idea and experience associated in some way with cats. More likely, one will be reminded of a cat one once knew, or recall that one has to pick up some cat food at the grocery store. If associations only partially determine the thoughts which follow a given thought, what are the other factors which select a *single* thought? How is this thought selected depending on the particular context?

Second, associationist theories lack a theory of representation of concepts. How, for example, does one form associations based on similarity (if such associations are viewed as distinct from those based on space and time contiguity)? When two ideas are claimed to be similar, what, exactly, is similar about those ideas? How are two ideas recognized to be similar? How are ideas represented in the first place?

Third, associationist theories of daydreaming do not account for its *generative* aspect. Thought does not consist merely of the reproduction of past ideas, but rather involves the construction of new ideas out of existing ideas and components of existing ideas. For example, it is possible for one to imagine a hypothetical sequence of events which one has never before imagined or experienced in that particular combination.

Associationism has been revived in recent years under the names of *spreading activation* (Quillian, 1968; J. R. Anderson, 1983) and *connectionism* (Rumelhart et al., 1986). This work has suggested solutions to some of the problems associated with classical associationism. Nonetheless, difficult problems remain.

Our view is that in order for associationist mechanisms to be useful, they must be extended with other processing mechanisms and with rich conceptual representations. As, for example, James (1890a, pp. 571-577) has proposed, other factors such as recency, vividness, and emotional congruity must be employed in order to select from among multiple associations. The important problem thus becomes that of *representation*: How are experiences represented? What representations enable recognition of salient, important, and interesting experiences?

9.1.2 James's Stream of Thought

It was James (1890a) who introduced the phrases "stream of thought" and "stream of consciousness" in a chapter of his *Principles of Psychology*. There he discusses five characteristics of thought, briefly summarized (without bothering to discuss possible objections) as follows:

- Thoughts are unique to a particular personal consciousness; one cannot directly know the thoughts of another consciousness.

- Thought is constantly changing; a thought state can never be identical to a past thought state; even a particular fact is seen from a new angle each time the same fact recurs in thought.

- Thought is a continuous stream; despite any time gaps (due to sleep, coma, and so on) or breaks in the quality of consciousness, there is continuity of the self; changes in the quality of consciousness are never "absolutely abrupt"; the stream of thought consists of a series of flights from one resting point to another.

- Thought need not distinguish between the object of thought and thought itself; thoughts are not composed out of other thoughts, but, rather, are undivided entities.

- At any time, thought chooses one part of its object over another; presented with many sensations, thought selects certain ones and rejects others.

9.1.3 Freud's Theory of Dreaming and Daydreaming

S. Freud's (1900/1965, 1908/1962) theory of daydreaming is based on his influential theory of night dreaming. The fundamental assumption of this theory is that dreams are significant—that is, rather than being random nonsense, dreams have a coherent meaning. Furthermore, this meaning is hidden: The dream recalled upon awakening—the "manifest dream" or "dream-content"—is distinguished from the underlying substance of the dream—the "latent dream" or "dream-thoughts." It is the operation of "censorship" which inhibits the direct expression of the unconscious dream-thoughts in consciousness and thus in the manifest dream. The dream-thoughts may be represented in the dream-content only once they have been appropriately distorted. At the core of the dream-thoughts is a repressed infantile wish. Each manifest dream involves the fulfillment, in disguise, of such an wish.

Freud proposes that dreams are instigated by significant recent events as well as significant recent internal experiences (such as a train of thought or memory experienced during the day). These significant experiences, which stir up repressed wishes, are concealed in the manifest dream.

The technique of "free association" is used to interpret a dream—that is, to uncover the latent dream corresponding to a manifest dream. In this technique, the patient is to notice and report to the psychoanalyst any and all thoughts which come to mind regarding elements of the dream. The patient must allow thoughts to enter into consciousness no matter how unimportant, absurd, or unpleasant those thoughts may seem. The purpose of dream interpretation is to discover pathological ideas (as part of the dream-thoughts) responsible for the patient's problems; it is theorized that these stuctures may be removed by unraveling them in psychotherapy.

The "dream-work" transforms the dream-thoughts into the dream-content through "condensation" and "dream-displacement." Condensation represents many elements of the dream-thoughts as a much smaller number of elements of the dream-content. The reduction does not occur, for example, by a process of omission, but through the selection of "'nodal points' upon which a great number of the dream-thoughts [converge] ..." (S. Freud, 1900/1965, p. 317) (Note the similarity of this mechanism to that of *spreading activation.*) Each element of the manifest dream is therefore *overdetermined*—determined by many elements of the latent dream. It is also possible for single elements of the dream-thoughts to be represented by several elements of the dream-content. In dream-displacement, the "center" of the dream-content—that is, the apparent nucleus of significance in the dream-content—is skewed so that it does not correspond with the "center" of the dream-thoughts—the true nucleus of significance.

The methods of "composition" and "identification" enable condensation and dream-displacement: In composition, several persons having some significant element in common in the dream-thoughts are represented in the dream-content as a single, composite figure derived from those persons. However, in this composite figure the significant element is deemphasized or dropped, and indifferent features unique to one or other of the persons are combined and emphasized. In identification, the several persons are represented as one of those persons and indifferent aspects of that person are emphasized. Identification and composition enable dream-displacement by representing insignificant elements as significant in the manifest dream, and by avoiding representation of significant elements.

Freud notes that emotions in dreams often do not correspond to those one might experience if one were to encounter the situation represented in the dream while awake—for example, one may feel fear when there appears to be nothing in the dream to fear, or one may fail to experience fear in a dream in which there is much to fear. In order to account for this, he proposes that although ideational material of the dream-thoughts is transformed, the emotions remain unchanged. Thus each emotion in the manifest dream may be traced to some dream-thought. In addition, he proposes that emotions associated with dream-thoughts may be suppressed and that emotions can be turned into their opposite—both consequences of the censorship of dreams.

"Secondary revision" is the process whereby a manifest dream is rendered more coherent—in a fashion similar to the way waking thought imposes order on perceptual material. Secondary revision, operating simultaneously with other dream processes, "seeks to mould the material offered to it into something like a day-dream." (S. Freud, 1900/1965, p. 530) Alternatively, if an appropriate, possibly unconscious, daydream has already been formed, secondary revision may adopt this daydream and incorporate it—in part or in whole—into the dream-content.[2]

Freud proposes that daydreams, like dreams, are fulfillments of an ungratified wish:

> Like dreams, [daydreams] are wish-fulfilments; like dreams, they are based to a great extent on impressions of infantile experiences; like dreams, they benefit by a certain degree of relaxation of censorship. If we examine their structure, we shall perceive the way in which the wishful purpose that is at work in their production has mixed up the material of which they are built, has re-arranged it and has formed it into a new whole ... (S. Freud, 1900/1965, p. 530)

> The motive forces of phantasies are unsatisfied wishes, and every single phantasy is the fulfilment of a wish, a correction of unsatisfying reality. These motivating wishes ... fall naturally into two main groups. They are either ambitious wishes ... or they are erotic ones. ... (S. Freud, 1908/1962, pp. 146-147)

S. Freud (1908/1962) also proposes that daydreaming arises as a substitute for childhood play:

> [T]he growing child, when he stops playing, gives up nothing but the link with real objects; instead of *playing*, he now *phantasies*. He builds castles in the air and creates what are called *day-dreams* ... (p. 145)

9.1.4 Bleuler's Autistic Thinking

Bleuler (1912/1951) discusses a form of thought similar to daydreaming called "autistic thinking." Autistic thinking occurs in the thought processes of schizophrenics who have withdrawn into a dream world, ignore reality, and fulfill their wishes—or believe themselves to be persecuted—in their delusions. Bleuler characterizes autistic thinking, which is present to some degree in every person, as follows:

- Autistic thinking is concerned with the fulfillment of wishes; it seeks pleasant ideas and avoids painful ones.

- Affects strive to perpetuate themselves in autistic thinking; ideas which are in accord with a given affect are favored over those which are not. For example, a depressed person tends to continue to have negative thoughts.

---

[2]S. Freud (1900/1965, pp. 533-537) proposes that such a mechanism accounts for dreams which (it is claimed) occur instantaneously or in compressed time (see, for example, Maury, 1878): The entire daydream does not have to be reviewed while asleep; it only must be "touched on." The daydream is then later elaborated when the dream is recalled while awake. This is similar to the "cassette theory of dreams" discussed by Dennett (1978, pp. 136-144). Incidentally, recent dream research seems to indicate that dream time is proportional (Dement & Kleitman, 1957) or identical (LaBerge, 1985, pp. 74-78) to real time, despite the myth of accelerated dreaming.

- Autistic thinking ignores the rules of logic, allows the impossible to be possible, overlooks obstacles, and assumes goals have been attained that have not.

- Autistic thinking permits contradictory wishes to be maintained simultaneously. For example, in autistic thinking one may wish to be a carefree child and a powerful executive at the same time.

- Autistic thinking disregards time relationships—an old memory can become real again.

Bleuler states that a degree of autism may be of some use in everyday life; however, an excessive amount can have negative consequences.

Bleuler explains the prevalence of autistic thinking in the play of children, in daydreaming, in dreams, in mythology, and in pathological phenomena as follows:

- Autistic thinking provides a means of achieving pleasure, which is an adaptive goal in humans: "the anticipated pleasure enforces consideration and preparation prior to an endeavor and enhances the energy of the striving." (p. 434)

- Autistic thinking enables the abreaction of (releasing of the repressed emotion resulting from) negative experiences.

- Autistic thinking provides exercise for thinking ability (especially in children).

- Autistic thinking enables one to overestimate the value of a goal and to overlook potential obstacles; this increases the desire to achieve the goal and decreases the possibility of becoming discouraged.

9.1.5 Varendonck's Psychology of Daydreaming

J. Varendonck published a book in 1921 called *The Psychology of Day-dreams* containing retrospective reports of over 30 of his own daydreams, detailed analyses of many of these daydreams, and conclusions regarding the process of daydreaming (or what he calls "fore-conscious thought" as distinguished from unconscious and conscious thought).[3] Based on my own tabulation, about one-third of the reported daydreams occurred before falling asleep at night, another

---

[3]Rapaport (1951) has pointed out that Varendonck's term "fore-conscious" stems from a misunderstanding of Freud's constructs: The "unconscious" refers to thoughts which have a resistance to becoming conscious, while the "preconscious" refers to thoughts, not currently conscious, which may become conscious at any time. Thus one does not say that dreaming is carried out through "unconscious thought," although the study of dreaming is useful in uncovering unconscious thoughts (S. Freud, 1900/1965). Similarly, daydreaming is not carried out preconsciously (or fore-consciously). We may assume that in using the term "fore-conscious," Varendonck simply wanted to suggest the special state of consciousness in which daydreaming appears to occur and its involuntary, undirected nature.

third while reading (during the day or in the morning or night in bed), and the remainder before getting up in the morning, while on a train, while walking, and during a conversation. Although many of Varendonck's reported daydreams occurred prior to sleep, few of them have the nonsensical or bizarre quality characteristic of the hypnagogic reveries common in the transition from wakefulness to sleep. Varendonck was probably in a wakeful state when the reported daydreams occurred—in many instances he admits that he was having trouble getting to sleep. We may thus consider Varendonck's presleep reveries as bona fide "daydreams" (see, however, Wallas, 1926, pp. 72-74, for an alternative view).

Varendonck (1921) postulates that daydream sequences originate with an emotionally-emphasized recollection, which is triggered through association with an external stimulus or with "ideas which are selected experimentally" (p. 40), or which "simply obtrudes itself upon" (p. 179) one's attention. Often the memory is a "day remnant" (or recent experience). For example, one of Varendonck's daydreams was triggered by the sound of an exploding shell outside, which reminded him of a previous shell that was dropped on the village a week earlier causing many casualties.

Once started, a daydream, according to Varendonck, is composed of a sequence of hypotheses and rejoinders which attempt to solve one or more problems or satisfy one or more wishes. Each hypothesis is a suggestion, supposition, or question (often of the form "What if ...?" or "How shall I ...?"). Each rejoinder is an objection to, rejection of, or acceptance of the suggestion, or an answer to the question. For example, Varendonck had the following experience when he was working in a Belgian military hospital: An orderly brought a request from the head nurse to Varendonck but "spoke and behaved so impolitely that [Varendonck] flatly refused to grant it" (Varendonck, 1921, p. 64). The orderly left threatening Varendonck with the vengeance of the head nurse. Varendonck then sent a letter to his commanding officer demanding punishment of the orderly. The same night while trying to get to sleep, Varendonck had a daydream concerned with ensuring the proper punishment of (or taking revenge on) the orderly. He analyzes a portion of this daydream into hypotheses and rejoinders as follows:

> What if I warned [the head nurse] (so as to dispose her in my favour and arouse her feelings against the culprit)? I start composing the letter.
>
> If I enclosed my visiting-card? No reply (suggestion accepted).
>
> If I asked Captain Y. to send his corporal with it? He is too busy.
>
> And if he went after office-hours? No reply (suggestion accepted).
>
> If I added a copy of my report? A great waste of time to copy it.
>
> If I tore my own report out of my notebook? I might still want it.
>
> If I asked her to return it? (Understood: she might not do so.)

If I went to see her myself? I shall put on my best uniform.

If I handed my card to an orderly and asked for an interview? (Suggestion accepted and visualized, as well as the interview itself.) ... (p. 75-76)

Varendonck concludes that such hypotheses and rejoinders which attempt to solve a problem (such as how to punish the orderly) are generated through the adaptation of previous memories: "The content of every single created thought, in the form either of a question or an answer, is nothing else than an element borrowed from the store of memory and applied to the present need as created by an affect." (Varendonck, 1921, p. 183)

According to Varendonck, a daydream terminates "at a moment of mental passivity under the influence of some affect which causes [streams of thought] to rise to the surface ... " (p. 180) or when another memory is recalled in response to an external stimulus.

9.1.6 Green's Study of Daydreaming and Development

G. H. Green (1923) explored the relationship of daydreaming to childhood development. He proposed several stages, each marked by daydreams of a particular theme: In the "age of imagination," beginning around the third year, the play and daydreams of the child are concerned with an imaginary companion. By the next stage, beginning around the tenth year, daydreams are no longer accompanied by external actions; the "gang or team fantasies" of this stage involve the daydreamer as the leader of a group engaged in some adventure or quest. In the "romantic fantasy" stage, beginning around the fourteenth year, the daydreamer imagines being with a companion of the opposite sex—sitting or walking together in beautiful surroundings.

Green views daydreams as fulfillments and temporary partial gratifications of repressed infantile wishes (as does Freud), and as representing an unconscious enduring attitude toward the world.

9.1.7 The Research of Singer and Colleagues

J. L. Singer and McCraven (1961) developed a daydream questionnaire in order to answer such questions as: How frequent is daydreaming in different samples of the general population? What is the typical content of daydreaming? Does the frequency and content of daydreaming depend on the age, sex, and other background factors of the individual? Do certain patterns of daydreaming go together? The questionnaire was composed of a large number of daydreams collected from various sources. Subjects were to report whether they had experienced a given daydream and with what frequency. The earliest versions of the questionnaire contained daydreams such as "I suddenly find I can fly, to the amazement of passers-by" and "I picture an atomic bombing of the town I live in" (J. L. Singer, 1975, p. 53).

Over the years, this questionnaire evolved into what is now called the Imaginal Processes Inventory (IPI) (J. L. Singer & Antrobus, 1963, 1972), consisting of 344 items divided into 28 groups. The IPI contains items representing common daydreams, as well as statements designed to assess other personality characteristics of the individual. Each item of the IPI is rated by the individual along a 5-point scale according to how true that statement is of that person. A few items from the IPI are as follows:

Daydreams or fantasies make up ___ % of my waking thoughts. (2)
In my daydreams, I have succeeded in becoming a respected figure in my field of work. (106)
I often wonder about the life of a person I happen to see standing at a window of an apartment building. (123)
While reading, I often slip into daydreams about sex or making love to someone. (244)
I see myself attaining revenge against someone who has deceived me. (342)

J. L. Singer (1975) concludes from the results of administering such questionnaires to many individuals that:

[D]aydreaming is a remarkably widespread occurrence when people are alone and in restful motor states. It is a human function that chiefly involves resort to visual imagery and is strongly oriented toward future interpersonal behavior. (p. 55)

Several studies (Cundiff & Gold, 1979; Giambra, 1977; J. L. Singer & Antrobus, 1972) have supported the reliability of the IPI. That is, when subjects are tested more than once they are consistent in their responses.

Factor analyses of responses to the IPI (J. L. Singer & Antrobus, 1963, 1972; Huba, Segal, & J. L. Singer, 1977; Giambra, 1980; Segal, Huba, & J. L. Singer, 1980) have yielded at least the following three major factors: The "positive-vivid" factor is associated with individuals who daydream frequently, whose daydreams are accompanied by positive emotions and vivid sensory imagery, and who employ daydreaming for constructive problem solving. The "guilty-dysphoric" factor is associated with individuals whose daydreams have a negative emotional character, and whose daydreams reflect guilt and self-doubt on the one hand, and a striving for achievement on the other. The "anxious-distractible" factor is associated with individuals having difficulty attending to tasks, and whose daydreams are characterized by fleeting thoughts and accompanied by anxiety and worry.

J. L. Singer and McCraven (1961) found cultural differences in the content and frequency of daydreaming; other individual differences have been found as well. Since the literature on this topic is quite extensive, the reader is referred to the reviews of J. L. Singer (1978) and McNeil (1985).

J. L. Singer (1974, pp. 247-248) views long-term memory as an active process, normally just outside our awareness, in which items are constantly replayed and rehearsed; when we are otherwise unoccupied, we become aware of this internal source of stimulation and make use of its contents in daydreaming. J. L. Singer and colleagues have conducted various studies to investigate the relationship of external and internal sources of stimulation. In these studies, the subject sits in a darkened, soundproof cubicle and either listens to tones via a pair of headphones or watches lights. The subject is to perform a "signal detection" task—to press a button in order to discriminate, for example, low from high tones or bright from dim light flashes.

One study (Antrobus & J. L. Singer, 1964) found that subjects engaged in counting out loud, rather than continuous free association, while performing the signal detection task were more likely to report themselves as feeling drowsy. Subjects engaged in counting rather than free association were also found to produce a greater degree of accuracy in the signal detection task. These results suggest that although daydreaming serves to maintain arousal while performing a boring task, it also reduces accuracy in that task.

In another set of experiments (Antrobus, J. L. Singer, & Greenberg, 1966), subjects were interrupted at regular intervals while performing a signal detection task and asked to report whether any task-irrelevant thoughts occurred since the last interruption. An increase in the rate of the signals was found to correspond with a decrease in the amount of task-irrelevant thought. Nonetheless, even when the signal rate was increased, a certain minimum amount of task-irrelevant thought was found to persist. When, while waiting to perform the experiment, subjects heard a fake radio broadcast telling of an escalation in the Vietnam war and draft calls, the amount of task-irrelevant thought was found to increase substantially. However, the error rate in the signal detection task was not significantly worse, suggesting to the researchers a greater "bandwidth" or capacity for processing both internal and external material than they had expected.

J. L. Singer (1975) proposes the following functions for daydreaming in humans: generating solutions to life problems; planning and rehearsing future actions; achieving positive mood states by diverting attention from negative thoughts to more positive fantasies (in contrast to *catharsis* theories of daydreaming); self-entertainment and maintaining arousal in boring situations; increasing the probability of recalling, and dealing with, unfinished business; and, in general, enriching our daily lives. He also proposes that adult daydreaming evolves out of childhood pretend play. As a child grows up, the overt play, by and large frowned upon by society, is gradually internalized until it is exclusively carried out in the form of private daydreaming.

9.1.8 Klinger's Theory of Fantasy

In *Structure and Functions of Fantasy*, Klinger (1971) proposes a psychological theory of fantasy. He defines fantasy as verbal reports of mental activity other than problem-solving in a goal-oriented situation, and other than the scanning of external stimuli or orienting responses to external stimuli. Klinger intends this definition to exclude thoughts concerned with explicit planning of future events (but to include thoughts about the possible course of future events).

Klinger proposes that fantasies are composed of a sequence of segments (called "main routines") consisting of homogeneous thematic content. These main routines may be further broken down into "subroutines," which may be further broken down into "sub-subroutines," and so on—fantasies may thus be represented in terms of a tree structure. High agreement between two judges segmenting fantasies into main routines was found and moderately high agreement was found at the subroutine level.

Klinger classifies thought segments into two types: operant and respondent (after Skinner, 1935). A parallel is drawn between respondent and operant behavior, and primary and secondary process thought (S. Freud, 1900/1965): respondent behavior and primary process thought are both initiated involuntarily and unconcerned with feedback from the external environment, while operant behavior and secondary process thought are both initiated voluntarily and directed and modified in response to feedback indicating whether the activity accomplishes a given objective. Operant behavior is accompanied by a greater sense of mental effort than respondent behavior. Fantasy, dreams, and play are all viewed as instances of primarily respondent sequences. Jung (1916) makes a similar distinction between directed thinking—primarily carried out in words and with effort—and dream or phantasy thought—primarily carried out in images. He argues that phantastic thinking in present-day humans retains a condensation of mythical material from the past of humankind; modern science, on the other hand, is an example of directed thinking.

Klinger defines a "current concern" as the state of having initiated a behavioral sequence for attaining an anticipated incentive or goal state. A current concern persists until the goal is either achieved or abandoned. Current concerns lead to respondent segments concerned with the means for achieving the associated goals, but only if the odds of success are in doubt. In part, Klinger's current concerns inspired the "concern" construct of DAYDREAMER.

Klinger argues that concepts such as motives and needs are too coarse-grained to account for moment-to-moment shifts in fantasy. Instead, he proposes the following account of transitions from one segment to another: A segment related to one current concern may contain an element or cue related to another concern. This induces a shift to a segment related to that other concern. He later called this the "induction principle." (Klinger, 1978)

9.1.9 Pope's Studies of the Stream of Thought

Pope (1978) conducted several experiments in order to investigate the effects of solitude, posture, and gender on daydreaming. He also compared think-aloud protocols with a nonverbal method of report: A key-press device was hooked up to a polygraph recorder and subjects were instructed to hold down a "present" or "absent" button depending on the current topic of thought.

Clear differences were found between the think-aloud and key-press methods of report. The number of segments was about five times greater in the key-press method than in either the subject-rated or judge-rated think-aloud reports. Frequency and cumulative duration of "present" segments were also greater for the key-press method. Pope reports that "many participants expressed the difficulty of putting their thoughts into words and doing so quickly enough so that they neither slowed down their thoughts nor left our [sic] parts of a description because they were in a hurry or because they could not find the right words." (Pope, 1978, p. 262)

The majority of the time seemed to be spent in activities such as recalling past experiences or future fantasies. Only about 35-55% of the segments and about 30-40% of the 5-minute period was found to be concerned with the present (percentages not varying with the method of reporting).

9.1.10 Shanon's Investigation of Thought Sequences

Shanon (1981) collected a corpus of daydream protocols and wrote a brief paper summarizing the patterns revealed in the investigation of those protocols. He posits several "local mappings" for generating the next thought in a sequence given the previous thought.

The "association" and "content" mappings produce what we would call re-mindings, through the use of stored or generated (inferred) links between perceptual objects, concepts, or episodes. Thus "hearing a tune" may be followed by "N. introduced me to this singer."

"Formal" mappings consist of structural transformations on thoughts such as negation, generalization, specification, and interrogation. So, for example, "to ask R. about it" may be mapped through the operation of generalization to "to ask adults about it." Although Shanon does not provide examples of these mechanisms, we assume that negation of "to ask R. about it" produces "not to ask R. about it" and that interrogation produces "why ask R. about it?" Such transformations may be related to the action mutations of DAYDREAMER discussed in Chapter 5 and to the transforms of Colby (1963), among other proposals.

"Symbol" mappings involve a "shift in the perspective" of a thought: "Eating spaghetti with a spoon" is followed by "why isn't there a tool that will keep the noodles and let the liquid pass," which in turn is followed by "fork." This example seems to be related to planning. In the other example of symbol mappings, "(reading) I draw my finger across her forehead" is followed by "An image of myself drawing my finger across L.'s forehead." If I understand this example correctly, it appears to consist of the analogical application of an episode involving a character in a story to a daydream involving the self.

Shanon also makes some interesting suggestions that thought sequences involve "shifts of levels," "interactions," and triggering of thoughts by "partially specified" constituents (but, unfortunately, does not elaborate on these mechanisms).

9.2 Related Work in Night Dreaming

As might be expected from its name, daydreaming[4] is similar in many respects to night dreaming: Both involve a withdrawal from the external environment, both involve a sort of "waking up," both consist of fanciful, somewhat unrealistic sequences of events, both involve imagery, and both are difficult to recall without practice.

However, there are also several differences: Dreaming is carried out while asleep, daydreaming while awake; dreaming thus involves near complete isolation from the external environment (Foulkes, 1966), while in daydreaming one is to varying degrees aware of the external environment and receptive to external stimuli (Klinger, 1978).

Like daydreams, the plots of dreams are coherent (Foulkes, 1985). However, dreams seem to involve a greater degree of bizarre juxtapositions of elements and transitions from one event to another. The events of a dream are less to be taken at face value than those of a daydream: Dreamed situations are often metaphors for current problems of the dreamer (Baylor & Deslauriers, 1986), whereas in daydreams current concerns are represented directly (Klinger, 1971).

The imagery in dreams is realistic and lifelike, whereas in daydreaming it is more vague. In dreaming we think we are experiencing real events, while in daydreaming we know the events are imaginary. Nonetheless, there are exceptions: First, in waking thought, one may sometimes believe the imagined events are actually occurring: For example, in a study reported by Foulkes and Fleisher (1975), subjects were interrupted at random while lying down in a state of relaxed wakefulness. In 19 percent of the content-producing interruptions, subjects described their thoughts as hallucinatory. Posey and Losch (1983) administered a questionnaire to 375 normal subjects; 71 percent reported having

---

[4]The earliest known occurrence of the English word "daydream" reported by the *Oxford English Dictionary* is from 1685: "And when awake, thy soul but nods at best, Day dreams and sickly thoughts revolving in thy breast." "Dream" meaning "thoughts, images, or fancies passing through the mind during sleep" seems to have appeared first in Middle English (first reported quotation circa 1250), while the first reported quotation for "dream" in the sense of indulging in "fancies or day-dreams" occurs in 1533. The first quotation for "fantasy" in the sense of "imagination" is from 1589.

experienced auditory hallucinations involving voices in wakeful situations. Second, sometimes one may be aware that one is dreaming (LaBerge, 1985; C. E. Green, 1968; S. Freud, 1900/1965, pp. 610-611). In such a case, however, the events of the dream still seem real. Third, some thought activity during sleep may lack vivid visual imagery and more resemble daydreaming than classical night dreaming:

> The impression is that NREM mentation resembles that large portion of our waking thought that wanders in a seemingly disorganized, drifting, nondirected fashion whenever we are not attending to external stimuli or actively working out a problem or daydream. (Dement, 1976, p. 44)

There are other interesting relationships between daydreaming and dreaming as well. The author has consistently observed, for example, that it is possible to think to oneself—that is, to daydream—about the current situation in a dream. Thus it appears that daydreaming is possible within dreaming.

Is there similarity in the content of the dreams and daydreams of a particular individual? Klinger (1971, p. 55) concludes that there is little evidence to this effect. However, experiments conducted by Starker (1982, pp. 62-68) demonstrated a correspondence between the IPI daydreaming factor of a subject (discussed earlier) and the emotional tone, length, and strangeness of that subject's dreams: Dreams of "guilty-dysphoric" daydreamers contained more negative emotions than those of "positive-vivid" daydreamers; dreams of "anxious-distractible" daydreamers were somewhat longer and stranger than those of "guilty-dysphoric" daydreamers; and dreams of "positive-vivid" daydreamers were shorter, less strange, and contained less negative emotions than those of "anxious-distractible" daydreamers.

Because of the relationship of night dreaming and daydreaming, in this section we review previous work in night dreaming which is relevant to the present research. We also briefly review two other forms of dreaming of potential relevance to our research—hypnagogic dreams and lucid dreams.

9.2.1 Foulkes's Scoring System for Latent Structure

Foulkes (1978) proposed an explicit dream analysis technique called the Scoring System for Latent Structure (SSLS) (which, unfortunately, was not implemented as a computer program). This technique is used to analyze the private meaning of a dream given transcripts of the dream and free associations on each part of the dream. The primary objectives of SSLS are: an understanding of *dream processes*—those mechanisms whereby dream-thoughts are transformed into dream-content—and a reliable method of *dream interpretation* for clinical use—determining the dream-thoughts from the dream-content.

In this method of dream analysis, each conceptualization contained in the dream and free association transcripts is coded as a subject-verb-object proposition (using a concise but cryptic set of special symbols). Two classes of propositions—"interactive" and "associative"—are used to represent all conceptualizations. Interactive propositions may employ the following verbs: *moving toward* (loving, helping, penetrating, and so on), *moving from* (withdrawing, neglecting, and so on), *moving against* (being hostile to, destroying, and so on), and *creating* (producing, discovering, nurturing, and so on). Associative propositions may employ the verbs of *with* (associated with, near to, and so on), *equivalence* (being identical to), and *means* (serving as an medium for sustaining some relationship).

Each segment of a dream report and each free association from a dream element is called an "association chain." From each association chain, an "association path" is formed as follows: The first element of the path is the first element of the chain; the second element of the path is the remaining element of the chain having the greatest number of nouns in common with the first element of the path; the third element of the path is the remaining element of the chain having the greatest number of nouns in common with the second element of the path; and so on.

A network of interconnected associated noun elements called a "structuregram" is then formed from one or more association paths. If a composite structuregram is formed from all of the dream and free association paths, the complete set of transformations from dream-thoughts to dream-content is exposed. Foulkes discusses how mathematical analysis may be applied to structuregrams in order to find "condensing carriers," "branch intersects," "trunk condensers," and the like. Although such constructs may be interesting from the standpoint of digraph theory, it is not clear how they enrich one's understanding of a particular dream. Nonetheless, such investigations are somewhat useful in the investigation of dream processes: For example, Foulkes provides a mathematical characterization of Freud's dream-work processes of condensation and displacement.

9.2.2 The Computer Analogy for Night Dreaming

Several researchers (C. Evans, 1983; C. Evans & Newman, 1964) have proposed theories of night dreaming based on an analogy with computers. These theories may be stated roughly as follows: The brain is controlled by a collection of programs, many of them innate. These programs must continually be modified in order to meet the changing demands of the world. However, this modification process may occur only when the brain is not receiving any new information from the external world—that is, while a person is asleep. Dreams are what is recalled of this activity of updating these programs in light of recent experiences. Evans suggests that "dreaming may—in humans at least—be concerned with running through and inspecting a social adaptation program, for it could be said that young children who have had little actual social experience in the outside world need to prepare for the interactions to come by staging their own nightly series of internal dramas." (p. 156) Note the similarity between this theory of dreaming and our hypothesis that daydreaming serves the functions of learning from past experiences and planning for the future. A. Maeder and Alfred Adler have also previously proposed a planning or "thinking ahead" function for dreaming (S. Freud, 1900/1965, p. 618, footnote 1).

9.2.3   The Dream Simulation of Moser et al.

Although DAYDREAMER is (to the author's knowledge) the first computer model of daydreaming, a computer simulation of *night* dreaming has previously been constructed by Moser, Pfeifer, W. Schneider, Von Zeppelin, and H. Schneider (1980, 1982). The simulation is based on psychoanalytic theory (S. Freud, 1900/1965), on a previous theory which views dreaming as attempting to deal with unresolved conflicts (French, 1952, 1953), and on previous theories which view dreaming as the deferred processing of daytime information (Ben-Aaron, 1975; Antrobus, 1977). The program (implemented in Lisp) simulates the production of two dreams originally reported during psychotherapeutic sessions. The output is in propositional (rather than English) form.

In the theory of dreaming of Moser et al., a "latent conflict" or, as we call it here, an "abstract wish," is triggered by an experience during the day. An abstract wish is the goal to perform an action which is prohibited by a rule of the "superego." An abstract wish consists of:

- an action

- desired attributes for the actor and objects of that action

- a rule of the superego prohibiting the performance of certain instances of this abstract wish (each rule of the superego is associated with a person from whom the rule is derived [usually a parent] and who will enforce this rule through punishment if it is violated)

- the extent to which the abstract wish has not been realizable in the past, or the strength of the abstract wish (called NEGEM for "negative emotions")

- the degree of the anxiety emotion to be expected upon fulfilling the abstract wish despite the superego rule (called ANX for "anticipated anxiety")

Dreaming, in this theory, is broken into two phases: the "involvement phase" and the "commitment phase." In the involvement phase, defensive transformations are performed on an instance of the abstract wish in order to render it acceptable to the superego. In the commitment phase, the transformed instance of the abstract wish is achieved in a sequence of dreamed situations. Each such situation consists of a setting, an action, an actor of the action, and objects of the action. The program implements the production of dreams given a list of abstract wishes.

The operation of the program will be described in the context of how it accounts for the production of the following "Manure Wagon Dream":

> I am fleeing with my brother in a wheat field. I don't know what I am fleeing from. I am in great fear. I am trying to hide. Suddenly I see a manure wagon with a white liquid mush coming out the front. I have the feeling that I absolutely have to have the mush. This gets me into the conflict of being seen. But finally I go to the wagon all the same and drink. The [sic] I wake up, I am depressed, but then I laugh and say: "But the mush was good." (Moser, Pfeifer, W. Schneider, Von Zeppelin, & H. Schneider, 1980, p. 10)

The program is started with a list of abstract wishes reactivated during the day. In the "Manure Wagon Dream," instances of the two abstract wishes are (a) to suck his brother's male organ and ingest the product and (b) to suck his mother's breast and drink the milk. These abstract wishes are assumed to have been reactivated by a "transference" situation: Although the analyst usually provided the patient with a cigarette, the day before the patient had the "Manure Wagon Dream" she refused to give him a cigarette.

The program is a loop which operates on one abstract wish at a time: The current abstract wish is instantiated different ways, depending on the value of the associated NEGEM. If NEGEM is small (as in the "Manure Wagon Dream"), the dreamer is selected as the actor. If NEGEM is large, the dreamer is selected as a "spectator." If NEGEM is very large, a "deanimated" (French, 1953) version of the dreamer is selected as the actor: The animate dreamer is transformed into an inanimate object having as many of the desired actor attributes (specified in the abstract wish) as possible. This is the case in the other dream modeled by the program—in this other dream, the patient dreams she is a big beautiful plum hanging from a tree. If NEGEM is very large, the action and setting are chosen as described below, the situation is "expanded by a detailed description of the [actor's] attributes" (p. 74), and the program terminates (which models waking up). Otherwise, if NEGEM is small or large, the program continues as follows:

If the current abstract wish does not have an associated superego rule, the commitment phase may begin. Otherwise, a suitable object for the action of the abstract wish is selected according to the desired object attributes (specified in the abstract wish) and the constraint that the object must enable the action of the next abstract wish. In the "Manure Wagon Dream," his brother's male organ is chosen as the object. If the superego rule does not prohibit the action with the given actor and objects, the commitment phase begins. Otherwise, the action is transformed as follows: If the object processor violates the superego rule, an acceptable object processor is substituted. If the type of action violates the rule, a new action type is substituted depending on the value of the ANX associated with the abstract wish: If ANX is large, the action of fleeing is selected; if ANX is small, an action of sitting, walking, or waiting is selected. The objects of the original action become actors in the new action along with the original actor. In the case of the "Manure Wagon Dream," the original action is prohibited by a rule stemming from the patient's mother, ANX is large, and so the action is transformed into fleeing with his brother.

Next, a setting for the action is chosen according to desired attributes common to all the abstract wishes, and the constraint that the action must be able to be performed in that setting. In the example dream, a wheat field is chosen as the setting because one may flee in a wheat field, and a wheat field is nourishing. At this point, the first dream situation has been generated. A dynamic emotion variable called DMA (for "dream manifest anxiety") is set depending on various conditions. If DMA is low, the commitment phase begins.

Otherwise, as is the case in the "Manure Wagon Dream" the following operations are performed: The transformed action is output (in the above dream as SUB [I] WITH [brother] PI [runaway] IN/ON/AT [wheat field]). An emotion (whose level corresponds to that of the DMA) is output (in the above dream as AFF-STATE [great fear]). A special dream operation called "omitting a segment" is performed (and output in the above dream as SUB [I] ACT [trying to hide]). Appropriate objects for the action of the next abstract wish are selected, and those objects are introduced into the dream by "seeing." Thus in the "Manure Wagon Dream," mush and a manure wagon are selected as appropriate objects for the drinking action of the following wish, so SUB (I) PI (see) OBJ (manure wagon) OB (PROD mush) is output. The next iteration of the loop is then performed (commitment phase is skipped in this case).

In the next iteration of the loop in the example dream, the action of the abstract wish is to drink. The dreamer is selected as the actor of the action (since NEGEM is small). The manure wagon and the mush have previously been chosen as objects of the action of drinking. Since no superego rule is associated with drinking, the program enters commitment phase. The commitment phase consists of performing an "approach" if it has not already been performed, producing the action as the next dream event, and performing the next iteration of the loop. In the above dream, the approach is first output as SUB (I) OBJ (manure wagon) PI (approach) and then the drinking action is output as SUB (I) PI (drink) OBJ (mush).

Although Moser et al. (1980) consider the transformation of abstract wishes to be a form of problem solving, the program contains no general-purpose planning and control mechanism (as in DAYDREAMER). Instead, the program is written directly in Lisp, according to a flowchart which appears to be specifically tailored to the two dreams which the program generates—in fact, the flowchart does not even achieve an economical generalization of the two dreams since each dream requires a separate branch! Because the control structure is so specialized, it is doubtful that it would be suitable for the generation of other dreams. Furthermore, the mechanisms invoked within the flowchart (such as to find appropriate settings, actors, objects, or transformations of actors, objects, and actions) are largely specific to the two dreams.

Like DAYDREAMER, the program incorporates emotions. However, the model of the initiation, modification, and influence on dreaming behavior of emotions, remains undeveloped: Both NEGEM and ANX are static values provided with input abstract wishes; their influence is limited to the selection of actions or actors of actions in an often arbitrary way. While DMA is a dynamic value, it merely invokes a particular sequence tailored to the "Manure Wagon Dream."

Although the dream simulation program of Moser et al. is not of the same scope and level of generality as DAYDREAMER, it was an interesting and worthwhile experiment.

9.2.4   Hypnagogic Dreams

The dreaming which occurs just before falling asleep is called *hypnagogic* dreaming (Maury, 1878; S. Freud, 1900/1965; Silberer, 1909/1951; McKellar, 1957; Foulkes, 1966). According to Foulkes (1966), sleep onset is defined as an alpha EEG (electroencephalogram or brain waves) and slow eye movements, or theta EEG with no eye movements (descending stage 1 sleep). This form of dreaming typically involves nonsensical concepts and bizarre visual or other imagery. There is also a related *hypnopompic* state which occurs during the return to wakefulness (Foulkes, 1966; McKellar, 1957).

Silberer (1909/1951) investigated a kind of hypnagogic dream—which he calls the "autosymbolic phenomenon"—occurring both while in a drowsy state and making an effort to think about something. This phenomenon involves the transformation of feelings or abstract thought contents into symbolic images (usually visual images). Silberer classifies autosymbolic phenomena into three classes: "material," "functional," and "somatic" phenomena. Material phenomena are autosymbolic experiences involving the transformation of thought contents into images: When thinking about rewriting part of an essay, Silberer saw himself planing a piece of wood. Functional phenomena involve the transformation of thoughts about the effectiveness or functioning of a thought process (i.e., "meta" thoughts) into images: After having lost his train of thought and thinking about this fact, he visualized a portion of typeset text with the last few lines missing. Somatic phenomena involve the transformation of external or internal bodily sensations into images: After taking a deep breath, he saw himself and another person lifting a table up high.

It turns out that bizarre thoughts such as those found in hypnagogic dreams occur in waking thought as well: Foulkes and Fleisher (1975) conducted a study in which subjects were interrupted while in a state of relaxed wakefulness. In 25 percent of the interruptions, the reported thoughts were judged to be "regressive." Klinger (1978) reports that 22 percent of the subjects sampling their thoughts out of the lab (n = 285) described their thoughts as strange or distorted.

9.2.5 Lucid Dreams

Lucid dreams (C. E. Green, 1968; Garfield, 1974; LaBerge, 1985) are those night dreams in which one is aware that one is dreaming and is able consciously to manipulate the content of the dream, as if one were awake, to varying degrees and for varying lengths of time. When one says to oneself "this is only a dream" while dreaming or forces oneself to awaken out of a nightmare, one has experienced a rudimentary lucid dream. While full-blown lucid dreams are rare for most persons, there are several techniques for increasing the frequency of this form of dreaming (Gackenbach, 1985).

In some ways, lucid dreaming resembles daydreaming—the lucid dreamer, like the daydreamer, is able to control the events of the dream or daydream to some degree and is aware that the events of the dream or daydream are not real. In other ways, the two forms of dreaming are quite different: While daydreaming consists of a stream of *thought* accompanied by vague imagery, lucid dreaming involves such extremely vivid and realistic imagery that it would more accurately be called a stream of *experience*!

A computer model of lucid dreaming might incorporate the following elements:

1. a mechanism for the production of conventional night dreams (a "dream generator")

2. a process for generating a sequence of thoughts which enables one to recognize that one is dreaming and thus to initiate voluntary thought

3. a voluntary thought process which attempts, not always successfully, to guide the operation of the dream generator according to various goals and previously formed intentions to carry out certain activities (as is encouraged by some techniques for lucid dreaming [LaBerge, 1985])

4. a process whereby voluntary thought is terminated or waking up occurs

Of course, such a model must take a stand on what "voluntary thought" is, in computational terms. This issue has been largely ignored in the present work for simplicity.

Chapter 10

Underpinnings of a Daydreaming Theory

This chapter addresses the philosophical, scientific, and methodological issues involved in the construction of a computational theory of daydreaming. First, we examine the philosophical justifications and implications of our underlying assumption that daydreaming can, at least in principle, be captured in computational terms. We address the apparent paradox in constructing an objective computer program to capture a subjective phenomenon such as the stream of consciousness. Second, we address the problem of semantics in computer programs. We examine the sense in which a computer made out of electronic components can be said to exhibit the same cognitive processes or mental states as a human. Third, we address the scientific issues involved in our theory: (a) the objectives—in particular, the explanatory goals—of the theory, (b) the methodology employed in constructing the theory, and (c) the testing and evaluation of the theory.

10.1 Philosophical Underpinnings

What is the true nature of the human mind? How is the mind (that is, the notion of an entity which thinks and is conscious) related to the brain (that is, the physical organ)? What, in reality, are mental entities such as thoughts, sensations, emotions, desires, and beliefs? How do such mental entities relate to the physical world? What do we mean by a computational theory of an inherently subjective phenomenon such as the stream of consciousness? These questions are aspects of what is known in philosophy as the *mind-body problem.*

10.1.1   Functionalist Approach to Mind-Body Problem

Theories which address the mind-body problem may be divided into two classes:[1] *dualist* theories and *materialist* theories. Dualist theories maintain that the mind is a nonphysical entity—that intelligence cannot be explained in physical terms alone. That is, part (or all) of intelligence is provided by something outside of the physical realm. Materialist theories maintain that the mind *can* be explained in purely physical terms—that mental states and processes correspond in some way, however complex, to the states and operations of the brain.

In the dualistic theory of René Descartes (Russell, 1945, p. 561), the mind (or soul) exerts an influence on the body and this influence is located in the pineal gland (a pinecone-shaped body attached to the base of the brain thought by Descartes to be unique to humans but later found to occur in all vertebrates [Asimov, 1963]). After the discovery by scientists of conservation of momentum, which would rule out a nonphysical influence on the physical world, his disciple Geulincx proposed (Russell, 1945, pp. 561-562) that the soul does not act on the body; rather, the body operates in exact synchronism with the mind so that it appears that there is an influence when in reality there is not. The problem with this theory is that if the body behaves according to rigid, physical laws, then the mind, operating simultaneously, would be characterizable in terms of those same laws (and thus dualism is lost).

Dualistic theories, which place the mind in the nonphysical realm, contend that a purely physical explanation (and therefore a scientific explanation) of the mind is not possible. In seeking a scientific, objective (at least partial) explanation of the phenomenon of human daydreaming, we make the assumption that such an explanation is possible. It is logically possible for this assumption to be false—it may eventually turn out, as in the theory of Descartes, that the mind, a nonphysical entity, is in fact able to influence the body in the physical world (for example, via some as yet unknown form of energy exchange [Churchland, 1984, pp. 9-10]). Nonetheless, cognitive science and the neurosciences have already achieved a degree of explanatory success operating under the assumption that an objective characterization of mind is indeed possible (see Churchland, 1984, pp. 18-21 for a more detailed version of this argument).

In a *functionalist* theory of mind—a special brand of materialist theory—the essential nature of a mental state is how that mental state is causally related to inputs (environmental stimuli), outputs (bodily responses), and other mental states (Fodor, 1981; Dennett, 1978). Mental states have a "life of their own" independent of any particular physical realization. In principle, it does not matter whether mental states with a given set of causal properties are realized on a digital computer, in the brain, or in some alien biology. See, for example, Hofstadter's (Hofstadter & Dennett, 1981, p. 376) discussion of how human

---

[1]The taxonomy presented here is based on those provided by Churchland (1984) and Fodor (1981).

mental processes might be simulated using a variety of media including toilet paper.

In computer science terms, mental processes can be carried out on any device providing a suitable *interpreter*. An interpreter for a given programming language L is a program P', written in a language L', which carries out the operations (including generation of output) specified by a program P written in L, given P (and data) as input. Interpreters enable programs in a given language to be run on computers with different hardware architectures and instruction sets, provided that an interpreter for the language exists for the particular computer. Interpreters may be *nested*: An interpreter for one language may be run by another interpreter, which in turn is run by another interpreter, and so on, until eventually there is an interpreter which is able to execute on the computer in question.

A computational theory of mind is a functionalist (and therefore materialist) theory. Computational theories are not specified in terms of neurophysiological processes, but in terms of an abstract computing device called the *Turing machine* (Turing, 1936).[2] Turing machines provide a hardware-independent language for describing mental processes in a functionalist theory of mind.[3] The modern digital computer may be characterized as a Turing machine.[4] Rather than specifying a computational theory directly in terms of a Turing machine, the theory is usually programmed on a computer running a high-level language such as Lisp (Winston & Horn, 1981).

10.1.2 Problems with Theories of Subjective Phenomena

The stream of consciousness is an internal phenomenon experienced subjectively, from a particular point of view which is unique to an individual. How can there be a computational, materialist theory of a subjective phenomenon such as the

---

[2]Some authors (for example, Johnson-Laird, 1983, pp. 9-10) define functionalist theories as those which characterize mental processes in terms of effective procedures (or algorithms) as distinguished from Turing machines. The Church-Turing thesis (which we adopt) states that the intuitive notion of an effective procedure is equivalent to a Turing machine.

[3]The Turing machine provides a deterministic model of human behavior. In order to admit indeterministic quantum mechanical effects we would have to add a true random number generator to our Turing machine. However, whether and how quantum effects at the subatomic level might cascade to the level of human behavior presents a very difficult problem (Dennett, 1984, pp. 77, 135-136).

[4]The digital computer is theoretically equivalent to the finite-state automaton (with a very large number of states) rather than the more powerful Turing machine with its infinite tape. Nonetheless, most researchers view computation in terms of Turing machines for (at least) the following reasons: First, the memory of modern computers is so large that for most purposes they can be considered as Turing machines (Lewis & Papadimitriou, 1981, p. 362). Second, the Turing machine is closer to the computer in its basic architecture than the finite-state automaton: A finite-state automaton is a large state-transition network and has no memory analogous to the tape of a Turing machine or the primary and secondary memories of a computer (consult Pylyshyn, 1984, pp. 69-74 for an elaboration of this point). However, see Kugel (1986) for an interesting proposal that thinking may be more than computing.

stream of consciousness?

First of all, what do we mean when we say that the stream of consciousness is "experienced subjectively"? Everyone is familiar with the feeling of consciousness: There seems to be an "I" at the center of everything which observes, experiences, and—to varying degrees—controls the events which make up consciousness. As Dennett (Hofstadter & Dennett, 1981, p. 6) has noted, we feel that the "I" is located roughly in back of our eyes and in between our ears. The "I" perceives the world through the senses, feels various emotional states, experiences the "I" feeling, thinks, imagines, decides to perform—and then performs—various actions in the world, and so on. However, this "I" feeling is sometimes lost—in unconscious states, in certain dream states, in certain daydreaming states (Watkins, 1976), or even when becoming very involved in a movie or novel.

The subjective rate of consciousness is not always linear with real time. There is the common observation that "time flies when you're having fun." On the level of milliseconds, one is often surprised, upon looking at a wristwatch, by how long the first click of the second hand seems to take (if one starts looking at the watch at the very instant of a tick)—thereafter, the second hand resumes its normal subjective rate.

We assume that other human beings have internal experiences similar to our own—that others, for example, have the "I" feeling. Others describe their experiences to us, and we can imagine—with varying degrees of apparent success—how it would feel to have those experiences. However, the exact subjective experience of another person is inaccessible to us (just as others do not have access to our subjective experiences). We will never know exactly how a particular experience felt to the other person. Even if someone describes a subjective experience which seems to coincide with one of our own subjective experiences, there is no guarantee that the experience is actually felt the same way. This problem is demonstrated in the classic philosophical conundrum of color perception: How does one person know that another person sees the same color when both look at a "red" object? It is possible that the second person subjectively perceives what the first person would call "blue," but calls it "red" because that is the language convention which the second person has learned. Even if similar physical events occur in the brains of two persons when viewing a red object, this does not necessarily imply that the color is subjectively experienced in the same way.

Nagel (1974) considers the problem of subjective experience in asking the reader to ponder "what it is like to be a bat."[5] Bats perceive the world in a way entirely foreign to humans: Instead of normal vision or hearing, bats use echolocation (bouncing high frequency sounds off of objects in the environment to determine their distance, velocity, size, and shape—a kind of natural sonar). How can a human ever hope to know what it is like to be a bat? You can try to

---

[5]The example of bat experience was used previously by Farrell (1950, pp. 183-184).

imagine being a mouselike animal with webbed wings who flies around at night to capture and eat insects; you "see" objects only using sonar; this is completely natural to you; and so forth. But to the degree that you are able to imagine this, you would only know what it would feel like *for you* to behave as a bat, not what it feels like *to the bat* to be a bat. The question of knowing what it is subjectively like to be another creature presents a difficult philosophical problem, since in order to know what it is like to be that creature, you would have to be that creature, in which case you would no longer be yourself, and hence it would no longer be *you* knowing what it is like to be that creature.

Narayanan (1983) asks "What is it like to be a machine?" in the context of an example involving a nuclear reactor controller which hypothesizes and prepares for potential future events—an instance of daydreaming. Narayanan proposes that it is necessary for one to take the subjective "viewpoint" of such a program in order to understand its operation.

The common, intuitive notion of subjective experience is difficult if not impossible to define in objective terms.[6] We should not then expect a computational theory to account for subjective experience in this sense. If a computational theory were constructed which completely accounted for the behavioral impact of consciousness (ignoring for the present discussion what constitutes a "complete account"), it appears that the theory would be logically compatible with the absence of this concept of subjective experience. Put another way, it seems that a human or computer which had such subjective experiences would behave identically to one which did not. This makes one wonder what the role of these subjective experiences could possibly be.

If the stream of consciousness is defined in terms of subjective experience, how can a computational theory be constructed to account fully for it—in all its subjective glory? Since pinning down the common notion of subjective experience is an insoluble (for the present, at any rate) philosophical problem, we must instead formulate a definition of the stream of consciousness which does not make use of this notion. Our definition must employ an objective, external frame of reference, rather than a subjective, internal one.

The stream of consciousness must therefore be defined operationally in terms of the input-output behavior—verbal and otherwise—of an organism having (the common notion of) a stream of consciousness. Since we have made the materialist assumption that human behavior is capturable in computational terms, the stream of consciousness, redefined as input-output behavior, is also capturable in computational terms.

At this point, the reader may well wonder if we have rendered our definition so unlike the intuitive conception of the stream of consciousness as to make it useless and irrelevant to those interested in this private, internal phenomenon.

---

[6]On the face of it, a dualist theory would seem to be more likely to account for subjective experience since in such a theory the "I" feeling could be located outside the physical world. However, as argued above, dualist theories fail to provide a satisfactory scientific account because such theories are not stated in objective terms.

But the purpose of this definition is merely to ensure that the phenomenon under investigation is in principle capturable by a computational theory. In fact, such a theory—if complete—will have to account for a rich set of behavioral phenomena, including practically every aspect of the intuitive notion of the stream of consciousness—that is, every aspect *except* this elusive notion of subjective experience. Most, if not all, aspects of our common conception of the stream of consciousness—the *contents* of the stream—will be accounted for by the theory, since most such aspects have a causal effect on behavior. For example, when we feel a negative emotional state, we tend to behave more negatively. When we decide to perform an action in the future, we often do perform that action. We are capable of producing rich, detailed descriptions of our internal experiences. Our verbal behavior is thus influenced in a very complex way by what we refer to commonly as the stream of consciousness. Accounting for this behavior is a sufficiently—in fact, extremely—difficult problem without also having to explain the true nature of subjective experience (whatever that might be).

The mere fact that the content of internal experiences can be described to others implies that those experiences are not epiphenomenal. If the experiences were epiphenomenal, they would not have a causal effect on verbal behavior. However, the *subjective sensation* of experience may indeed be epiphenomenal. We do not deny the existence or reality of this sensation. From an internal point of view, this sensation certainly seems to exist. However, we do assume that this subjective sensation has no causal effect on behavior. This is a reasonable assumption as argued above, since its falsehood would seem to imply influence on the brain from outside the objective, physical realm. If, possibly, the assumption is false, a stumbling block will eventually be reached. However, we are a long way from reaching such a stumbling block. In the meantime, we can investigate what appears to be a large component of our intuitive notion of the stream of consciousness which does have an effect on behavior. If we are eventually able to construct a complete theory of the behavioral impact of the stream of consciousness, which addresses all the important issues regarding our intuitive notion of the stream of consciousness except for the subjective sensation, we may decide that the question of subjective sensation is no longer an important question for us.

Thus our theory of the stream of consciousness will attempt to account only for its objective, empirically observable aspects. That does not mean that we must ignore the many varied aspects of consciousness with which we are all familiar, since these aspects are observable in the descriptions we receive from people. DAYDREAMER is the computer program embodiment of a computational theory of the stream of consciousness as defined operationally in terms of the impact of the intuitive notion of the stream of consciousness on behavior: verbal and otherwise.

Why focus on the stream of consciousness? How do we know that the stream of consciousness is a relevant object of study? Some psychologists (see, for example, Nisbett & Wilson, 1977; Watson, 1924) have criticized the use of introspective reports of consciousness (James, 1890a; Titchener, 1912) in inferring cognitive processes. Although the *subjective* aspect of consciousness may have no causal effect of behavior, the contents of consciousness influences verbal behavior, and, furthermore, reflects the useful activities of creative problem solving, learning from imagined experiences, and emotion regulation (as argued in Chapter 1). In addition, Ericsson and Simon (1984, pp. 24-30, 48-61, 220) present extensive arguments—which need not be repeated here—that verbal reports of the stream of consciousness are not epiphenomenal but highly relevant to the study of cognitive processes and memory structures.

If our theory is to account for the effect of the intuitive notion of stream of consciousness on behavior, our theory must include some construct, or architectural element, corresponding to the stream of consciousness. In fact, it does: the current planning context discussed in Chapter 7 is such an element.

10.2   The Problem of Semantics

What is the nature of the meaning of concepts? How are concepts programmed in a computer program? How can a data structure in a computer program be said to represent a concept? What do we mean when we say a computer program has semantics? What is semantics, anyway? How can we represent the subjective aspects of consciousness in a computer program? How could a mental state such as *embarrassment* be programmed into a computer?

10.2.1   Structure-Content Distinctions

A distinction is often made between claims regarding the *content* of a system and claims regarding the *structure* of the system. Newell (1982) distinguishes between the "knowledge level" and the "symbol level" (and several other lower levels). The knowledge level is a system which consists of a set of goals, actions, and a body of knowledge. The behavior of the knowledge level is specified by the "principle of rationality": If the system has knowledge that an action will achieve one of its goals, that action is selected. (Knowledge may thus be viewed as whatever a system has that enables its behavior to be computed according to the principle of rationality.) However, this principle does not specify which action will actually be performed, for example, in the event that several actions are selected or if multiple, conflicting goals are present in a given situation. Thus the knowledge level provides only a partial specification of behavior. The remainder of the specification is given by the symbol level, which provides mechanisms for control, encoding of knowledge, memory storage and retrieval, and so on.

Pylyshyn (1984) distinguishes three levels: the "physical level" (or biological level), the "symbol level," and the "semantic level." These correspond respectively to the device, symbol, and knowledge levels of Newell. Symbol level generalizations are expressed in terms of the "functional architecture" of the system: A "virtual machine" for cognitive processes providing a set of primitive operations. Pylyshyn proposes the criterion of "cognitive penetrability" for determining whether a given function should be located within the symbol or semantic levels: If a given function is alterable in a semantically regular way through the use of mental representations such as goals and beliefs, then that function is cognitively penetrable and should be located in the semantic level.

10.2.2   Semantics of the Stream of Consciousness

In order for DAYDREAMER to model the stream of consciousness, it must contain representations for the contents of the stream of consciousness. As we have noted above, the stream of consciousness contains a variety of phenomenological events. How can we represent events which are inherently subjective: valid only from a single point of view?

Pekala and Levine (1981) have developed a methodology for assessing subjective experience. Subjects are asked to sit quietly with their eyes open for several minutes and then complete a 60-item questionnaire. Items are drawn from the following aspects of consciousness: body integrity, time, state of awareness, attention, volition, self-awareness, perception, positive affect, negative affect, imagery, internal dialogue, rationality, memory, meaning, and alertness. The subject rates each item along a 7-point scale, where extremes are provided by statements such as "I experienced no strong feelings of anger" and "I experienced very strong feelings of anger" (p. 43). The questionnaire thus provides one representation for the phenomenology of consciousness. However, this representation is too broad for our purposes since it provides too low a level of detail (e.g., why is the subject angry, and at whom?), and does not measure the instantaneous change of consciousness over time (e.g., was the subject angry during the entire period, or merely for an instant?).

Although stream of consciousness events are subjective, it is possible to make generalizations about them. It is reasonable to expect that human beings have similar subjective experiences. Such a set of generalizations is provided by human language. Through use of such generalizations it is possible to describe subjective experiences to others. Another possibility is to provide a "key" which maps the program's representations to descriptions of what subjective experiences they involve: The key could state, for example, that a particular representational structure corresponds to the feeling of embarrassment in humans. This is in effect what the English generator of DAYDREAMER accomplishes.

One might argue that this is begging the question—we have substituted one ill-defined description language for another. This problem of *absolute meaning* is not unique to the phenomenology problem, however. An identical problem occurs for any meaning representation scheme. No matter what syntax is used to implement a representation—whether it be node and links, strings over a finite alphabet, or positive integers—the meaning of that representation is completely dependent upon what interpretation is placed on the syntax. That is, syntactically identical structures may be used to represent different concepts, just as syntactically different structures may be used to represent identical concepts. If we attempt to formalize the interpretation process, we are forced to invent yet another representation which itself is subject to interpretation. Thus we can never have an absolute representation whose meaning is specified objectively. In denotational semantics (Stoy, 1977), for example, the meaning of a computer program in one language is specified in terms of another language—a metalanguage. The meaning, however, of the metalanguage is left unspecified. Usually the metalanguage consists of abstract mathematical concepts whose meaning is assumed to be well understood.

That the same problem arises for both phenomenological and conceptual representations hints at the possibility that they are really instances of the same problem. In the case of music, L. B. Meyer (1956) argues that "there is no diametric opposition, no inseparable gulf, between the affective and the intellectual responses made to music" (p. 39).

10.2.3 The Functionalist Approach to Semantics

If representations can have no inherent, absolute meaning, then just how *do* they acquire meaning? In functionalist theories, representational entities exist at the semantic level (Newell, 1982; Pylyshyn, 1984), and correspond (at least to a first approximation) to various concepts from folk psychology such as beliefs, goals, emotions, and so on. A functionalist theory of mind specifies how these representational entities causally relate to input, output, and other representational entities.

Thus a representational entity alone does not have semantics. Rather, a representational data structure acquires meaning only in conjunction with processes that operate on that representation and which have a causal relation with external objects, events, or concepts of other intelligent entities, such as humans. So, for example, in the knowledge level of Newell, goals, actions, and other knowledge elements have meaning only in combination with the "principle of rationality," which relates the knowledge elements to processes carried out in the external world.

10.2.4 The Elusiveness of Concepts

Finding suitable representations for concepts, however, is a very difficult task. Concepts have an elusive, open-ended quality which makes complete and consistent characterization difficult. For any representation of a concept, a counterexample may be found which demonstrates the incompleteness or inconsistency of that concept.

First, there is *representational incompleteness*. Suppose, for example, we wish to represent the concept of buying something. We might define buying as two instances of the ATRANS primitive in conceptual dependency (Schank,

1975), which refers to abstract transfer of possession of an object from one person to another. Buying thus defined consists of an ATRANS of money from the buyer to the seller and an ATRANS of the object being sold from the seller to the buyer. But this definition is obviously incomplete. What if a credit card is used for payment? In this case, there is an ATRANS of the credit card from the buyer to the seller, an ATRANS of the card back to the buyer, an ATRANS of a pen from the seller to the buyer, an ATRANS of the charge sheet, a signing of the charge sheet, an ATRANS back of the charge sheet and pen, and so on. Then there is an ATRANS of the charge sheet from the seller to a third party, the credit card company. Then there is an ATRANS of a bill from the credit card company to the buyer, followed by an ATRANS of a check from the buyer to the credit card company, and so on. What if the transaction takes place by mail order? What if the sold item is delivered to the seller's home by another party? But these are mere details—in each of the above cases, money and the sold object are transferred in some way, and these facts are captured by the definition.

However, a more serious form of incompleteness may arise. Suppose a buyer purchases an item by check from a mail order company. Two months later, the buyer receives the cancelled check, but the item still has not arrived. At this point, the seller is obligated to deliver the item or to refund the buyer's money. We see now that the original definition of buying is but a component of a larger picture involving higher-level notions of *obligation* and implicit *contracts*. Note, however, that a mere ATRANS of money to someone does not automatically invoke such obligations—if the ATRANS were accidental, such as if money were found on the ground, there is no obligation to return the money to its original owner (this is the case of "finder's keepers losers weepers.")

Incompleteness of representations may continue indefinitely if one tries hard enough. Of course, this phenomenon is paralleled in humans who always continue to learn more about concepts. Schank (1986, pp. 1-24) argues that understanding of concepts may be characterized along a spectrum from "making sense" at one end, to "cognitive understanding" in the middle, to "complete empathy" at the other end (the domain of present work in artificial intelligence being between "making sense" and "cognitive understanding" on this scale [p. 9]). Crook (1983) contends that attribution of consciousness to another requires empathy, analogical sympathy, and concordance.

Another problem is that of *representational conflict*. In this situation, two representations or rules are postulated which appear correct when considered in isolation from each other. However, when those representations are incorporated into a system, it is noticed that the two representations contradict each other in a certain situation. For example, suppose there is one rule which states that "if a person is friends with another person, the former person will help the latter person," and another rule which states that "if a person is angry at another person, the former person will not help the latter person." These two rules conflict when a person is angry at a friend. In order to cope with a problem such as this, one may order the rules so that one rule has priority over another, introduce further rules to resolve conflicts, modify the conditions of the rules (e.g., "if a person is friends with and not angry at another person ...''), or otherwise rethink and reorganize the representations.

Naturally occurring concepts are very rich entities. Computer programs will not display intelligence approaching that of humans unless they contain a collection of multifarious concepts. The importance of this has been recognized by researchers in narrative comprehension engaged in the construction of "in-depth" representational schemes (Dyer, 1983a). Representations derived initially from folk psychological concepts will have to be enriched. Our view is that if sufficiently complex representational systems are built, and augmented through the incorporation of both real and daydreamed experiences, the concepts contained in that system will exhibit the same elusive, vague character.

10.2.5 Connectionist Approaches to Semantics

The elusive quality of concepts has led some researchers toward a holistic approach to representation known as *connectionism* (Rumelhart et al., 1986; Hofstadter, 1983; J. A. Feldman & Ballard, 1982; Hinton & J. A. Anderson, 1981; Grossberg, 1976; Rosenblatt, 1962). In this approach, specific nodes in a representation do not refer to particular entities of the knowledge level. Rather, semantics is distributed among the nodes of the system in a "holographic" manner. It is argued (McClelland, Rumelhart, & Hinton, p. 12) that various proposed knowledge structures—such as scripts, frames, and schemata—are in fact approximate descriptions of the "emergent" (Dennett, 1978, p. 107; Neisser, 1963, p. 194) properties of an appropriate connectionist network.

The *holistic* connectionist approach is characterized by the assumption that high-level conceptual representations can be generated automatically starting from low-level (or even random [Rumelhart & Zipser, 1986]) initial representations. Thus researchers who work in this paradigm are typically not very interested in the explicit construction of higher-level knowledge structures. Not all connectionist models are holistic (see, for example, Waltz & Pollack, 1985; Rumelhart & Norman, 1982). However, our emphasis in this section is on holistic connectionism, since it provides an alternative to knowledge-level representations.

A connectionist system consists of a set of nodes connected by directional links (Rumelhart, Hinton, & McClelland, 1986). These links are of two types—excitatory and inhibitory. Each node has an activation level and each link has a strength. Activation spreads from a node to connected nodes. The amount of activation which is spread depends on the strength of the link from one node to the other. Positive activation is spread if the link is excitatory; negative activation is spread if the link is inhibitory. Short-term information is stored in the activation levels of nodes, while long-term information is stored in link strengths. Thus a connectionist system must learn through the alteration of the strengths of connections between nodes. Some nodes, called input and output nodes, communicate with an external environment. The remaining nodes are called hidden nodes. In 1890, William James, inspired by physiological research of his day, outlined a similar connectionist model. This model incorporated a learning rule for modification of link strengths, and the ideas of inhibition and spread of activation (James, 1890a, p. 567).

Representation in a connectionist model may be *local, distributed,* or a combination of the two. In a local representation scheme, single nodes represent single concepts. In a distributed scheme, each concept is spread across many nodes and furthermore each node represents many concepts. In distributed representations "it is impossible to point to a particular place where the memory for a particular item is stored" (Hinton et al., 1986, p. 80). Emergent, holistic properties are possible in both local and distributed representational schemes. It is often difficult to distinguish clearly between distributed and local representations in a holistic connectionist system (Hinton et al., 1986, p. 85).

Connectionist models take their inspiration from the physiology of the brain, which is estimated (in humans) to contain tens of billions of neurons. Like the nodes and links of connectionist models, neurons (nodes) have dendrites (links to a node) and an axon (the link from a node) connected by excitatory and inhibitory synapses. Neurons are quite slow (on the order of milliseconds) compared to the electronic circuits of computers (on the order of nanoseconds). J. A. Feldman and Ballard (1982) have therefore proposed the "100-step program" constraint—for those tasks which take about a second, a program should require only about that many steps of a parallel system. Connectionist models are materialist theories of cognition which can, in fact, be simulated by a Turing machine, albeit at a very slow rate.

Researchers differ in their view of the correspondence between connectionist nodes and neurons. As Smolensky (1986b) points out, there are several possibilities: Connectionist models can be given either local or distributed neural interpretations, and either local or distributed conceptual interpretations. These mappings are independent. In a local neural interpretation of a model, for example, each node corresponds to a neuron. It is possible for such a model to have a local conceptual interpretation (in which each node and neuron represents a single concept) or a distributed conceptual interpretation (in which concepts are represented by a set of nodes or neurons). In a distributed neural interpretation of a model, each node corresponds to a set of neurons. If such a model has a local conceptual interpretation, each node represents one concept which is in turn represented by many neurons. If the model has a distributed conceptual interpretation, concepts are represented by a set of nodes, each of which is represented by a set of neurons.

Some of the tasks which connectionist models are able to perform are as follows (Rumelhart & Zipser, 1986, p. 161):

- *Auto association.* The model is trained with a set of concepts. Given part of a concept, the system will then retrieve the remainder of that concept.

- *Pattern association.* The model is trained to associate concepts. Given one concept the system will retrieve the associated concept.

- *Classification.* The model is trained with a set of concepts and classifications for those concepts. Given a concept similar to those already seen, the system will return the best classification for that concept.

- *Regularity detection.* The model discovers salient features or central tendencies in a collection of given concepts.

In each case above, the system is first trained. Later, tasks are accomplished by activating appropriate input-output nodes, waiting for the network to reach a stable pattern of activity, and then reading the results from the remaining input-output nodes. Most applications of connectionist networks may be viewed as instances of one or more of the above cases. For example, the storage and retrieval of information in human memory is modeled as a form of auto association (McClelland & Rumelhart, 1986).[7]

In another application of connectionist models, schema acquisition is modeled as regularity detection (Rumelhart, Smolensky, McClelland, & Hinton, 1986, pp. 22-38). Although the system exhibits what might be called schemata or prototypes for rooms in a house (kitchen, living room, bathroom, and so on),[8] it should be noted that these schemata are not as conceptually rich as, say, restaurant scripts (Schank & Abelson, 1977).

One difficult problem with holistic connectionist representations is as follows: Since concepts are represented as a stable pattern of activation of the nodes in the network, only one concept can be represented at a time (Rumelhart et al., 1986, p. 38). That is, one cannot have multiple copies of a concept. Therefore, it is difficult to represent even simple relationships among multiple concepts (Hinton et al., 1986, pp. 82-84, 105). One solution is to have a separate copy of the network for each concept, and "gateway" networks for relating those concepts (Hinton, 1981). First, this seems a bit costly to someone who is used to creating a new concept by allocating a few bytes of storage. Second, and more importantly, if this approach is adopted, many of the advantages originally claimed for connectionist models are lost: instead of semantic relationships emerging from the "microstructure" of the system, the semantic relationships

---

[7]See Chapter 8 for a critical discussion of the memory model of McClelland and Rumelhart (1986).

[8]Tarnopolsky (1986) independently chose a similar example involving objects in a living room and bedroom for his simple connectionist model of "spontaneous thinking." Currently, his model is able to produce thought streams such as:

There are THINGs like the PICTURE which are SMALL. The THING which is SMALL may be a PICTURE. The THINGs such as a CURTAIN and a PICTURE are in the LIVING ROOM. (p. 18)

are determined by the "macrostructure" of the system of modules as designed by the programmer. One might think that a "scrambling" function could be run on the result in order to gain back the potential emergent properties lost in the modularization process. However, emergent properties occur in both local (modularized) and distributed holistic connectionism schemes, so it may not be necessary to rescramble the result. Furthermore, the fact remains that the programmer is forced to make semantic-level design decisions.

It would be difficult to construct a theory of daydreaming in the holistic, connectionist paradigm as currently conceived, for several reasons: First, knowledge structures such as schemata are important in the generation of daydreaming behavior; although connectionist models have recently been able to achieve certain aspects of schemata in an emergent fashion, it is difficult then to employ those schemata in an interesting way. Second, even operations which are staples in artificial intelligence programs, such as instantiation, become a major obstacle in a connectionist scheme. Most generally, we do not see why we should limit ourselves to what a given low-level mechanism may happen (or not happen) to be able to do.

It seems to be difficult to implement instantiation without abandoning the holistic paradigm. Touretzky (1986) has developed an implementation of Lisp cons cell allocation in connectionist networks; such a mechanism could be used to implement instantiation. However, it must be stressed that this is *classical* instantiation, which requires that the programmer specify the representation of concepts in advance. Thus the system cannot discover its own concepts as in the holistic paradigm. At present, the only advantages of redeveloping traditional artificial intelligence tools for connectionist networks are the "blurring" and fault tolerance properties provided by those networks. If a holistic, connectionist network could be made to discover its own instantiations and relationships among concepts, that would be interesting indeed.

We do not deny the importance of emergent phenomena. It is our hope, for example, that the elusive property of concepts will arise once a sufficiently complex knowledge level has been constructed. However, any approach that seeks to achieve intelligence through emergent phenomena must give careful consideration to the following problem: If intelligent behavior is finally produced by the model, will we be satisfied that we understand how that behavior is achieved? Will the model serve as an explanation of the phenomenon? Will the explanation be at a level which is of use to us? Will we be able to explain or predict phenomena of interest? For example, will the model be useful in diagnosing and treating depression?

Of course, in any artificial intelligence endeavor, lack of understanding of the behavior of a running program is a potential source of difficulties. Simply because one can "name" nodes does not mean one understands the operation of the system (McDermott, 1976; Hinton et al., 1986, p. 85)—the semantics of *any* computer program resides, not in any single node or data structure, but rather in the processes which operate on those data structures, and how a subset of those data structures are eventually related to the external world.

Although holistic connectionism presents several difficult problems, it may in fact be fruitful to investigate *semantic* connectionist models—those in which knowledge-level information is communicated between active processing elements in parallel (Charniak, 1983a; Small & Rieger, 1982; Minsky, 1977, 1981; Hewitt, 1977) or in which knowledge-level information is represented by nodes and links where processing is carried out by other mechanisms (J. R. Anderson, 1983). A knowledge-level connectionist theory of daydreaming is beyond the scope of the present work; however we do consider it an interesting topic for future research.[9]

10.3 Scientific Underpinnings

What are the scientific objectives of our research? Why do we want to understand how the mind works? According to what methodology will the construction of the theory proceed? How do we go about constructing the theory? How do we test and evaluate the theory? This section addresses these issues.

10.3.1 Scientific Objectives and Methodology

In constructing a computational theory of daydreaming there are two primary objectives: an understanding of human daydreaming—a *scientific* objective—and the application of daydreaming in order to make computer systems more useful—an *engineering* objective. In this section, we consider the scientific objectives of the present research (applications are discussed in Chapter 11): What constitutes an understanding of a phenomenon? What constitutes a successful theory?

Our theory must satisfy the following requirements: First, it must be stated in physical terms—that is, as a computer program. This ensures that no hidden or unknown operations are required in order to apply the theory. Second, the theory must be stated at a useful level of abstraction. That is, the theory should provide generalizations which are useful in making predictions (Pylyshyn, 1984, ch. 1).

We explore how a theory may satisfy these requirements, and how understanding of a phenomenon benefits from imposing these requirements on the theory, in the context of a detailed example—that of an electronic telephone switching system.

The electronic switching system is contained in a sealed box which is plugged into a power outlet and connected by wires to four telephones. Most everyone is familiar with the operation of the telephone. After a bit of thought, one may propose an initial concrete theory of the telephone switching system. For simplicity, we will not actually describe the theory as a computer program;

---

[9]We outline a preliminary version of such a theory in Chapter 11.

however, a demon-based implementation (Dyer, 1983a) is possible directly from our English description.

The theory is as follows: First of all, the system has an internal *representation* of the time-varying physical state of each of its phones, including the current hook state (on-hook or off-hook), current phone number being dialed, and instantaneous air pressure level (i.e., sound) at the phone's transmitter. The system can also influence the physical state of each of its phones in the following ways: It can make the phone ring, and it can modify the air pressure level over time at the phone's receiver.

A bit of terminology will simplify the description of the remainder of the theory: When the system begins the continual ringing of a phone, we will say that the system *starts* ringing the phone. While this process is being performed we will say the phone *is ringing*. When this process terminates we will say that the system *stops* ringing the phone. When the system begins the continual modification of the instantaneous air pressure level of the receiver of a phone to produce the sound of a particular tone, such as a dial tone or audible ring tone, we will say that the system *gives* the phone that tone. While this process is being performed we will say that the phone *has* the tone. When this process terminates we will say that the system *removes* the tone. When the system begins the continual modification of the instantaneous air pressure level of the receiver of one phone in proportion to the sensed instantaneous air pressure level of the transmitter of another phone, and vice versa, we will say that the system *connects* those two phones. While this process is being performed we will say the the two phones are *connected*. When this process terminates we will say that the system *disconnects* the two phones.

A few constraints which the system obeys are as follows: A phone may only have one tone at a time. A phone may be connected to at most one other phone. A phone may be in, at most, one of the following conditions at any given time: It may have a tone, be connected to another phone, or be ringing.

The processes which the telephone switching system performs may then be described as follows:[10]

- When the hook state of a phone changes from on-hook to off-hook, the system gives that phone a dial tone.

- When a phone number is dialed by a phone that has a dial tone, the system removes the dial tone from this *calling* phone, gives the phone an audible ring tone, and starts ringing the phone corresponding to the dialed number—the *called* phone. The system notes the calling phone associated with the called phone.

---

[10]It is assumed that the multiple actions necessary to handle a given phone state change are carried out as an atomic (uninterruptable) action. That is, the system finishes handling one state change before it handles the next.

- When a phone is ringing and the hook state of that called phone changes from on-hook to off-hook, the system stops ringing the phone, removes the audible ring tone from the associated calling phone, and connects the calling and called phones.

- When two phones are connected and the hook state of one of the phones changes from off-hook to on-hook, the system disconnects the two phones.

One reason for stating a theory as a computer program is to ensure the concreteness of the theory. Constructing a program forces the theory designer to consider cases that might have otherwise been overlooked. Another reason is to facilitate testing of the theory. By executing the programmed theory on a computer, one can easily try out various example situations in order to:

- Find bugs in the program/theory. Cases which were not considered when constructing the program/theory are likely to be exposed by running the program. Incomplete and inconsistent aspects of the program/theory are discovered when the program bombs.[11] These problems may then be fixed.

- Discover differences in the behavior of the program/theory and the known behavior of the actual system. This leads to appropriate revisions of the theory/program.

- Suggest experiments to be carried out on the actual system. When a certain behavior is produced by the program in some situation, and the theory designer does not know how the actual system behaves in that situation, an experiment may be conducted in order to find out. If the prediction of the theory holds in reality, confidence in the theory is strengthened. Otherwise, the theory must be revised.

Our initial theory of the telephone switching system may have several faults: It may be internally inconsistent or incomplete, and it may not correspond to the actual behavior of the system. However, now that we have an explicit theory, we can begin testing the theory. We consider various possible sequences of hook states and dialed numbers. If problems are discovered, the theory will be modified in order to fix those problems.

What if a number is dialed by a phone when that phone is connected to another phone? What happens in the program? Nothing. This behavior is consistent with the known behavior of the telephone switching system. What if the hook state of the phone associated with a ringing phone changes from off-hook to on-hook? That is, what if there is no answer and the caller hangs up? The program does nothing. In this case, the behavior is inconsistent with the

---

[11] The branch of computer science sometimes known as *verification* is concerned with discovering such problems in a program automatically or semiautomatically. However, these techniques are not yet able to cope with programs as intricate as most artificial intelligence programs. See the criticisms of DeMillo, Lipton, and Perlis (1978) of this approach.

known behavior of the telephone switching system. Therefore, the program must be appropriately modified: In this situation, the system should stop ringing the called phone and remove the audible ring tone from the calling phone. What if a phone which has a dial tone goes on-hook before a number is dialed? The program must be modified to disconnect the dial tone.

What if a dialed phone number corresponds to a phone that currently has a tone, is connected to another phone, or is ringing? That is, what if the called phone is busy? The program in this case will violate the constraint that a phone may not be ringing and connected to a phone at the same time—the program will bomb. Thus the program must be modified to check whether the called phone is busy, and if so, to give the calling phone a busy signal (and to remove the busy signal when the calling phone goes on-hook).

Once the above modification has been performed, a particular special case of busy phones may be considered: What if a dialed phone number corresponds to the very same phone that dialed that number? That is, what happens when a phone calls itself? The theory predicts that the phone will be given a busy signal. In this case, however, one may not know what the actual behavior of the telephone switching system is. Consequently, an experiment must be carried out on the actual system. If a busy signal is in fact given to the calling phone, this prediction of the theory is shown to be correct. Otherwise, the theory must be appropriately modified.[12]

Another case involving busy signals might be considered: If the caller hangs on the line after receiving a busy signal, will the system remove the busy signal, give the caller an audible ring tone, and start ringing the called phone when the called phone eventually becomes free? The theory predicts that this is not the case—the busy signal will remain until the caller hangs up. This turns out to be true when tried out on the actual system.

Other situations cause the program to bomb: When a ringing phone is picked up, the system also gives that phone a dial tone, violating the constraint that a phone may not have a tone and be connected to another phone at the same time. The program must be fixed to give a phone a dial tone only if it is not ringing. What happens if an unassigned number is dialed? The program again bombs in this case. The program must be fixed to give the calling phone a recording.

Still other predictions of the theory may be found to be incorrect: For example, suppose when two phones are connected, one phone goes on-hook for an instant and then goes off-hook again. The program would disconnect the two phones, while most real telephone systems keep the phones connected.

Instead of simply asserting that the telephone switching system somehow lets one "dial the number of a person and then talk to that person," we constructed

---

[12]The behavior of actual phone systems varies: Some give a busy signal and some give a recording. Still others, such as my old local exchange in Brentwood, California (a Western Electric No. 1A Electronic Switching System), give a sequence of clicks; when one then hangs up, the phone rings.

an explicit, computational theory of the system. In the process, we discovered many unanticipated complexities. We were forced to ask and answer questions that we might have otherwise neglected. Although there are probably other problems with our theory of the telephone switching system, it now comes quite close to reality. The theory may be used to understand, explain, and predict the operation of the system.

The theory presented above is a high-level theory (in terms of abstract states and processes) of the telephone switching system. This theory was constructed based on the external behavior of the system. Suppose, now, that we are allowed to take apart the sealed box containing the switching system. Now we can examine and catalog each and every electronic component and connection in that system—we can produce a complete low-level description of the system in terms of an electronic circuit diagram.

The high-level theory may be related to the low-level description via a certain transformational sequence: If the circuit incorporates a traditional microprocessor and memory, the transformations are those of a compiler which converts the high-level language program into a machine language program for the microprocessor. Otherwise, the transformations must convert the high-level language program directly into a hard-wired circuit (see, for example, the proposal of Mostow, 1983).

The two levels have their own advantages and disadvantages. The low-level description is useful if we wish to construct another identical electronic switching system. However, constructing a modified version of the system would be extremely difficult, since one has *only* a low-level electronic description of the circuit and not a more abstract understanding of the functional modules or programs contained in the circuit (a more abstract understanding would in fact be a high-level theory). In contrast, the high-level theory facilitates modification of the system, provided that one has a way to implement that high-level theory. A high-level theory also has the advantage that it can be implemented in any medium for computation.

The high-level theory is more useful for explanation. Suppose we are given the question, Why does a phone get a busy signal when the number of that same phone is dialed? Our high-level theory provides the following answer: Whenever one calls a number corresponding to a phone that is in use, one receives a busy signal. It does not matter if that phone happens to be the same phone one is calling from. The need for a busy signal derives from the constraint that a phone may not be connected to two things at once. In constrast, the low-level description (for a hard-wired logic circuit) would only give us an explanation something like: "Phone 4 is given a busy signal when phone 4 is dialed because when input lines 11 and 12 are both high, and pass through AND gate 62, the resulting high value is ANDed with input line 4 by gate 112, producing a high value which is sent to OR gate 19, resulting in a high value which is ANDed by gate 442 with a high value from flip-flop 45, producing a high value which is ANDed by gate 32 with a high value from AND gate 48 whose inputs are the outputs of flip-flops 12 and 13, resulting in a high value which sets flip-flop 21, whose high output is connected to driver 24, whose 12 volt output is in turn attached to relay 24 that connects phone 4 to the busy tone generator."

The high-level theory is also more useful for making predictions. For example, suppose one wishes to know what happens if two parties place a call to each other at the same time. Our high-level theory states that a busy signal is given to the caller whenever the called phone has a tone, is connected to another phone, or is ringing. Since both parties have a dial tone while dialing the other party, the answer is that both parties receive a busy signal. It is more difficult to answer this question using the low-level description: One would have to specify the appropriate time-varying input values (corresponding to the hook state changes and dialed numbers of the two phones of interest), simulate the operation of the circuit for those inputs, and then interpret the resulting output levels (which determine what connections the system makes). The simulation would then show that the system gives the two phones a busy signal. However, what is true of two particular phones may not be true of all pairs of phones. Therefore, the simulation would have to be carried out for every possible pair of phones. Furthermore, in general, the simulations would have to be carried out for every possible initial system and input state, and every possible sequence of state changes of the remaining inputs—those not associated with the two phones of interest. This adds up to an awful lot of simulations—one is probably better off just trying out the various cases on the actual system.

The purpose of a theory is to provide generalizations which are useful in explaining and predicting the behavior of the system. The high-level theory provides relevant generalizations, whereas the low-level description does not. Of course, there may be certain phenomena which the high-level theory does not address, but which the low-level description does. For example, suppose one bit of the memory cell that holds the state information for a given phone is damaged. This may result in behavior that is difficult to characterize in terms of the high-level theory. Thus some information is lost in the transformation from the low-level description to the high-level theory. But the advantage of the high-level theory is that it provides generalizations which are useful in predicting most behaviors. Of course, the ultimate understanding may consist of high- and low-level theories, and a mapping between the two levels. Indeed, one may feel that one understands a system best if one understands the operation of the system at a high level (its logical specification), the operation of the system at a low level (its physical implementation), and the connection between the two.

The case of human cognition, however, is more difficult. Constructing a complete low-level account of cognition faces the following obstacle: We cannot take apart the brain in order to analyze its operation the way we can take apart and analyze the electronic telephone switching system. It would destroy the pattern of interest to us. That is, there is no way to, say, insert a sensor at every neuron and synapse in order to gain a complete picture of the operation of the brain.

Using partial data or even complete data (if, perhaps, a way of scanning the brain with sufficient time and space resolution were somehow eventually realized), a low-level theory may be developed. However, in order for a low-level account to be useful, the constructs of the lower level must be related to the more familiar high-level constructs—this was demonstrated even in the simple case of the telephone switching system. In order for a low-level account of cognition to succeed as an explanatory theory, high-level abstractions will still have to be developed.

The transformations which map the high-level theory of the telephone switching system to the low-level description of that same system were seen to be fairly difficult. The transformations involved in mapping a knowledge-level theory to a neuroscientific theory are likely to be orders of magnitude more complex than this. (However, if materialism is true, there must be *some* mapping.)

We emphasize a top-down approach toward the explanation of human behavior, while the neurosciences and connectionism emphasize a bottom-up approach. Using either approach, the task is extremely difficult. We start with concepts from folk psychology, such as emotions, goals, and beliefs, and processes which operate on these concepts. These concepts and processes are continually refined to provide progressively better accounts of human behavior. We hope that at each incremental modification of the theory, a useful level of explanation can be retained. The neurosciences and connectionism, however, attempt to bootstrap human behavior starting from the structure of the human brain and neurons or neuron-like units. But in the connectionist paradigm, for example, when a certain high-level behavior is produced, it is not always clear how that behavior is actually generated. Saying that the schema is an "emergent property" is not enough. How exactly does the schema emerge and why?

Perhaps in the end neither approach will succeed at providing an explanation, or perhaps one will and the other will not. Probably, as Churchland (1984) points out, the end result will lie somewhere between a folk psychological explanation and a neuroscientific account. It will most likely involve the construction and conceptualization by humans of an entirely new, as yet unanticipated taxonomy—what may become the folk psychology of future generations. Perhaps one or both of the approaches will succeed in achieving truly intelligent behavior *without* retaining a useful explanation of that behavior. Even this, of course, would be a significant accomplishment.

10.3.2   Goals and Limits of Artificial Intelligence

Why attempt to understand the mind? There are those who would argue that explaining human intelligence is either impossible or, if possible, a misguided endeavor: By succeeding at a complete mechanistic explanation of the mind, all that is most unique and valued about our own humanity would be eliminated. First of all, such complete demystification is not probable in the near (or even not-so-near) future. Furthermore, by pursuing the enterprise of artificial intelligence, our fascination with human intelligence and creativity will only be strengthened. Because of the near intractability of human creativity, artificial intelligence researchers will have no difficulty finding interesting problems to look at for plenty of years to come. Artificial intelligence can be thought of as another creative human endeavor—one which presents a difficult problem to solve, a fun exercise, a new approach to discovering the many ways in which humans are so amazing.

What are the limits of artificial intelligence? Will our efforts, if continued over many years, eventually converge upon a complete theory of human intelligence? Will we reach a true thinking machine? Might we reach a limit point after which no more progress could be made?

If functionalism is true, not only is it possible for a theory of intelligence to be constructed in terms of a computer program, but, in principle, a computer program could actually *be* intelligent. Obviously, DAYDREAMER is a long way from having a true, human stream of consciousness. Computer programs are not likely to approach the level of human intelligence for many years to come, since the processing mechanisms, knowledge, and experiences which humans have are so much more complex and rich than those which so far have been put into computer programs.

Perhaps materialism is true and there will come a time when a fully human, intelligent computer program will be developed. However, even if materialism is true, we may not necessarily be able to construct such a program. It may be logically impossible for the mind to understand itself, since, as Johnson-Laird (1983) states the paradox, perhaps "the mind must be more complicated than any theory proposed to explain it: the more complex the theory, the still more complex the mind that thought of it in the first place" (p. 1). There are several responses to this problem. Perhaps such a theory, while not constructible by a single person, may be achieved by many persons working together. Perhaps such a theory, while not understood by any single person, may be understood by a collection of persons.

Sometimes it is argued that a mind understanding itself is impossible on the grounds that infinite storage would be required: If a complete model of the human mind were contained in a human mind, then this model would have to contain yet another complete model, which in turn would have to contain another complete model, and so on *ad infinitum*.[13] However, through economical processing and representation of recursion, this problem may be avoided: Instead of having a separate copy of the model embedded within each copy of the model, the top-level model need only *refer* to itself recursively and for recursive processes to be carried out on a demand basis; how to accomplish this is suggested by such programming language constructs as *pointers* and *lazy evalu-*

---

[13]But, as K. M. Colby (personal communication, February, 1987) points out, if the mind is infinite, it could contain itself as a subset (just as the integers contain the even integers as an isomorphic subgroup).

ation (Sussman & Steele, 1975). As Minsky (1968) suggests (but using different terminology), if an interpreter for a programming language is written in that same language (assuming, of course, that the interpreter is eventually executed by another hardware or software interpreter for that language with equivalent behavior), the interpreter can be run on itself to see how it would operate on an arbitrary program in that language (which itself could be the interpreter operating on yet another program in that language, and so on). There is no problem with infinite storage since there need only be one copy of the interpreter. In addition, several computer programs which can be said to introspect about their own operation—in a way more interesting than simple self-interpretation—have been constructed (B. Smith, 1982; Doyle, 1980).

Even if it were not logically impossible for the mind to understand itself, the mind might still turn out to be too complex for a person or group of people (even aided by partially intelligent computers!) to figure out. Perhaps the experimental methods which are available for examining the mind will prove insufficient.

Physiological mechanisms interact with psychological mechanisms and contribute much to the experience of being human (see, for example, Bloch, 1985). Even if materialism is true, a truly human artificial intelligence (if we wanted to construct one) might have to be embedded within a biological system (or within a simulation of a biological system).

Programs that seem human will be constructed way before those programs actually come close to achieving a level of intelligence comparable to that of humans. Indeed, the programs PARRY (Colby, 1973, 1975, 1981) and ELIZA (Weizenbaum, 1966) have already fooled some people, that is, passed weak versions of the Turing test (discussed in the following section).

As our computer programs become more intelligent, we might find humans becoming even more clever in order to demonstrate their superiority over their mechanical counterparts. As R. E. Mueller (1963) puts it:

> Perhaps the human mind has only been creating in low gear up to now. If a computer was invented that outinvented man, imagine how inventively man would try to outinvent the computer! (p. 152)

Materialism may turn out to be false—the phenomenological aspect of consciousness may turn out to be inexplicable in purely physical terms. Perhaps after trying so hard to design a computer program with artificial consciousness, and achieving a device with equivalent behavior to a human with true consciousness, we may find out that true phenomenological experiences are of a nonphysical nature. Of course, if the behavior of the artificial system is identical to the natural one, this might not make any difference to us. Another possibility is that the behavior of the program will not be equivalent to human behavior in very subtle ways—ways that depend on the true nature of mental phenomena. For example, there might be something "not quite right" about the program's descriptions of its internal experiences.

As we construct closer and closer computer approximations to human intelligence, and as we run into difficulties, it may be very difficult to pinpoint the cause of these difficulties: We are not smart enough, the systems lack physiology, materialism is false, and so on.

Even if it eventually turns out that materialism is false—that a full computational account of intelligence is not possible—we will have gained insight into those aspects of human cognition which *are* characterizable in these terms. We will also have better knowledge of which aspects of intelligence are capturable and which are not. If it were to turn out that *no* part of intelligence is capturable in computational terms, then we still will have gained insight into *artificial* intelligence.

Whether or not all of our theoretical assumptions (such as materialism) turn out to be true, and whether or not we ever converge upon a complete explanation of human behavior, we will have uncovered a whole range of previously uninvestigated human cognitive phenomena and previously unasked questions. We will thus have learned much about human thought processes.

10.3.3   Theory Construction, Testing, and Evaluation

What is the methodology for the construction of our computational theory of daydreaming? How is such a theory tested? How do we evaluate such a theory? What constitutes a successful theory? In what senses might the processes posited by the theory be considered equivalent to human mental processes?

Theory construction begins by examining one or more human daydream protocols. An initial theory is devised and implemented as a computer program which is able to exhibit the behavior of these protocols and other, similar, protocols. The theory is not developed in a vacuum. Rather, previous related theories from psychology, artificial intelligence, and philosophy are studied. Experiments from the psychological literature are consulted in choosing among alternative explanations of the same behavior. Because our objective is an explanatory theory of daydreaming at an appropriate level of abstraction, the theory is stated in terms of knowledge-level representations which correspond roughly to folk psychological notions such as emotions, goals, beliefs, attitudes, and so on.

A rudimentary test of the theory occurs in its implementation as a computer program. At the very least, the fact that the theory can be implemented demonstrates that the theory is physically possible. Moreover, it ensures that the theory does not depend on any hidden or mysterious mechanisms.

The theory is next tested by running the computer program. The program is presented with various input states and actions in performance mode, and the resulting external and daydreaming behaviors are observed. This exposes various problems—the program may crash or it may otherwise not exhibit the desired output behavior. Modifications are continually made to the initial theory and program in response to problems discovered from running the program. In order to increase the generality of the theory and program, additional human daydream protocols are considered and incorporated into the theory and program.

Since everything must be made concrete in a computer program, there are certain aspects of the program which might not be considered a part of the theory. As Colby (1981) writes:

> All the statements of a theory are intended to be taken as true, or as candidates for truth, whereas some statements in the programming language in the model, features necessary to get the model to run, are intended to be read as false, or at least of uncertain truth value. (p. 516)

That is, since a given theory must concentrate on some aspects of intelligence while overlooking others, those parts of the program which deal with issues not addressed by the theory must either accept input from a human, or be implemented in a simplistic fashion (analogous to the routines called "stubs" in software engineering [Kernighan & Plauger, 1976]). Indeed, one of the advantages of implementing a theory as a computer program is the discovery of such loose ends. At first, one may devise a quick, specific solution or "kludge" to enable the program to handle such a situation. However, upon closer examination of the situation, substantial theoretical issues may be revealed. One may then update the theory and program in order to handle that situation (and others) in a more general fashion. Alternatively, that situation may justifiably be deemed beyond the scope of the theory. A theory is best developed in conjunction with the writing of the program.

Once the behavior of the program has achieved rough face validity—that is, the output more or less resembles the modeled daydreaming behavior—more exacting tests may be employed: Turing (1950) proposed the "imitation game" as a test for computer intelligence: Suppose there are three persons—a woman, a man, and an interrogator of either sex. The interrogator is in a separate room from the woman and man, and communicates with them via a teletype connection. The task of the interrogator is to determine which of the two persons is the woman and which is the man. The man attempts to fool the interrogator into thinking he is the woman, and the woman attempts to convince the interrogator that she is the woman. Suppose now that a computer takes the place of the man. Turing contends that the computer may be considered intelligent if it is able to fool the interrogator as often as a man is able to. What is often called the *Turing test* is a modified version of this game: If an interrogator is allowed to converse freely with a computer program, and after some long length of time the interrogator is fooled into thinking the program is a human, that computer program is said to have passed the Turing test. The Turing test therefore ascribes importance to the input-output behavior of the human or computer program (rather than to other attributes). A *Turing-like* test may be used to evaluate a program which models some aspect of human intelligence: If the output of the program cannot be distinguished from the output of a human engaged in the activity modeled by that program, the program passes the test. Other authors (Hofstadter & Dennett, 1981; Colby, 1981; Heiser, Colby, Faught, & Parkison, 1980) treat Turing-like tests in greater detail.

How might a Turing-like test be applied to DAYDREAMER? One might argue that such a test is insufficient because daydreaming is an output-less activity—any program could pass such a test for daydreaming! As Weizenbaum (1974) pointed out in a letter to the *Communications of the ACM*, a program which responds to each input of an interrogator in the same way as an autistic patient—that is, not an iota—hardly constitutes an explanation of autism. However, daydreaming as defined here—as a verbal report of the stream of consciousness—is not an output-less activity. The daydreams produced by the current DAYDREAMER resemble those of a human daydreamer to a large degree; however, we suspect that the output of the program can be distinguished from human daydream protocols. This is due in part to the current English generator—perhaps if daydreams generated by a human and daydreams generated by the program were both edited for style by a human, the resulting daydreams would be indistinguishable. However, we have not yet attempted such an experiment.

10.3.4   Pylyshyn's Strong Equivalence

Pylyshyn (1984) argues that if a computer program is to be taken literally as a model of human behavior, the program must generate behavior in the *same* way as a human. In other words, he argues that a Turing-like test is too weak a criterion for an explanatory theory—rather, up to some level of granularity, the processes carried out by the program and by the human should be equivalent. He proposes the notion of "strong equivalence." Informally, two programs are strongly equivalent if they carry out the "same" algorithm. If two programs carry out the "same" algorithm, it should be possible to run both programs *directly* on the same functional architecture: Consider the *random access machine*, which has an infinite number of memory locations, accessible in a single step by address. Although Turing machines have the same computational power (are able to compute the same functions) as random access machines, there are certain algorithms—such as hashing—which can be directly executed on a random access machine, but not on a Turing machine. In order to run a hashing algorithm on a Turing machine, random access memory would have to be simulated. This can be accomplished by using the Turing machine's tape to store the contents of memory and accessing locations by scanning the tape until the desired location is reached. But in the process of simulation, the *computational complexity* of the program (measure of space and time necessary for a computation as a function of the input) is increased—on a Turing machine the number of steps to retrieve an item is dependent upon the total number of stored items, whereas on a random access machine the number of steps is independent of the number of items.

Consequently, complexity is a way of defining what is meant by direct execution of a program on a given functional architecture—if the basic operations of a program are not realized by the functional architecture in a number of steps independent of the input, then, by definition, the program is not executed directly by that architecture. A program which employs random access as a basic operation may therefore not be executed directly on a Turing machine. Strong equivalence of two processes requires that both processes run directly on the same functional architecture. Of course, a major task in constructing a cognitive model is that of finding the unknown functional architecture of cognition. The use of complexity measures and reaction-time experiments using humans may assist in this process. The strongest requirement for equivalence might be *phenomenological equivalence*—that the behavior be experienced subjectively in the same way. We have already discussed the paradoxes involved in such a notion.

There are several problems with Pylyshyn's arguments: First, he argues that Turing (input-output) equivalence is too weak and proposes the notion of "strong equivalence" which makes use of complexity equivalence and reaction times. But Turing equivalence does not necessarily free a theory from accounting for reaction times: For example, reaction times may be reflected in verbal behavior simply by requesting that a subject count out loud while not verbalizing, or by reading elapsed time off a stopwatch before each verbalization. Even if we do not require reaction-time equivalence, there are many important features of the functional architecture which can be correct without satisfying the criteria for strong equivalence. There are significant aspects of the functional architecture serving to determine behavior which can implemented using different algorithms. For example, failing (rather than delaying) at recalling something because it has not recently been primed is typically explained at the level of functional architecture (e.g., in terms of a spreading activation mechanism). And yet, this mechanism could be implemented using very different algorithms (for example, one on a Turing machine, one on a random access machine) not satisfying complexity equivalence. Thus even weak Turing equivalence is really not so weak.

Second, Pylyshyn appears to require that primitive operations of the functional architecture have a time complexity independent of input. A problem with this is that the primitives of most cognitive science architectures, present and future, hardly have this property: Spreading activation, for example, might have a time complexity dependent upon the lengths of connections in memory. There is no *a priori* reason for the functional architecture of cognition not to be implemented using various levels of interpretation. Perhaps there is theoretical justification for this view; however, Pylyshyn does not address the basic problem that functional architectures with the same primitive operations can be, and often are, implemented in radically different physical systems with resulting radically different complexity properties. For example, a random access machine can be implemented directly in hardware, or it can be implemented in microcode on a very fast Turing machine. If this requirement for primitive operations is enforced, then it is likely that the primitives of the functional architecture will not be very complex. When the primitives are thus constrained, all the ways of implementing them at the physical level will probably turn out to be the same algorithm. Thus this requirement has the side effect that two strongly equivalent programs will not simply be equivalent up to the level of the functional architecture, but since the functional architectures on which the programs are based are also implemented in the same way, the two programs will be equivalent all the way down to the physical level. This dilutes the notion of strong equivalence as an "abstract sameness" of algorithms only up to the symbol level. The problem is probably that degrees of "sameness" cannot be quantized.

If, on the other hand, primitive operations are not required to have time complexity independent of input, then reaction time experiments no longer constrain the functional architecture; instead, they constrain the entire implementation down to the physical level. This destroys the utility of reaction time tests for strong equivalence, which is only equivalence up to the level of functional architecture. However, reaction times could be used to test for a "stronger" equivalence requiring that the two programs execute the same algorithm all the way down to the physical level. Thus we have argued that either strong equivalence, as defined, ends up being all the way down to the physical level, or reaction times are useless in assessing strong equivalence.

Third, Pylyshyn is not clear on the issues of: (a) how complexity equivalence serves to locate the symbol-semantics boundary (although it is clear how the criterion of cognitive penetrability accomplishes this, even if it is just a restatement of the definition of the symbol-semantics distinction [see also Johnson-Laird, 1983, p. 152]), (b) how finding an appropriate symbol-semantics boundary (via any method) implies there is strong equivalence, (c) whether the program which runs on the functional architecture is identical to the semantic level (although the answer to this is most probably yes, it is far from clear, and there are problems: it appears that a semantic level program could not contribute complexity effects, because any such effects would be the result of cognitively impenetrable operations; yet paradoxically, strong equivalence is defined as equivalence up to, but not including, the functional level; that is, the programs which run on the functional level [semantical-level programs] must be equivalent; but this then means that complexity equivalence is meaningless as a test for strong equivalence), and (d) how the cognitive penetrability criterion ensures that the division of levels in the program corresponds to the division of levels in the person (this might be the case if the program has already been shown to be weakly equivalent to the person; however, if the program is weakly equivalent to the human, the levels are already correct).

10.3.5 Discussion

In the body of this book, we have specified our theory in English, with the major propositions typeset in italics. A detailed description of the DAYDREAMER computer program itself is given in Chapter 7.

Our theory of daydreaming represents one possible way in which daydreaming behavior may be produced. How might we evaluate whether and to what degree this behavior is achieved in the same way in humans? Both the major propositions of the theory and the detailed algorithms may be tested empirically. Although conducting appropriate psychological experiments is beyond the scope and resources of the present work, it will be useful to conduct such experiments in the future.

From one or more propositions of the theory, further consequences or side effects of the theory may be inferred. Nonobvious implications of the theory may also be discovered through observation of the behavior of the program—some consequences arise only when the dynamic behavior of a theory is explored in a running program. If predictions generated by the theory and program are validated experimentally, confidence in the theory is reinforced. This is the case whether or not the experimental evidence existed previously—provided that this evidence was not consulted in constructing the theory. Checking a theory against the data from which it was derived hardly serves to validate that theory (although it may serve as a check that we have not lost sight of the data). Rather, a theory must be tested through independent and indirect means. Whenever aspects of the theory have been derived from existing experimental evidence, relevant literature is cited in the text. If existing evidence for a given proposition of the theory is sufficiently conclusive, there is no need to test that proposition further. It may, however, be necessary to test that proposition in interaction with other propositions of the theory.

In addition, there are potential problems in validating a theory through its ability to explain (new) behaviors after the fact—rather than predicting behaviors before the fact. Rapaport (1960, p. 18, footnote) gives suggestions on the proper handling of such "postdiction" by a theory: It is necessary to make explicit what information is given in the behavioral data and what information can only be inferred (in a nonobvious manner) by postdiction from the given information.

Predictions regarding computational complexity may be generated by the theory and program. One example is the algorithm for serendipity which is presented in Chapter 5. The way the program carries out a given task might then be shown to be "strongly equivalent" to how humans carry out a similar task—this would be accomplished by measuring the reaction times of humans in a certain set of situations and comparing these times with the reaction times of the program in those situations.

However, if a prediction generated by the program turns out to be incorrect—sometimes this is obvious, as when the program crashes—there are two possibilities: Either the program does not correspond to the theory, or the program does correspond to the theory and the theory is (at least in part) incorrect. At this point, one must examine the theory and program and devise appropriate modifications. Testing the program and theory is thus important not only to validate the program and theory, but to suggest improvements in the program and theory.

Another means of evaluating DAYDREAMER is in terms of the proposed functions of daydreaming discussed in Chapter 1. In previous chapters, we have already discussed how the program achieves the functions of learning and creative problem solving.

DAYDREAMER as presented in this book is the result of many revisions (E. T. Mueller & Dyer, 1985a; E. T. Mueller & Dyer, 1985b). The behavior of the program resembles that of a human daydreamer—to what degree remains a question for future research. DAYDREAMER appears to account for much of the daydreaming behavior of humans, although a number of aspects of the theory are still to be tested in the psychological laboratory. Most likely, DAYDREAMER will succeed at some future tests and fail at others. Considering the magnitude of the problem, we consider it to be an achievement simply to have constructed a plausible computational theory of daydream production. Although this theory may not represent *the* way that daydreams are generated, it represents *one* *possible* way that daydreams may be produced—for now this must suffice since, to our knowledge, no alternative computational theories of daydreaming have yet been proposed. Thus DAYDREAMER is really a point of departure for future theories of daydreaming. We expect this theory to be continually refined in the years to come.

Chapter 11

Future Work and Conclusions

Capturing daydreaming computationally is a difficult task. Nonetheless, this book has shown how behavior resembling protocols of human daydreaming may be produced using a few simple but powerful mechanisms.

In response to external and internal circumstances, the program activates, processes, and terminates multiple *concerns*. Each concern is an instance of a *personal goal* or ongoing life objective of the program.

Concerns are motivated by data structures called *emotions* which determine which concern to process at any given time. The strengths of these emotions, while initially set according to the intrinsic importances of particular personal goals, are subject to dynamic modification as unexpected consequences of a concern are recognized.

The sequences of both fanciful and realistic events which make up daydream scenarios are produced using the building blocks of *planning and inference rules* and *episodes*. Planning rules specify methods of varying degrees of plausibility for breaking down a subgoal into further subgoals, while inference rules specify consequences of various situations.

Daydream scenarios are generated by the *planning* mechanism which repeatedly applies planning and inference rules to a selected concern. The planning mechanism is employed to generate possible behaviors of the daydreamer as well as the possible behaviors of others.

A mechanism for modifying existing daydream scenarios is the *mutation* of the objectives of unsuccessful action subgoals.

Episodes are aggregates of rule instances applied in a concrete real or imagined situation. Some episodes are hand-coded and provided to the program as input, while others are generated as the program daydreams and interacts in the simulated real world. Episodes are *indexed* in *episodic memory* under subgoal objectives, emotions, persons, and other features. Once *retrieved*, past episodes may be applied, in whole or in part, to a new concern by the *analogical planning* mechanism. Episodes reduce search in planning and enable scenario details to be filled in.

The *serendipity* mechanism recognizes the unexpected applicability of some possibility (external or internal situation or retrieved episode) related to one concern, to another active concern. The serendipity mechanism conducts an intersection search, from a point associated with a new possibility to a point associated with an active concern, through the space of currently accessible planning rules. Found paths are then verified by progressive unification and employed through analogical planning. Once a serendipity occurs, the resulting plan is stored in episodic memory so that in the future a similar plan can be generated without having to chance upon a similar serendipity.

A collection of *daydreaming goals* augment the program's personal goals and initiate useful daydreaming activity. The daydreaming goals of *rationalization* (generating scenarios to rationalize a failure), *roving* (shifting attention from an unpleasant failure), and *revenge* (generating scenarios in which revenge is attained) model the daydreaming which humans perform in order to reduce negative emotional states. The daydreaming goals of *reversal* (generating scenarios in which a past or imagined future failure is avoided), *recovery* (generating scenarios in which a goal which failed in the past succeeds in the future), *rehearsal* (generating possible scenarios for achieving an active goal), and *repercussions* (exploring and planning for hypothetical future situations) enable the program to improve its future external behavior—to *learn*.

Learning through daydreaming is accomplished by adding daydreamed episodes to memory: Various alternative scenarios involving a given situation are generated, evaluated as to their *realism* and *desirability*, and stored as episodes. When a similar situation arises in the future, the best and most similar retrieved episode is applied to that situation through analogical planning. Generation of hypothetical future scenarios improves the behavior of the program since negative consequences of various courses of action can be detected in advance and thus avoided.

Learning is also accomplished by adding new planning and inference rules: In response to a side effect real or imagined failure, the *reversal* daydreaming goal determines what actions might have been taken in order to avoid that failure. Rules are then created to anticipate similar failures in the future and carry out appropriate actions to prevent those failures.

In this chapter, we present: (a) possible extensions to, and applications for, DAYDREAMER, (b) the limitations of DAYDREAMER and how they arise, (c) how these limitations might be overcome in the future, and (d) final closing remarks.

11.1 Eventual Applications and Extensions

The components of DAYDREAMER described above are largely independent of the particular *interpersonal* domain of the program. Thus DAYDREAMER provides a framework for the construction of daydreaming systems in other domains. The components of DAYDREAMER which *are* dependent upon the domain are: (a) planning and inference rules, (b) personal goals, and (c) episodes. For other domains it may be necessary to add new daydreaming goals to the system. However, the existing set enables the system to daydream—which provides useful functions such as learning from past mistakes and anticipation of future experiences.

In this section, we consider the following domains for the potential future application of DAYDREAMER: autonomous robots, operating systems, creative writing, conversation, art and music, psychotherapy and psychiatry, and education and games.

11.1.1 Autonomous Robots

Currently, DAYDREAMER interacts in a *simulated* world whose behavior is determined in part by a human user who types in English phrases describing actions and states in that world. Eventually, however, DAYDREAMER could be employed as the controlling program for an autonomous robot interacting in the *real* world (or some part of the real world). A self-contained robot needs to decide what to do next. Multiple personal goals and motivating emotions provide this function. A household robot, for example, might have such ongoing objectives as: recharging batteries, pleasing owner, preparing meals, keeping house clean, watering plants, and so on. Emotions would be tuned to the intrinsic importance as well as dynamic urgency of a given objective or need. Such an autonomous robot would have to cope with situations, and improve at coping with various situations, without explicit human assistance. If the owner requested Shao Mai for breakfast the following morning, the robot might start daydreaming about preparing them and realize that some ingredient is missing and needs to be purchased. If the owner were about to leave for a week, the robot might daydream about one of its batteries failing and as a result tell the owner to leave it a spare. Although such a household robot is beyond what can be achieved with current technology, a rudimentary daydreaming robot with simple motion and object sensors and a limited set of spoken input commands could be constructed today.

11.1.2 Operating Systems

The notion of spending free time planning for future events can be applied to computer operating systems. When otherwise unoccupied with processing, the system could daydream about future operations of the user in order to perform those operations in part or otherwise prepare for those operations. For example, while unoccupied with processing in the early morning hours, the system might anticipate two of its heavy users creating large data files the following day whose size would exceed the remaining space in a certain file system. It might then relocate one of the users to another file system, or send messages to the users informing them of the potential problem. Even during normal daytime processing, the system might anticipate the use of a particular file (such as a mail file upon logging in) and start loading that file. Any operations with side effects (other than those involving performance) carried out during daydreaming would be tentative and subject to rollback using a mechanism such as nested transactions (E. T. Mueller, 1983). In general, maintaining consistency of the system in the face of daydreaming is an interesting topic for future research.

11.1.3   Creative Writing

Since human daydreaming often involves the generation of creative possibilities, daydreaming will be a useful component of a creative computer program. Constructing a program which assists the user in solving a problem by suggesting and organizing the application of various *general* brainstorming (Osborn, 1953), lateral thinking (de Bono, 1970), or conceptual blockbusting (J. L. Adams, 1974) techniques is relatively easy. However, a version of DAYDREAMER equipped with suitable representations of the problem domain would be able to suggest *specific* solutions to a problem.

The plot-oriented aspect of daydreaming (Klinger, 1971, p. 142; S. Freud, 1908/1962) suggests the potential application of DAYDREAMER to the creative domain of story invention.[1] The plots of RATIONALIZATION1 and REVENGE1, for example, resemble those of several recent motion pictures. DAYDREAMER could provide initial ideas for stories, animations, or live-action films, and once begun, suggest continuations, modifications, or further possibilities.

The program would have to be extended with appropriate personal and daydreaming goals for story invention. As input, the program would take any constraints of the user, such as characters and their personality traits, initial situations, desired outcome situations, points to convey, and the like. As output, the program would produce several stories satisfying—or even not satisfying— those constraints (and a modified episodic memory of the program for use in generating future stories). The output stories could be generated in English or even fed to a visual invention system for graphical animation.

DAYDREAMER could also be used at an entirely different level in story invention: as a means of generating accounts of the internal thought sequences and daydreams of story characters.

---

[1]Metz (1982), however, likens daydreaming to the *viewing* of a film.

11.1.4 Conversation

J. L. Singer (1981) has suggested that daydreaming allows more creative conversation and social interaction. An example of this may be found in a protocol of a conversation recorded by Hobbs and D. A. Evans (1980) (although such creative aspects were not the point of their work). The subject, discussing her daughter's entry to an art contest, interjects the following daydream-like scenario into the conversation:

> You send it to a P.O. box. ... What happens if you have dishonest mailmen ... and they see all these things going to an art contest, so they open it up and change it so that it's being sent from them? (p. 357)

Daydreaming, both prior to and during a conversation, will be useful in conversational computer systems to liven up the quality of the interaction. Furthermore, it has been observed (Schank, 1977) that what one may say in a conversation is a subset of the things that come to mind. Reik (1948) conversely observes that "thoughts are speeches not made, or condemned to silence" (p. 208). Although modeling "what comes to mind" in conversational contexts involves processes which not addressed by DAYDREAMER, a system which is able to daydream in conversational situations may indeed provide one component of a conversational computer system.

11.1.5 Art and Music

Artists, composers, choreographers, novelists, and poets daydream in images of their respective media. For example, Housman (1952, p. 91) reports new lines of verse flowing through his mind while taking a walk. According to Copland (1980), composers have "the ability to imagine sounds in advance of their being heard" (p. 24). Shapero (1946/1952) writes that "if [a composer] focuses his attention on a definite key and beats mentally in a chosen meter, musical images will be set in motion in his mind, and the entire musical texture generated in this way" (p. 52).

DAYDREAMER may be adapted for applications in various art forms by the provision of an appropriate set of personal goals, daydreaming goals, planning and inference rules, and episodes. This task is naturally accomplished by the artist, who would probably start by providing the program with representations of the artist's own style and techniques, including appropriate visual (Arnheim, 1974), musical, or other principles, works of past artists, and so on. The artist would run the program, examine its products, and revise the rules as appropriate. Of course, in constructing and revising representations, the artist need not be limited to the existing style and technique of that artist. Rather, the artist may experiment with new ideas whose expression or execution in conventional media would be impractical or impossible. Thus the artist might evolve a new style possible only in the *computational* medium.

A version of DAYDREAMER equipped with suitable representations might be applied in the following ways: First, the program could function as an *artist's apprentice*: It could be used to generate ideas which the artist might not otherwise have thought of, or to execute ideas which the artist might not otherwise have been able to execute. These ideas could then be incorporated into artistic works or discarded as the artist saw fit. Second, if the program evolved to a point where the artist considered its output as valid works of art, the program could function as an *artist's proxy*: Works produced by the program would be fully authorized by the artist. In this case, one might adopt the view that it is not the output of the program which is the work of art, but the program itself and its entire range of potential output. Thus such a program might function as *art object*.

Is it possible to create art in the medium of computation? Can one internalize and habituate oneself to this medium to the extent necessary for artistic expression? There are certainly obstacles to algorithmic art at present: The languages available for expressing computational ideas are generally too rigid, too concerned with details, too fragile in the face of incompleteness and inconsistency. Not every artist will wish to set aside years of training in traditional media in order to pursue computation. Perhaps we will have to wait until future generations of artists who as children learned about frames, production rules, and unification along with the three Rs.

Many artists would probably choose to withhold their programs from distribution in order to prevent others from learning how their program operates, constructing derivative versions, or producing further works using the program. Nonetheless, there is a necessity to avoid reduplication of effort; some artists and programmers would probably choose to make their basic representations of art and music theory and history available to others. A potential danger, of course, with the wildfire-like spread of artistic computer programs, is that everyone's work would become homogeneous (not to mention the problems of copyrights and royalties). Artists would therefore avoid basing their work on previous programs—except for a basic set of tools sufficiently rich to enable many artists to construct unique and diverse works (just as the tools of conventional media enable the production of new and different works).

If an artist's program were distributed to others, that program and its outputs would be more vulnerable to analysis than are conventional works of art. After all, the program provides an explicit, complete model of how its outputs are generated. Would examination of such a program reduce the apparent creativity of its outputs? First of all, the program might be of such a level of complexity that it would not be possible to understand exactly how it produces a given output. But even if one were able to understand how the program worked, one would have to realize, as with any creative product, that it is much easier to understand a product after the fact than to have thought of it in the first place.

If a new artistic program is derived from a program written by another artist, is that program then a joint work? When, if ever, does a derivative work cease to be the work of the original artist? Is computational art culture best developed through the sharing of programs or should interaction among artists be restricted to observation of program output behavior?[2]

Although the productions of an artistic program should not consist of mere "variations," there would nonetheless appear to be fundamental limits on the quantity of creative products which could be generated. That is, the more we ask a program to generate outputs, the less creative those outputs are likely to seem. Does generating a large quantity of outputs automatically negate the value of those outputs (or might the outputs taken as a collection still be considered creative)? Can it be that artistic value is subject to some sort of law of conservation—that only a certain number of objects or "chunked" classes of objects may be considered creative at a time? We leave questions such as these to art theorists and to future experience with artistic computer programs.

Actually, artists and composers have already begun to experiment with such programs: Hiller and Isaacson (1959) constructed a program which was used to compose the *Illiac Suite for String Quartet*; Xenakis (1971) employed computers to assist in his stochastic compositions; H. Cohen (H. Cohen, B. Cohen, & Nii, 1984) wrote a program able to create its own drawings based on heuristic rules; Whitney (1980) has used computers to execute a form of animated visual art based on an analogy with harmonic relationships in music; and so on (see, for example, the reviews and discussions of Hiller, 1984, R. E. Mueller, 1983, Sofer, 1981, Leavitt, 1976, Cope, 1976, pp. 77-114). What DAYDREAMER adds to this previous work is a framework for the construction of creative programs; this framework consists of the following components: multiple creative tasks directed by emotions and daydreaming goals, recognition and exploitation of serendipitous relationships among tasks, fanciful possibility generation through arbitrary mutation and other methods, storage and later application of previous experiences, learning from mistakes, and knowledge representation in the form of both generic rules and episodes. Whereas in DAYDREAMER, episodes are employed in generating new daydreams, in an advanced artistic program, episodes might function in a contrary fashion: The system might actually try to *avoid* creating new works similar to episodes representing previous works.

11.1.6 Psychotherapy and Psychiatry

DAYDREAMER could eventually be employed as a model to test various strategies for treating depression (see, for example, Beck, 1967) and for investigating the processes that might lead to depression. In addition, certain modifications

---

[2]R. E. Mueller (1967, pp. 273-274) discusses similar issues.

to DAYDREAMER might lead to behavior resembling various pathological conditions of humans.

What is the long-term impact on the behavior of DAYDREAMER of various sets of daydreaming strategies? Which combinations are more advantageous and which are less so? The **RATIONALIZATION**, **ROVING**, and **REVENGE** daydreaming goals serve the function of reducing negative emotional states. What would happen to the behavior of DAYDREAMER in the long run if one or more of these goals were removed? What would happen if these strategies were applied excessively? If, for example, **RATIONALIZATION** were removed, but **ROVING** were retained, the program would reduce negative emotions but never alter the negative emotions associated with episodes. It might, therefore, remain depressed as a result of "repressing" its negative emotions—immediately diverting attention from negative emotions and not coping with them through long-term modifications such as that provided by **RATIONALIZATION**.

If all daydreaming goals were removed, DAYDREAMER would be unable to improve its behavior in response to performance mode experiences, plan for future ones, or reduce negative emotional states. It would most likely end up in a negative state. If performance mode were removed, DAYDREAMER would never be able to achieve its personal goals and would therefore also end up in a negative state.

Perhaps it is when certain daydreaming strategies get out of hand that personality disorders such as paranoia begin to develop, as suggested by the following anecdote (reported by Dyer, 1983c):

> A man was driving along a lonely road late at night when his car ran out of gas. He remembered seeing a farm house a mile back, and so started walking toward it. As he walked along he thought: "It's pretty late. If someone were to awaken me at this hour, I might be pretty annoyed." He kept on thinking along these lines: "The farmer will have to get dressed and siphon gas out of his tractor for me. He may find that very inconvenient..." As he approached the farmhouse he became more and more annoyed.
>
> The farmer was awakened by loud knocking at his door. When he went down to answer it, there stood a man who barked: "Who needs your stupid gas anyway!" and then stomped off into the night. (Dyer, 1983c, p. 76)

Daydreaming can result in the excessive raising or lowering of expectations; without appropriate controls, daydreaming can cause one to lose touch with reality. In DAYDREAMER, such an effect could be achieved through the removal of reality assessment: The program would then generate external behavior in accordance with unrealistic daydreams; its expectations of the world's behavior in response to its own would be too high, and it would fail at achieving its personal goals in the short and long term.

Although worrying has its disadvantages, as one can see from the above story, worrying also serves the useful function of anticipating possible future events and planning for them. There is a subtle line between adaptive and maladaptive instances of daydreaming. J. L. Singer (1975, pp. 180-204) discusses the relationship between daydreaming and psychopathology in some detail.

11.1.7 Education and Games

There are potential applications of DAYDREAMER in education and games for children and adults. Just as LOGO (Papert, 1980) enables one to experiment with the consequences of a few graphics primitives, a suitably simplified DAY-DREAMER (perhaps using a representation based on English phrasal patterns instead of slot-filler objects) would enable one to enter a few daydreaming rules, and run the program to see what it generates. DAYDREAMER could eventually be used as part of a *shared daydreaming* game in which the user and computer would take turns in expanding out possible scenarios in some domain.

11.2 Shortcomings of the Program

DAYDREAMER cannot daydream for a long time and cannot generate many novel sequences. Unexpected behaviors of the program were more useful in understanding possible daydreaming pathologies and flaws in the details of the mechanism, than they were for generating truly novel solutions to interpersonal problem situations.

One unexpected behavior was an *infinite daydreaming loop.* Before a mechanism was added to inhibit the activation of multiple top-level goals with the same objective, the following cycle occurred: The **REVERSAL** daydreaming goal would generate an alternative scenario in which the first of two personal goal failures resulting from an earthquake was avoided. (**REVERSAL** is only able to plan to avoid one failure at a time in the current version of DAYDREAMER.) Since the second personal goal still failed in this scenario, a new **REVERSAL** daydreaming goal would be activated to avoid the second failure. This in turn resulted in a scenario in which failure of the second personal goal was avoided, but failure of the first still occurred. Thus another **REVERSAL** daydreaming goal would be activated for the first personal goal and the cycle would repeat indefinitely.

A flaw that was discovered in the system was its failure to terminate a **RE-HEARSAL** daydreaming goal once the activity to be rehearsed was being carried out in reality. Thus the system might continue to rehearse an action after it had already been completed. This flaw in fact led to an unanticipated, but valid, serendipity: when *the waiter serves Guy* was received as input during an actual **M-DATE** experience, this action was detected as serendipitously applicable to a still active **REHEARSAL** concern for that **M-DATE**.

Whether the program's various mechanisms will be sufficient for the generation of truly novel daydreams as well as long-term learning and emotion regulation remains to be seen. The length of continuous nonredundant daydreaming and its degree of novelty is limited in the current program by several interacting factors:

- *Set of rules and episodes*: This set is less than some "critical mass" (if there is such as thing) necessary for the generation of an endless variety of daydreams.

- *Susceptibility to minor bugs*: If a rule or episode is coded incorrectly, daydreams which could be generated by that rule of episode are not generated, or the future behavior of the program is adversely affected. Unification, the basis of most of the program's operations, is very sensitive to the exact form of representations.

- *Speed of the program*: By its very nature, daydreaming eats processing time and a supercomputer was not available. Although many optimizations have been employed, such as rule chaining and indexing of facts in contexts, the program is still quite slow. It takes many hours to find out the results of a given addition to the program. In conjunction with the above problem involving minor bugs, this makes it difficult to add rules and episodes.

- *Degree of program tailoring*: Whenever the program fails to produce a desired daydream, an incorrect rule or mechanism is fixed. Although it is important for the program mechanisms to function properly, it is less clear to what extent one should be permitted to modify the set of rules. As one continually adjusts rules to attain some desired behavior of the program, one gets the feeling that it is the programmer, rather than the program, who is carrying out the process of search. When program execution is slow, this advance search by the programmer speeds debugging of the essential components of the program. Unfortunately, the more the program is tailored in advance by the programmer, the less likely it is to generate novel products.

The speed of the program could be improved to some degree through various modifications. For example, contexts never to be used again in the future could be destroyed and the storage reclaimed. Such contexts consume time during garbage collection and may also slow the program down through increased paging because the contexts are scattered in virtual memory. Ultimately, DAYDREAMER must be reimplemented in a machine with at least the power of a Connection Machine (Hillis, 1985).

The program's susceptibility to minor bugs could be lessened through the use of *relaxed unification*. Techniques for relaxed unification include: ignoring variable type restrictions, ignoring a small number of slots which fail to unify, ignoring minor differences of type, and so on. Relaxed unification could also be used as a component of *relaxed analogical planning*—the application of an episode to goals which match the goal of the episode only in a relaxed fashion. In exchange for a reduced number of rules failing to fire when they should (errors of *omission*), however, this will increase the number of rules which fire when they should not (errors of *commission*). Thus "incorrect" daydreams might result, and strategies for pruning incorrect sequences would have to be devised. In addition, the system would become even slower as a result of having to generate and test these additional possibilities.

Techniques will have to be devised for automatically acquiring the rules and hand-coded episodes[3] necessary to generate novel daydream scenarios. This would ease the task of extending the program and might at the same time provide a solution to the problems of program tailoring and bug susceptibility. A future DAYDREAMER should be able not only to learn through daydreaming, but also to learn how to daydream.

In the following section, we present an alternative architecture for daydreaming which may help overcome the above shortcomings of the current system. This architecture is presented as one possible topic for future research in machine daydreaming.

11.3 Overdetermination of Daydreaming

What factors are responsible for the production of the daydream REVENGE1 (see page 4)? In the current version of DAYDREAMER, this daydream results from a **REVENGE** daydreaming goal activated in response to a failed **LOVERS** goal in LOVERS1: Achieving a position of prestige, that of being a famous movie star, results in the movie star having a positive attitude toward DAYDREAMER so that he will ask her out on a date, so that she can turn him down, so that she can obtain revenge. On the other hand, this daydream can also be seen to fulfill a **RECOVERY** daydreaming goal resulting from the **SOCIAL ESTEEM** goal which also failed in LOVERS1: Achieving a position of prestige results in the star having a positive attitude toward DAYDREAMER, resulting in success of the same social esteem goal which previously failed.

How might the above daydream have been generated in a human? There are three logical possibilities (if our assumption that daydreaming goals are the source of daydreams is correct): First, it may be that the daydream is generated in response to the **REVENGE** daydreaming goal and the daydream simply happens to satisfy the **RECOVERY** daydreaming goal. Alternatively, it may be that the daydream is generated in response to the **RECOVERY** daydreaming goal and the daydream simply happens to satisfy the **REVENGE** daydreaming goal. The third possibility is that the daydream may have been the result of

---

[3]Of course, DAYDREAMER already acquires episodes through interactions in performance mode and also through storage of previous daydreams.

combined efforts to satisfy both the **REVENGE** and **RECOVERY** daydreaming goals simultaneously.

The latter case is an example of the principle of *overdetermination* introduced by Breuer and S. Freud (1895/1937, pp. 156, 219). This principle states that a behavior (such as a symptom or dream element) generally stems, not from a single cause, but from a combination of causes. That is, while any single cause is insufficient to result in the given behavior, several causes taken together *are* sufficient to result in that behavior.

There are numerous examples of overdetermination in the products of cognition: S. Freud (1900/1965, pp. 311-339) shows how events represented in a dream may be traced to multiple underlying causes. Reik (1948, pp. 28-35) provides examples of overdetermination in a daydream protocol. A single conversational utterance may be overdetermined in the sense that it achieves several goals of the speaker simultaneously (Hobbs & D. A. Evans, 1980; P. N. Johnson & Robertson, 1981). Sentences with double meanings are another instance of overdetermination: In some contexts, an idiom may be employed and understood in both its literal and figurative sense. Readers are able, for example, to recognize both meanings of *Kathy always lands on both feet* in the context of whether or not she will succeed at a risky parachute jump (R. A. G. Mueller & Gibbs, 1987).

The psychoanalytic conception of overdetermination emphasizes unconscious sources of behavior. In general, one may or may not be conscious of the various elements which contributed to a given behavior. This section considers the possibility that the stream of consciousness might result from the *merged* products of many processes, rather than from any single process as in DAYDREAMER. We propose a basic model for the generation of overdetermined behavior, describe some associated problems and possible solutions, and present some initial ideas toward a system for the overdetermined production of daydreams.

11.3.1   An Overdetermination Mechanism

We propose a mechanism for the production of overdetermined behavior consisting of the following components: (a) a large collection of nonconscious, concurrent *generative processes* which provide the underlying source material for behavior, and (b) a collection of nonconscious, concurrent *merging processes* which produce the final behavior of the system through merging of the results of several generative processes. Thus the only elements which rise to the conscious surface are those which are multiply determined—participants in several nonconscious generative processes. (The term *nonconscious* is used to avoid confusion with S. Freud's [1900/1965] constructs of the "unconscious," which encompasses thoughts which have a high resistance to becoming conscious, and the "preconscious," which encompasses thoughts which can become conscious at any moment.)

In one way, this model parallels Chomsky's (1965) formulation of generative grammar: Generative processes correspond to the phrase structure grammar which generates a deep structure; merging processes correspond to the transformations which map the deep structure into the surface structure; the surface structure corresponds to conscious elements. In other ways, however, it is quite different: Chomsky does not postulate a large number of generative processes. Therefore, there is not much opportunity for overdetermination on a large scale. (A case, however, could be made for such a grammar introducing minor overdeterminations into the surface structure—one surface structure element could very well correspond to more than one deep structure element.) In addition, whereas the merging processes might introduce distortions or new elements of meaning, transformations in Chomsky's standard theory "cannot introduce meaning-bearing elements ..." (p. 132)

What psychological evidence is there for nonconscious, semantic processing? Habitual behaviors, such as driving to work, become automatic and hence performed with little or no conscious awareness (see, for example, Norman, 1982; J. R. Anderson, 1983). There is some evidence for perception without awareness and the nonconscious priming of concepts (Pötzl, 1917/1960; Dixon, 1981; Marcel, 1980). What are apparently conscious, voluntary behaviors may be influenced by nonconscious factors: In experiments conducted by Libet (1985) and colleagues, electrophysiological "readiness potentials" (negative shifts in electric potential recorded on the scalp) were found to precede by 400 milliseconds the conscious intention to perform a spontaneous and voluntary motor act (as measured by subjects' recall of the position of a revolving spot upon the first awareness of intention to act, and corrected by a control experiment involving the timing of subjects' reported awareness of skin stimuli). Libet concludes that initiation of such acts begins nonconsciously.

11.3.2 Potential Benefits

If an overdetermination mechanism could be made to work, it would provide solutions to two instances of the problem of elusiveness of concepts. First, overdetermination provides a potential solution to the *cognitive modeler's dilemma*: The processes and data entities hypothesized by the cognitive modeler to account for one or more human protocols are merely one set of processes and data entities which might have resulted in those protocols. In fact, the modeler often discovers more than one potential model. While choices can often be made in accordance with experimental data, theoretical assumptions, or principles of parsimony, at other times the modeler is forced to make an arbitrary choice.

Within the limits of the given theoretical framework, overdetermination would enable the modeler, when confronted with several possible sources of a behavior, to incorporate each of those sources into the model. This does not free the modeler, however, from: (a) finding each of these possible sources in the first place, and (b) eventually showing that these sources are valid (either that they have analogues in humans or that they are useful from an engineering or other standpoint).

Second, overdetermination provides a potential solution to the *scaling problem* in artificial intelligence programs: Once a given rule has been constructed, the programmer will frequently discover exceptions which render that rule incorrect in certain situations (see, for example, Michalski & Winston, 1986). Typically, the rule is then modified or the rule is discarded and a new rule is constructed which operates properly. However, when the system consists of a large number of rules, this process has a tendency to diverge: Whenever one problem is fixed, other problems are created; it also becomes very difficult to add further rules to the system without introducing new problems. (See the related discussion by D. E. Smith, Genesereth, and Ginsberg, 1986, pp. 347-350.)

Overdetermination would enable many conflicting rules to exist simultaneously. Exceptions would be handled, not by modifying existing rules, but by adding new rules. The behavior of the system in a given situation would be defined not by the products of single rules, but by the merged products of many rules. The rules would in effect *vote* on the final behavior. As a simple example, if, in a particular situation, a rule of attitudinal congruity (Heider, 1958) indicated a positive attitude toward some person while two defense mechanism rules (A. Freud, 1937/1946) indicated a negative attitude toward that person, the negative attitude would win out.

Overdetermination would thus enable the constructions of larger and more complete systems, because the inconsistency problems associated with large collections of rules in classical systems would no longer apply; it would no longer be necessary to ensure that each rule is wholly correct. Overdetermination thus adopts the "scruffy" approach to cognitive science (Abelson, 1981).

11.3.3   Problems for the Mechanism

There are many problems which arise in attempting to construct an overdetermination model of the stream of thought: If conscious sequences of imagined events are overdetermined by several nonconscious sequences of events, then how are those sequences of events merged together? Interlacing events from the two sequences is not likely to result in a coherent sequence. Yet daydreams are coherent.[4] How are coherent and continuous merged sequences generated? If two sequences are merged only if they are identical, there is the risk that two processes would never happen to generate identical sequences of events and that no conscious scenarios would ever be produced. If two sequences were merged when they differed only in, say, the persons and physical objects involved, similar sequences might still be rare occurrences.

In general, it is difficult to merge two scenarios since those scenarios may involve orthogonal world states. When the world states of scenarios do not clash

---

[4]S. Freud (1900/1965) proposed a process of "secondary revision" to render the manifest content of a dream more coherent and "like a day-dream." (p. 530)

entirely, how can they be reconciled? Merging might require alteration of one or both of the event sequences. How much alteration is permitted? Freud's dreamwork mechanisms, for example, permit a very distortive transformation from nonconscious to conscious elements. The task of merging might be facilitated if processes were allowed to interact as events were being generated.

How might emotions be modeled in an overdetermined processing scheme? Are there nonconscious emotions? Emotional *arousal* (Schachter & J. E. Singer, 1962), at least, is associated with conscious thought. Need emotions be handled any differently than other concepts?

One sometimes employs the technique of "counting sheep" in order to fall asleep. An overdetermination mechanism has no executive process for guiding the many nonconscious processes. How, then, does the "voluntary" or "directed" aspect of the stream of consciousness arise? How is it that we can "consciously direct" our thoughts? Overdetermination assumes that this aspect of thought is an emergent property of the many nonconscious processes. It is not at all clear how this property might be achieved. This problem presents an apparent paradox: Conscious control arises out of nonconscious processes, and yet it seems able to direct those very processes.

To what extent are the multiple causes of a behavior accessible to consciousness? S. Freud (1901/1960) argues, for example, that slips of the tongue originate in repressed (unconscious) thoughts. On the other hand, the utterance in which such a slip is embedded may have been consciously planned. Sometimes one utters a sentence with full knowledge of a double meaning, while other times one is not conscious of any second meaning. What beliefs do people have about the causes of their own behavior? To what extent are these beliefs accurate?

If conscious elements are overdetermined by nonconscious elements, then nonconscious elements might be *underdetermined* by conscious elements. (This is not necessarily so: Multiple conscious elements might share single nonconscious elements.) That is, it may be difficult or impossible for the modeler to figure out what the nonconscious elements are, given the conscious elements.[5] Whenever more remote potential sources for a given behavior are found in addition to more direct sources, how does the modeler know whether these more remote sources actually had any influence? Is it necessary to consider *when* the influence comes or *how much* influence there is from a particular source?

If the stream of consciousness merely arises out of nonconscious processes, then why is daydreaming—or the stream of consciousness—an interesting object of study? Should we not be studying nonconscious processes instead? First, the surface level of consciousness (as reported to others and as reflected in other actions) corresponds to human behavior and that is a primary object of study

---

[5]Even given further information, such as free associations, the nonconscious elements are not fully determined and the interpreter can hypothesize arbitrary nonconscious elements to suit the occasion. This has been one of the major criticisms of S. Freud's (1900/1965) wishfulfillment theory of dreams (see, for example, Grünbaum, 1984, pp. 235-239; Foulkes, 1978, pp. 45-46).

in the field of artificial intelligence. Second, consciousness is a window into nonconscious processes. By examining the results of nonconscious processes we may be able to figure out what those nonconscious processes are.

11.3.4 Initial Solutions

We now present initial solutions to some of the problems described above. In production systems (see, for example, Langley & Neches, 1981), conflicts are resolved through techniques such as selecting rules whose antecedent concepts have the highest activation, giving priority to more specific rules, and others. In an overdetermination mechanism, however, conflicting rules are permitted to fire. The outputs of those rules must then be merged into an appropriate result.

One way to merge *identical* concepts is to compile each new concept into a discrimination net (Feigenbaum, 1963; Charniak, Riesbeck, & McDermott, 1980). Concepts are merged whenever a new concept is compiled into the same location in the network as an existing concept. A number is associated with each unique concept indicating its degree of determination, or the number of times a concept was compiled into that location. Concepts with a degree of determination above a certain threshold would be the conscious results of the nonconscious processes.

In order to merge *similar* concepts, a simple similarity metric based on the syntactic form of slot-filler objects and embedded objects could be used. This would in fact be a general way of implementing the mechanism of "composition" (S. Freud, 1900/1965) in which several persons are represented in the surface content of a dream as a single person.

The content of daydreaming, however, is not single concepts but sequences of events. In order to merge two event sequences, each intended to achieve a different goal, transformational rules or optimization techniques such as the "critics" proposed by Sacerdoti (1977) could be employed. These transformations would find and eliminate redundancies in a plan consisting of the simple composition of the two plans (performing one sequence and then the other). For example, if one plan is to go to the grocery store to get groceries, and the other is to go to a newsstand to get a newspaper, the combined plan would consist of getting both the groceries and newspaper at the grocery store, paying not twice but once, and so on. Alternatively, transformations could be applied as the plans are constructed, rather than afterwards.

Another event merging process is suggested by the serendipity detection mechanism of DAYDREAMER: Whenever a subgoal of one plan under construction is recognized as applicable—directly or through the application of further rules—to some subgoal of another plan under construction, those two plans may be merged.

Some techniques have been developed for constructing overdetermined conversational utterances. The MAGPIE program (P. N. Johnson & Robertson, 1981) constructs utterances which achieve multiple goals as follows: An utterance is produced to achieve the first goal; the remaining goals are then achieved through incremental modification of the utterance. For example, MAGPIE models the generation of an utterance by a wife who has three goals: to obtain information about her husband's location the previous night, to express anger toward her husband, and to regain control over the relationship. Depending on the ordering of goals, the system is able to generate "Damn you, you were out too late last night. Where were you?" and "You were out so late last night, why?" (Anger, for example, is expressed in the former via the initial utterance "Damn you!" and in the latter through modification of the existing question utterance in order to emphasize the source of anger.) Hobbs and D. A. Evans (1980) suggest a similar approach in which goals compete to fill in "slots" of an utterance; material may be added to a given slot provided that (a) a stronger goal has not already filled that slot, and (b) the material is suitable given the contents of other slots.

11.3.5 DAYDREAMER*

One possible future direction for research is to construct a system for the overdetermined production of daydreams, called DAYDREAMER*. To handle the large number of independent processes, DAYDREAMER* would ideally be implemented on a massively parallel machine. In addition, several components of the current version of DAYDREAMER are suited to parallel implementation: The rule intersection search performed by the serendipity mechanism and investigation of alternative possibilities resulting from generic rules and episodes.

Instead of the stream of consciousness resulting from a shifting of processing from concern to concern, as in DAYDREAMER, the stream of consciousness in DAYDREAMER* would result from the merged results of processing on behalf of multiple concerns. Specifically, concerns would start out as independent processes executing simultaneously.[6] Then two processes would be *merged* upon detection of applicability of a combining transformation, upon detection of a common subgoal, or upon detection of potentially common subgoal (as in the serendipity mechanism of DAYDREAMER). This process could be applied repeatedly, resulting in the merging of what were once three or more independent processes. Processes would *split* again if the merged process was unable to achieve all (or some percentage) of its concern goals (or those processes would simply terminate).

Furthermore, why limit daydreaming on behalf of a given concern to a single event stream at a time? While alternative plans for achieving a given subgoal are investigated in serial order in DAYDREAMER, n alternative plans would result in a process being split into n parallel processes in DAYDREAMER*. Thus many alternative processes for each concern would be subject to merging with many processes for other concerns.

---

[6]This discussion applies only to daydreaming goal concerns. Personal goal concerns would be carried out in a single process corresponding to the simulated real world.

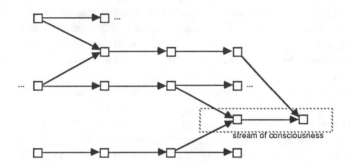

Figure 11.1: Overdetermined Stream of Consciousness

Processing in DAYDREAMER* can thus be viewed as a continuous activity of process splitting and merging. How is the stream of consciousness extracted from this activity? Perhaps it could be based on a "good" or "best" path through the sequence of merged processes. Figure 11.1 shows on a small scale what this might look like. (These mechanisms could be implemented on a serial system, but at a much slower rate.) Mechanisms for the "voting" of multiple conflicting planning and inference rules, and for pruning many possibilities (especially those generated through relaxed unification), will also have to be devised. It will be interesting to see to what degree the stream of consciousness resulting from such a system resembles the human stream of consciousness.

11.3.6   Relationship to Previous Work

J. L. Singer (1974, pp. 247-248) has previously proposed that daydreaming makes use of active long-term memory processes which take place just outside our awareness. Minsky (1977, 1981) has proposed a "society of mind" theory in which the mind is composed of many communicating and sometimes conflicting agents; in this theory, also, self-awareness or consciousness arises out of the complex pattern of agents rather than being contained in any one executive agent. A related idea is Hewitt's (1977) model in which computations are performed by agents called "actors" which send and receive messages to and from other actors and create new ones.

Several holistic connectionist researchers have proposed to model human intelligence using large networks inspired by the physiology of the brain (Rumelhart et al., 1986; Hofstadter, 1983; J. A. Feldman & Ballard, 1982; Hinton & J. A. Anderson, 1981). Connectionist networks provide a natural implementation for overdetermined behavior, since, as Norman (1986, p. 546) points out, there is no one reason for any given state of such a network; any state results from a conjunction of many factors. However, at present, connectionist networks lack an ability which is fundamental to daydream construction: The ability to instantiate and combine concepts. The problem is that a concept is represented in a connectionist network as a stable pattern of activation of the nodes in the net. As a result, only one concept can be represented at a time by a single network (Rumelhart, Smolensky, McClelland, & Hinton, 1986, p. 38), making it difficult to combine several concepts into a new one. Some researchers have therefore proposed keeping several connected copies of a network, one for each concept (Hinton, 1981). What is instead needed is some way, in a *single* network, to compose together previous substates of the network to form a new state. Minsky (1981) provides a suggestion as to how this might be accomplished via his mechanism of "K-lines" and the "K-recursion principle." Otherwise, the problem of constructing network architectures becomes equivalent to the problem of representing concepts by conventional, knowledge-level means.[7] We look forward to future experimentation with connectionist systems.

How does the view of overdetermined processing presented here compare to (nonholistic) spreading activation architectures for cognition such as that proposed by J. R. Anderson (1983)? Such an architecture consists of nodes connected by links, a process for spreading an analog level of activation from one node to connected nodes, and production rules which operate only on nodes above a certain level of activation.

If it is assumed that nodes above a certain threshold level of activation are conscious,[8] then this mechanism can describe the production of a conscious result by several nonconscious sources: Nodes whose activation is below the threshold of consciousness may spread activation to a node causing its level to rise above the threshold of consciousness. However, nonconscious processing in addition to this automatic process of spreading activation could not be described, because production rules may only access conscious nodes.

Alternatively, one threshold could be equated with consciousness, while another, lower, threshold could be used to determine whether processing might occur for a given node. This would enable modeling of nonconscious processing as well as the production of overdetermined elements. However, this would not enable overdetermination on a massive scale: Because activation level is intended to model the limited capacity of human short-term memory (Miller, 1956), only a relatively small number of nodes are permitted to be active, and accessible to processing, at any time. While in a spreading activation architecture only a small portion of the network is involved in processing, massive overdetermination requires that a large component of the network be involved in processing.

Thus in contrast to the view of spreading activation architectures that processing occurs only on short-term memory elements, overdetermination takes the view that processing occurs on many elements not present in short-term memory. In fact, overdetermination suggests that short-term memory (and

---

[7]See Chapter 10 for a more detailed discussion of this point.

[8]Although J. R. Anderson (1983) equates nodes above a certain activation level with human short-term memory, he does not assume such nodes correspond to consciousness (note, p. 309).

consciousness) may arise out of a large number of nonconscious processes. It remains to be determined whether and how this might be the case.

11.4   Was It Worth It?

Why was such a large problem as daydreaming chosen? Why was not a small, circumscribed topic instead chosen? There are several advantages to taking on a large problem: One is less likely to form a peephole view of things. One can integrate what were once seen as different problems. Work on one component often leads to progress on other components. For example, at the time the serendipity mechanism in DAYDREAMER was first designed, two other mechanisms were being used to recognize the applicability of input states to an active concern and to recognize the utility of an mutated action. It was then realized that the serendipity mechanism could accomplish both of these tasks. Similarly, the analogical planning mechanism has proved far more useful and general than was originally expected: It, for example, is used equally for carrying out previously daydreamed plans in the simulated real world, for generating daydream scenarios from past experiences, and for fleshing out a serendipity. If any one of the components of daydreaming had been studied alone, its relationship to other components might never have been noticed.

Taking on a large problem has its disadvantages as well: There is the danger of one's energy being spread too thin on too many problems; it is easy to become overwhelmed with the magnitude of the task; one might be so intent on doing everything that one ends up doing nothing. But any *true* artificial intelligence problem is by its very nature a large problem: How can we study natural language independently of other aspects of cognition? The same could be said for vision, problem solving, creativity, and so on. Daydreaming is but one more *AI-complete* problem: If we could solve any one artificial intelligence problem, we could solve all the others. We can benefit in the future from both comprehensive and more restricted approaches to these problems.

Despite its limitations, DAYDREAMER provides an explicit, running model of the seemingly elusive, but pervasive, phenomenon of daydreaming. The program demonstrates how the generation of various creative past, present, and future events enables beneficial modification of future behavior. Daydreaming is a useful capability for any computer system, and a necessary capability for a truly intelligent one.

References

Abelson, R. P. (1963). Computer simulation of "hot" cognition. In S. S. Tomkins & S. Messick (Eds.), *Computer simulation of personality.* New York: Wiley.

Abelson, R. P. (1981). Constraint, construal, and cognitive science. In *Proceedings of the Third Annual Conference of the Cognitive Science Society* (pp. 1-9), Berkeley, CA.

Adams, J. L. (1974). *Conceptual blockbusting: A guide to better ideas.* New York: W. W. Norton.

Anderson, J. R. (1983). *The architecture of cognition.* Cambridge, MA: Harvard University Press.

Antrobus, J. S. (1977). The dream as metaphor: An information-processing and learning model. *Journal of Mental Imagery, 2,* 327-337.

Antrobus, J. S., & Singer, J. L. (1964). Visual signal detection as a function of sequential variability of simultaneous speech. *Journal of Experimental Psychology, 68,* 603-610.

Antrobus, J. S., Singer, J. L., & Greenberg, S. (1966). Studies in the stream of consciousness: Experimental enhancement and suppression of spontaneous cognitive process. *Perceptual and Motor Skills, 23,* 399-417.

Arnheim, R. (1974). *Art and visual perception: A psychology of the creative eye* (the new version). Berkeley, CA: University of California Press.

Asimov, I. (1963). *The human brain: Its capacities and functions.* New York: Signet.

Barber, G. (1983). Supporting organizational problem solving with a work station. *ACM Transactions on Office Information Systems, 1*(1), 45-67.

Bartlett, F. C. (1932). *Remembering: A study in experimental and social psychology.* Cambridge: Cambridge University Press.

Baylor, G. W., & Deslauriers, D. (1986). Dreams as problem solving: A method of study—Part I: Background and theory. *Imagination, Cognition, and Personality, 6*(2), 105-118.

Beare, J. I. (1931). De memoria et reminiscentia. In W. D. Ross (Ed.), *The works of Aristotle* (vol. 3). Oxford: Clarendon Press.

Beck, A. (1967). *Depression: Clinical, experimental, and theoretical aspects.* New York: Harper & Row.

Ben-Aaron, M. (1975). The Poetzel effect: Corroboration of a cybernetic hypothesis. In R. Trappl & F. Pichler (Eds.), *Progress in cybernetics and systems research* (Vol. 1, pp. 247-252). Washington, DC: Hemisphere Publishing.

Bexton, W. H., Heron, W., & Scott, T. H. (1954). Effects of decreased variation in the sensory environment. *Canadian Journal of Psychology, 8*(2), 70-76.

Bleuler, E. (1951). Autistic thinking. In D. Rapaport (Ed.), *Organization and pathology of thought.* New York: Columbia University Press. (Original work published 1912).

Bloch, G. (1985). *Body and self: Elements of human biology, behavior, and health.* Los Altos, CA: William Kaufmann.

Bower, G. H., & Cohen, P. R. (1982). Emotional influences in memory and thinking: Data and theory. In M. S. Clark & S. T. Fiske (Eds.), *Affect and cognition: The 17th Annual Carnegie Symposium on Cognition.* Hillsdale, NJ: Lawrence Erlbaum.

Bower, G. H., Monteiro, K. P., & Gilligan, S. G. (1978). Emotional mood as a context for learning and recall. *Journal of Verbal Learning and Verbal Behavior, 17,* 573-585.

Breuer, J., & Freud, S. (1937). *Studies in hysteria.* Boston: Beacon. (Original work published 1895)

Bundy, A., Silver, B., & Plummer, D. (1985). An analytical comparison of some rule-learning programs. *Artificial Intelligence, 27,* 137-181.

Cannon, W. B. (1927). The James-Lange theory of emotions: A critical examination and an alternative theory. *American Journal of Psychology, 39,* 106-124.

Carbonell, J. (1980). Towards a process model of human personality traits. *Artificial Intelligence, 15,* 49-74.

Carbonell, J. G. (1983). Learning by analogy: Formulating and generalizing plans from past experience. In R. S. Michalski, J. G. Carbonell, & T. M. Mitchell (Eds.), *Machine learning* (pp. 137-161). Palo Alto, CA: Tioga.

Charniak, E. (1983a). Passing markers: A theory of contextual influence in language comprehension. *Cognitive Science, 7,* 171-190.

Charniak, E. (1983b). The Bayesian basis of common-sense medical diagnosis. In *Proceedings of the National Conference on Artificial Intelligence* (pp. 70-73). Los Altos, CA: Morgan Kaufmann.

Charniak, E., & McDermott, D. V. (1985). *Introduction to artificial intelligence.* Reading, MA: Addison-Wesley.

Charniak, E., Riesbeck, C. K., & McDermott, D. V. (1980). *Artificial intelligence programming.* Hillsdale, NJ: Lawrence Erlbaum.

Chomsky, N. (1965). *Aspects of the theory of syntax.* Cambridge, MA: MIT Press.

Churchland, P. M. (1984). *Matter and consciousness: A contemporary introduction to the philosophy of mind.* Cambridge, MA: MIT Press.

Clark, M. S., & Isen, A. M. (1982). Toward understanding the relationship between feeling states and social behavior. In A. Hastorf & A. M. Isen (Eds.), *Cognitive social psychology.* Amsterdam: Elsevier.

Clocksin, W. F., & Mellish, C. S. (1981). *Programming in Prolog.* Berlin: Springer-Verlag.

Cohen, H., Cohen, B., & Nii, P. (1984). *The first artificial intelligence coloring book.* Los Altos, CA: William Kaufmann.

Colby, K. M. (1963) Computer simulation of a neurotic process. In S. S. Tomkins & S. Messick (Eds.), *Computer simulation of personality.* New York: Wiley.

Colby, K. M. (1973). Simulations of belief systems. In R. C. Schank & K. M. Colby (Eds.), *Computer models of thought and language.* San Francisco: W. H. Freeman.

Colby, K. M. (1975). *Artificial paranoia: A computer simulation of paranoid processes.* New York: Pergamon Press.

Colby, K. M. (1981). Modeling a paranoid mind. *The Behavioral and Brain Sciences, 4*(4), 515-560.

Collins, A. M., & Loftus, E. F. (1975). A spreading activation theory of semantic processing. *Psychological Review, 82,* 407-428.

Cope, D. H. (1976). *New directions in music* (2nd Edition). Dubuque, IA: Wm. C. Brown.

Copland, A. (1980). *Music and imagination.* Cambridge, MA: Harvard University Press.

Crook, J. H. (1983). On attributing consciousness to animals. *Nature, 303*(5), 11-14.

Cundiff, G., & Gold, S. R. (1979). Daydreaming: A measurable concept. *Perceptual and Motor Skills, 49,* 347-353.

Dallin, L. (1974). *Techniques of twentieth century composition: A guide to the materials of modern music* (3rd Edition). Dubuque, IA: Wm. C. Brown.

Darwin, C. (1872). *The expression of the emotions in man and animals.* Chicago: University of Chicago Press.

de Bono, E. (1970). *Lateral thinking: Creativity step by step.* New York: Harper & Row.

DeJong, G. (1981). Generalizations based on explanations. In *Proceedings of the Seventh International Joint Conference on Artificial Intelligence* (pp. 67-69). Los Altos, CA: Morgan Kaufmann.

Dement, W. C. (1976). *Some must watch while some must sleep.* San Francisco: San Francisco Book Company.

Dement, W., & Kleitman, N. (1957). Cyclic variations in EEG during sleep and their relation to eye movements, body motility, and dreaming. *Electroencephalography and Clinical Neurophysiology, 9,* 673-690.

DeMillo, R. A., Lipton, R. J., & Perlis, A. J. (1978). *Social processes and proofs of theorems and programs* (Revised version, GIT-ICS-78/04). Atlanta: Georgia Institute of Technology.

Dennett, D. C. (1978). *Brainstorms.* Cambridge, MA: MIT Press.

Dennett, D. C. (1984). *Elbow room: The varieties of free will worth wanting.* Cambridge, MA: MIT Press.

Desoille, R. (1966). *The directed daydream.* New York: Psychosynthesis Research Foundation.

Dietterich, T. G., & Michalski, R. S. (1983). A comparative review of selected methods for learning from examples. In R. S. Michalski, J. G. Carbonell, & T. M. Mitchell (Eds.), *Machine learning* (pp. 41-81). Palo Alto, CA: Tioga.

Dixon, N. F. (1981). *Preconscious processing.* New York: Wiley.

Doyle, J. (1980). *A model for deliberation, action, and introspection* (Technical Report 581). Cambridge, MA: Massachusetts Institute of Technology, Artificial Intelligence Laboratory.

Duda, R. O., Hart, P. E., and Nilsson, N. J. (1976). Subjective Bayesian methods for rule-based inference systems. In *Proceedings of the 1976 National Computer Conference*. Montvale, NJ: AFIPS Press.

Dyer, M. G. (1983a). *In-depth understanding*. Cambridge, MA: MIT Press.

Dyer, M. G. (1983b). The role of affect in narratives. *Cognitive Science, 7,* 211-242.

Dyer, M. G. (1983c). Understanding stories through morals and remindings. In *Proceedings of the Eighth International Joint Conference on Artificial Intelligence*. Los Altos, CA: Morgan Kaufmann.

Ericsson, K. A., & Simon, H. A. (1984). *Protocol analysis: Verbal reports as data*. Cambridge, MA: MIT Press.

Ernst, G., & Newell, A. (1969). *GPS: A case study in generality and problem solving*. New York: Academic Press.

Evans, C. (1983). *Landscapes of the night: How and why we dream*. New York: Viking.

Evans, C. R., & Newman, E. A. (1964). Dreaming: An analogy from computers. *New Scientist, 419,* 577-579.

Faletti, J. (1982). *PANDORA—A program for doing commonsense planning in complex situations*. In *Proceedings of the National Conference on Artificial Intelligence* (pp. 185-188). Los Altos, CA: Morgan Kaufmann.

Farrell, B. A. (1950). Experience. *Mind, 54,* 170-198.

Feigenbaum, E. A. (1963). Simulation of verbal learning behavior. In E. A. Feigenbaum & J. Feldman (Eds.), *Computers and thought*. New York: McGraw-Hill.

Feigenbaum, E. A., & Feldman, J. (Eds.). (1963). *Computers and thought*. New York: McGraw-Hill.

Feigenbaum, E. A., & Simon, H. A. (1984). EPAM-like Models of Recognition and Learning. *Cognitive Science, 8,* 305-336.

Feldman, J. A., & Ballard, D. H. (1982). Connectionist models and their properties. *Cognitive Science, 6,* 205-254.

Feshbach, S. (1956). The catharsis hypothesis and some consequences of interaction with aggressive and neutral play objects. *Journal of Personality, 24,* 449-462.

Festinger, L. (1957). *A theory of cognitive dissonance*. Stanford, CA: Stanford University Press.

Fikes, R. E., Hart, P. E., & Nilsson, N. J. (1972). Learning and executing generalized robot plans. *Artificial Intelligence, 3,* 251-288.

Fikes, R. E., & Nilsson, N. J. (1971). STRIPS: A new approach to the application of theorem proving to problem solving. *Artificial Intelligence, 2,* 189-208.

Flavell, J. H. (1979). Metacognition and cognitive monitoring: A new area of cognitive developmental inquiry. *American Psychologist, 34,* 906-911.

Fodor, J. (1981). The mind-body problem. *Scientific American, 244,* 114-123.

Foulkes, D. (1966). *The psychology of sleep.* New York: Scribner's.

Foulkes, D. (1978). *A grammar of dreams.* New York: Basic Books.

Foulkes, D. (1985). *Dreaming: A cognitive-psychological analysis.* Hillsdale, NJ: Lawrence Erlbaum.

Foulkes, D., & Fleisher, S. (1975). Mental activity in relaxed wakefulness. *Journal of Abnormal Psychology, 84,* 66-75.

French, T. M. (1952). *The integration of behavior, volume I: Basic postulates.* Chicago: University of Chicago Press.

French, T. M. (1953). *The integration of behavior, volume II: The integrative process in dreams.* Chicago: University of Chicago Press.

Freud, A. (1946). *The ego and the mechanisms of defence.* New York: International Universities Press. (Original work published 1937)

Freud, S. (1936). *The problem of anxiety.* New York: W. W. Norton. (Original work published 1926)

Freud, S. (1960). *The psychopathology of everyday life.* New York: W. W. Norton. (Original work published 1901).

Freud, S. (1961). *Beyond the pleasure principle.* New York: W. W. Norton. (Original work published 1920)

Freud, S. (1962). Creative writers and day-dreaming. In J. Strachey (Ed. and Trans.), *The standard edition of the complete psychological works of Sigmund Freud* (Vol. 9). London: Hogarth Press. (Original work published 1908)

Freud, S. (1965). *The interpretation of dreams.* New York: Avon. (Original work published 1900)

Friday, N. (1973). *My secret garden.* New York: Pocket Books.

Gackenbach, J. I. (1985). A survey of considerations for inducing conscious awareness of dreaming while dreaming. *Imagination, Cognition, and Personality, 5*(1), 41-55.

Garfield, P. (1974). *Creative dreaming.* New York: Ballantine.

Giambra, L. M. (1977). Adult male daydreaming across the life span: A replication, further analyses, and tentative norms based upon retrospective reports. *International Journal of Aging and Human Development, 8*(3), 197-228.

Giambra, L. M. (1980). A factor analysis of the items of the Imaginal Processes Inventory. *Journal of Clinical Psychology, 36,* 383-409.

Gibbs, R. W., & Mueller, R. A. G. (1988). Conversational sequences and preference for indirect speech acts. *Discourse Processes, 11,* 101-116.

Goffman, E. (1959). *The presentation of self in everyday life.* Garden City, NY: Doubleday.

Goldman, N. (1975). Conceptual generation. In R. C. Schank (Ed.), *Conceptual information processing.* New York: American Elsevier.

Goldman, N. (1982). *AP3 reference manual.* Unpublished report. Marina del Rey, CA: USC/Information Sciences Institute.

Gouaux, C. (1971). Induced affective states and interpersonal attraction. *Journal of Personality and Social Psychology, 20*, 37-43.

Green, C. E. (1968). *Lucid dreams.* Oxford: Institute of Psychophysical Research.

Green, G. H. (1923). *The daydream: A study in development.* London: University of London Press.

Griffin, D. R. (1981). *The question of animal awareness* (Revised and Enlarged Edition). Los Altos, CA: William Kaufmann.

Grossberg, S. (1976). Adaptive pattern classification and universal recoding: Part I. Parallel development and coding of neural feature detectors. *Biological Cybernetics, 23*, 121-134.

Grünbaum, A. (1984). *The foundations of psychoanalysis: A philosophical critique.* Berkeley, CA: University of California Press.

Guilford, J. P. (1967). *The nature of human intelligence.* New York: McGraw-Hill.

Hariton, E. B., & Singer, J. L. (1974). Women's fantasies during sexual intercourse: Normative and theoretical implications. *Journal of Consulting and Clinical Psychology, 42*, 313-322.

Hayes-Roth, B., & Hayes-Roth, F. (1979). A cognitive model of planning. *Cognitive Science, 3*(4), 275-310.

Heider, F. (1958). *The psychology of interpersonal relations.* Hillsdale, NJ: Lawrence Erlbaum.

Heiser, J. F., Colby, K. M., Faught, W. S., & Parkison, R. C. (1980). Can psychiatrists distinguish a computer simulation of paranoia from the real thing? The limitations of Turing-like tests as measures of the adequacy of simulation. *Journal of Psychiatric Research, 15*, 149-162.

Hewitt, C. (1975). How to use what you know. *Advance papers of the Fourth International Joint Conference on Artificial Intelligence* (Vol. 1). Los Altos, CA: Morgan Kaufmann.

Hewitt, C. (1977). Viewing control structures as patterns of passing messages. *Artificial Intelligence, 8*(3), 323-364.

Hiller, L. (1984). The composer and the computer. *Abacus, 1*(4), 9-31.

Hiller, L. A., & Isaacson, L. M. (1959). *Experimental music.* Westport, CT: Greenwood Press.

Hillis, W. D. (1985). *The connection machine.* Cambridge, MA: MIT Press.

Hinton, G. E. (1981). Implementing semantic networks in parallel hardware. In G. E. Hinton & J. A. Anderson (Eds.), *Parallel models of associative memory* (pp. 161-188). Hillsdale, NJ: Lawrence Erlbaum.

Hinton, G. E., & Anderson, J. A. (Eds.). (1981). *Parallel models of associative memory.* Hillsdale, NJ: Lawrence Erlbaum.

Hinton, G. E., McClelland, J. L., & Rumelhart, D. E. (1986). Distributed representations. In D. E. Rumelhart, J. L. McClelland, & PDP Research Group (Eds.), *Parallel distributed processing: Explorations in the microstructure of cognition. Volume 1: Foundations* (pp. 77-109). Cambridge, MA: MIT Press.

Hobbes, T. (1968). *Leviathan.* Harmondsworth, England: Penguin. (Original work published 1651)

Hobbs, J. R., & Evans, D. A. (1980). Conversation as planned behavior. *Cognitive Science, 4,* 349-377.

Hofstadter, D. R. (1983). The architecture of jumbo. In *Proceedings of the Second Machine Learning Workshop,* Urbana, IL.

Hofstadter, D. R. (1985). *Metamagical themas: Questing for the essence of mind and pattern.* New York: Basic Books.

Hofstadter, D. R., & Dennett, D. C. (1981). *The mind's I: Fantasies and reflections on self and soul.* Toronto: Bantam.

Housman, A. E. (1952). The name and nature of poetry. In B. Ghiselin (Ed.), *The creative process.* New York: Mentor.

Huba, G. J., Segal, B., & Singer, J. L. (1977). Consistency of daydreaming styles across samples of college male and female drug and alcohol users. *Journal of Abnormal Psychology, 86,* 99-102.

Izard, C. E. (1977). *Human emotions.* New York: Plenum.

Jakobson, R. (1978). *Six lectures on sound and meaning.* Cambridge, MA: MIT Press.

James, W. (1890a). *The principles of psychology* (Vol. 1). New York: Dover.

James, W. (1890b). *The principles of psychology* (Vol. 2). New York: Dover.

Janis, I., & Mann L. (1977). *Decision-making.* New York: Free Press.

Johnson, M. K., & Raye, C. L. (1981). Reality monitoring. *Psychological Review, 88*(1), 67-85.

Johnson, P. N., & Robertson, S. P. (1981). *MAGPIE: A goal-based model of conversation* (Research Report 206). New Haven, CT: Yale University, Computer Science Department.

Johnson-Laird, P. N. (1983). *Mental models: Towards a cognitive science of language, inference, and consciousness.* Cambridge, MA: Harvard University Press.

Jones, E. (1908). Rationalisation in everyday life. *Journal of Abnormal and Social Psychology, 3,* 161-169.

Jung, C. (1916). *Psychology of the unconscious.* New York: Dodd, Mead, and Company.

Kahn, K. M. (1978). *Director guide* (AI Memo 482). Cambridge, MA: Massachusetts Institute of Technology, Artificial Intelligence Laboratory.

Kahneman, D., & Tversky, A. (1982). The simulation heuristic. In D. Kahneman, P. Slovic, & A. Tversky (Eds.), *Judgment under uncertainty: Heuristics and biases.* New York: Cambridge University Press.

Kedar-Cabelli, S. T. (1985). Purpose-directed analogy. In *The Seventh Annual Conference of the Cognitive Science Society* (pp. 150-159), Irvine, CA.

Kelley, H. H. (1967). Attribution theory in social psychology. In D. Levin (Ed.), *Nebraska Symposium on Motivation.* Lincoln, NB: University of Nebraska Press.

Kernighan, B. W., & Plauger, P. J. (1976). *Software tools*. Reading, MA: Addison-Wesley.

Klahr, P. (1978). Planning techniques for rule selection in deductive question-answering. In D. A. Waterman & F. Hayes-Roth (Eds.), *Pattern-directed inference systems* (pp. 223-239). New York: Academic Press.

Klinger, E. (1971). *The structure and function of fantasy*. New York: John Wiley.

Klinger, E. (1978). Modes of normal conscious flow. In K. S. Pope & J. L. Singer (Eds.), *The stream of consciousness*. New York: Plenum.

Koestler, A. (1964). *The act of creation: A study of the conscious and unconscious in science and art*. New York: Macmillan.

Kolodner, J. L. (1984). *Retrieval and organizational strategies in conceptual memory: A computer model*. Hillsdale, NJ: Lawrence Erlbaum.

Kolodner, J. L., Simpson, R. L., & Sycara-Cyranski, K. (1985). A process model of cased-based reasoning in problem solving. In *Proceedings of the Ninth International Joint Conference on Artificial Intelligence* (pp. 284-290). Los Altos, CA: Morgan Kaufmann.

Kornfeld, W. A., & Hewitt, C. (1981). The scientific community metaphor. *IEEE Transactions on Systems, Man, and Cybernetics, SMC-11*(1).

Kowalski, R. (1975). A proof procedure using connection graphs. *Journal of the ACM, 22*, 572-595.

Kris, E. (1952). *Psychoanalytic explorations in art*. New York: International Universities Press.

Kugel, P. (1986). Thinking may be more than computing. *Cognition, 22*, 137-198.

LaBerge, S. (1985) *Lucid dreaming: The power of being awake and aware in your dreams*. Los Angeles: Jeremy P. Tarchner.

Laird, J. E. (1984). *Universal subgoaling*. Doctoral dissertation, Department of Computer Science, Carnegie-Mellon University, Pittsburgh, PA.

Langley, P., Bradshaw, G. L., & Simon, H. A. (1983). Rediscovering chemistry with the Bacon system. In R. S. Michalski, J. G. Carbonell, & T. M. Mitchell (Eds.), *Machine learning* (pp. 307-329). Palo Alto, CA: Tioga.

Langley, P., & Neches, R. (1981). *PRISM user's manual* (Technical report). Pittsburgh, PA: Carnegie-Mellon University, Department of Psychology.

Lazarus, R. S. (1968). Emotions and adaptation: Conceptual and empirical relations. In W. J. Arnold (Ed.), *Nebraska Symposium on Motivation*. Lincoln, NB: University of Nebraska.

Leavitt, R. (1976). *Artist and computer*. New York: Harmony Books.

Lebowitz, M. (1980). *Generalization and memory in an integrated understanding system* (Research Report 186). New Haven, CT: Yale University, Computer Science Department.

Lehnert, W. G. (1982). Plot units: A narrative summarization strategy. In W. G. Lehnert & M. H. Ringle (Eds.), *Strategies for natural language processing*. Hillsdale, NJ: Lawrence Erlbaum.

Lenat, D. B. (1976). *AM: An artificial intelligence approach to discovery in mathematics as heuristic search* (Report No. STAN-CS-76-570). Stanford, CA: Stanford University, Computer Science Department.

Lenat, D. B. (1983). The role of heuristics in learning by discovery: Three case studies. In R. S. Michalski, J. G. Carbonell, & T. M. Mitchell (Eds.), *Machine learning* (pp. 243-306). Palo Alto, CA: Tioga.

Lenat, D. B., & Brown, J. S. (1984). Why AM and EURISKO appear to work. *Artificial Intelligence, 23*(3), 269-294.

Leuner, H. (1969). Guided affective imagery: A method of intensive psychotherapy. *American Journal of Psychotherapy, 23*, 4-22.

Lewin, K. (1951). Intention, will and need. In D. Rapaport (Ed.), *Organization and pathology of thought*. New York: Columbia University Press. (Original work published 1926).

Lewis, H. R., & Papadimitriou, C. H. (1981). *Elements of the theory of computation.* Englewood Cliffs, NJ: Prentice-Hall.

Libet, B. (1985). Unconscious cerebral initiative and the role of conscious will in voluntary action. *The Behavioral and Brain Sciences, 8*, 529-566.

Linton, M. (1982). Transformations of memory in everyday life. In U. Neisser (Ed.), *Memory observed: Remembering in natural contexts*. San Francisco: W. H. Freeman.

Loeb, A., Beck, A. T., & Diggory, J. (1971). Differential effects of success and failure on depressed and nondepressed patients. *The Journal of Nervous and Mental Disease, 152*, 106-114.

Loftus, E. F. (1975). Leading questions and the eyewitness report. *Cognitive Psychology, 7*, 560-572.

Luthe, W. (1969). Autogenic training: Method, research, and application in medicine. In C. T. Tart (Ed.), *Altered states of consciousness*. New York: John Wiley.

Mandler, G. (1975). *Mind and emotion.* New York: John Wiley.

Marcel, A. J. (1980). Conscious and preconscious recognition of polysemous words: Locating the selective effect of prior verbal context. In R. S. Nickerson (Ed.), *Attention and performance* (vol. 8). Hillsdale, NJ: Lawrence Erlbaum.

Maslow, A. H. (1954). *Motivation and personality* (2nd Edition). New York: Harper & Row.

Maury, L. F. A. (1878). *Le sommeil et les rêves.* Paris: Didier.

McCarthy, J., Abrahams, P. W., Edwards, D. J., Hart, T. P., & Levin, M. I. (1965). *LISP 1.5 programmer's manual.* Cambridge, MA: MIT Press.

McCarthy, J., & Hayes, P. J. (1969). Some philosophical problems from the standpoint of artificial intelligence. In B. Meltzer & D. Michie (Eds.), *Machine intelligence 4* (pp. 463-502). Edinburgh, Scotland: Edinburgh Univeristy Press.

McClelland, J. L. (1986). The programmable blackboard model of reading. In J. L. McClelland, D. E. Rumelhart, & PDP Research Group (Eds.), *Parallel distributed processing: Explorations in the microstructure of cognition. Volume 2: Psychological and biological models* (pp. 122-169). Cambridge, MA: MIT Press.

McClelland, J. L., & Rumelhart, D. E. (1986). A distributed model of human learning and memory. In J. L. McClelland, D. E. Rumelhart, & PDP Research Group (Eds.), *Parallel distributed processing: Explorations in the microstructure of cognition. Volume 2: Psychological and biological models* (pp. 170-215). Cambridge, MA: MIT Press.

McClelland, J. L., Rumelhart, D. E., & Hinton, G. E. (1986). The appeal of parallel distributed processing. In D. E. Rumelhart, J. L. McClelland, & PDP Research Group (Eds.), *Parallel distributed processing: Explorations in the microstructure of cognition. Volume 1: Foundations* (pp. 3-44). Cambridge, MA: MIT Press.

McClelland, J. L., Rumelhart, D. E., & PDP Research Group (Eds.). (1986). *Parallel distributed processing: Explorations in the microstructure of cognition. Volume 2: Psychological and biological models.* Cambridge, MA: MIT Press.

McDermott, D. V. (1976). Artificial intelligence meets natural stupidity. *SIGART Newsletter*, 57. New York: Association for Computing Machinery.

McDougall, W. (1923). *Outline of psychology.* New York: Scribner's.

McKellar, P. (1957). *Imagination and thinking: A psychological analysis.* New York: Basic Books.

McNeil, J. (1981). [Daydream diaries (retrospective reports) collected from subjects]. Unpublished raw data.

McNeil, J. (1985). *Interpersonal problem resolution in narrated imagery and verbal thought.* Unpublished doctoral dissertation, Department of Psychology, University of California, Los Angeles.

Meacham, J. A., & Leiman, B. (1982) Remembering to perform future actions. In U. Neisser (Ed.), *Memory observed: Remembering in natural contexts.* San Francisco: W. H. Freeman.

Mednick, S. (1962). The associative basis of the creative process. *Psychological Review, 69*, 220-232.

Meehan, J. (1976). *The metanovel: Writing stories by computer* (Research Report 74). New Haven, CT: Yale University, Computer Science Department.

Metz, C. (1982). *The imaginary signifier.* Bloomington, IN: Indiana University Press.

Meyer, L. B. (1956). *Emotion and meaning in music.* Chicago: University of Chicago Press.

Michalski, R. S., & Winston, P. H. (1986). *Variable precision logic. Artificial Intelligence, 29*, 121-146.

Miller, G. A. (1956). The magical number seven, plus or minus two: Some limits on our capacity for processing information. *Psychological Review, 63*, 81-97.

Miller, G. A., Galanter, E., & Pribram, K. H. (1960). *Plans and the structure of behavior.* New York: Holt, Rinehart, and Winston.

Minsky, M. (1968). Descriptive languages and problem solving. In M. Minsky (Ed.), *Semantic information processing* (pp. 419-424). Cambridge, MA: MIT Press.

Minsky, M. (1975). A framework for representing knowledge. In P. H. Winston (Ed.), *The psychology of computer vision.* New York: McGraw-Hill.

Minsky, M. L. (1977). Plain talk about neurodevelopmental epistemology. In *Proceedings of the Fifth International Joint Conference on Artificial Intelligence* (pp. 1083-1092). Los Altos, CA: Morgan Kaufmann.

Minsky, M. L. (1981). K-lines: A theory of memory. In D. A. Norman (Ed.), *Perspectives on cognitive science* (pp. 87-104). Norwood, NJ: Ablex.

Mischel, W. (1979). On the interface of cognition and personality: Beyond the person-situation debate. *American Psychologist, 34*, 740-754.

Mitchell, T. M., Kellar, R. M., & Kedar-Cabelli, S. T. (1986). Explanation-based generalization: A unifying view. *Machine Learning, 1*, 47-80.

Moser, U., Pfeifer, R., Schneider, W., Von Zeppelin, I., & Schneider, H. (1980). *Computer simulation of dream processes* (Technical Report 6). Zürich: Soziologisches Institut der Universität Zürich, Interdisziplinären Konfliktforschungsstelle.

Moser, U., Pfeifer, R., Schneider, W., Von Zeppelin, I., & Schneider, H. (1982). Experiences with computer simulation of dream processes. *Sleep 1982: 6th European Congress on Sleep Research* (pp. 30-44). Basel, Switzerland: S. Karger.

Moss, J. E. B. (1981). *Nested transactions: An approach to reliable distributed computing* (Technical Report 260). Cambridge, MA: Massachusetts Institute of Technology, Laboratory for Computer Science.

Mostow, D. J. (1981). *Mechanical transformation of task heuristics into operational procedures.* Unpublished doctoral dissertation, Department of Computer Science, Carnegie-Mellon University, Pittsburgh, PA.

Mostow, J. (1983). Program transformations for VLSI. In *Proceedings of the Eighth International Joint Conference on Artificial Intelligence* (pp. 40-43). Los Altos, CA: Morgan Kaufmann.

Mueller, E. T. (1983). *Implementation of nested transactions in a distributed system* (Technical Report CSD-831115). (Master's thesis). Los Angeles: University of California, Computer Science Department.

Mueller, E. T. (1987a). *GATE user's manual* (2nd edition, Technical Report UCLA-AI-87-6). Los Angeles: University of California, Artificial Intelligence Laboratory.

Mueller, E. T. (1987b). *Daydreaming and computation: A computer model of everyday creativity, learning, and emotions in the human stream of thought* (Technical Report UCLA-AI-87-8). Doctoral dissertation, Computer Science Department, University of California, Los Angeles, CA.

Mueller, E. T., Moore, J. D., & Popek, G. J. (1983). A nested transaction mechanism for LOCUS. In *Proceedings of the Ninth ACM Symposium on Operating Systems Principles* (pp. 71-89). New York: Association for Computing Machinery.

Mueller, E. T., & Zernik, U. (1984). *GATE reference manual* (Technical Report UCLA-AI-84-5). Los Angeles: University of California, Artificial Intelligence Laboratory.

Mueller, E. T., & Dyer, M. G. (1985a). Towards a computational theory of human daydreaming. In *The Seventh Annual Conference of the Cognitive Science Society* (pp. 120-129), Irvine, CA.

Mueller, E. T., & Dyer, M. G. (1985b). Daydreaming in humans and computers. In *Proceedings of the Ninth International Joint Conference on Artificial Intelligence* (pp. 278-280). Los Altos, CA: Morgan Kaufmann.

Mueller, R. A. G., & Gibbs, R. W. (1987). Processing idioms with multiple meanings. *Journal of Psycholinguistic Research, 16*(1), 63-81.

Mueller, R. E. (1963). *Inventivity: How man creates in art and science.* New York: John Day.

Mueller, R. E. (1967). *The science of art: The cybernetics of creative communication.* New York: John Day.

Mueller, R. E. (1983, January). When is Computer Art Art? *Creative Computing*, pp. 136-144.

Nagel, T. (1974). What is it like to be a bat? *Philosophical Review, 83*, 435-445.

Narayanan, A. (1983). *What is it like to be a machine?* (Research Report R-116). Exeter, England: University of Exeter, Department of Computer Science.

Neisser, U. (1963). The imitation of man by machine. *Science, 139*, 193-197.

Neisser, U. (1967). *Cognitive psychology.* New York: Appleton.

Neisser, U. (1982a). John Dean's memory: A case study. In U. Neisser (Ed.), *Memory observed: Remembering in natural contexts.* San Francisco: W. H. Freeman.

Neisser, U. (Ed.). (1982b). *Memory observed: Remembering in natural contexts.* San Francisco: W. H. Freeman.

Neisser, U. (1982c). Memory: What are the important questions? In U. Neisser (Ed.), *Memory observed: Remembering in natural contexts.* San Francisco: W. H. Freeman.

Newell, A. (1982). The knowledge level. *Artificial Intelligence, 18*(1), 87-127.

Newell, A., & Simon, H. A. (1972). *Human problem-solving.* Englewood Cliffs, NJ: Prentice-Hall.

Newell, A., Shaw, J. C., & Simon, H. A. (1957). Empirical explorations of the logic theory machine. *Proceedings of the West Joint Computer Conference* (vol. 15, pp. 218-239).

Nilsson, N. J. (1980). *Principles of artificial intelligence.* Los Altos, CA: Morgan Kaufmann.

Nisbett, R. E., & Wilson, T. D. (1977). Telling more than we can know: Verbal reports on mental processes. *Psychological Review, 84*, 231-259.

Norman, D. A. (1981). Twelve issues for cognitive science. In D. A. Norman (Ed.), *Perspectives on cognitive science* (pp. 265-295). Norwood, NJ: Ablex.

Norman, D. A. (1982). *Learning and memory.* San Francisco: W. H. Freeman.

Norman, D. A. (1986). Reflections on cognition and parallel distributed processing. In J. L. McClelland, D. E. Rumelhart, & PDP Research Group (Eds.), *Parallel distributed processing: Explorations in the microstructure of cognition. Volume 2: Psychological and biological models* (pp. 531-546). Cambridge, MA: MIT Press.

Ornstein, R. E. (Ed.). (1973). *The nature of human consciousness.* New York: Viking.

Osborn, A. F. (1953). *Applied imagination.* New York: Scribners.

Osgood, C. E., & Tannenbaum, P. (1955). The principle of congruity and the prediction of attitude change. *Psychological Review, 362,* 42-55.

Papert, S. (1980). *Mindstorms: Children, computers, and powerful ideas.* New York: Basic Books.

Paton, R. (1972). *Fantasy content, daydreaming frequency and the reduction of aggression.* Unpublished doctoral dissertation, City University of New York.

Pearl, J. (1982). Reverend Bayes on inference engines: a distributed hierarchical approach. In *Proceedings of the National Conference on Artificial Intelligence* (pp. 133-136). Los Altos, CA: Morgan Kaufmann.

Pekala, R. J., & Levine, R. L. (1981). Mapping consciousness: Development of an empirical-phenomenological approach. *Imagination, Cognition and Personality, 1*(1), 29-47.

Pfeifer, R. (1982). *Cognition and emotion: An information processing approach* (CIP Working Paper 436). Pittsburgh, PA: Carnegie-Mellon University, Department of Psychology.

Plutchik, R. (1980). A general psychoevolutionary theory of emotion. In Plutchik, R. (Ed.), *Emotion: Theory, research, and experience. Volume 1: Theories of emotion* (pp. 3-33). New York: Academic Press.

Pohl, I. (1971). Bi-directional search. In B. Meltzer & D. Michie (Eds.), *Machine intelligence 6* (pp. 127-140). Edinburgh, Scotland: Edinburgh Univeristy Press.

Poincaré, H. (1952). Mathematical creation. In B. Ghiselin (Ed.), *The creative process.* New York: Mentor. (Original work published 1908)

Polya, G. (1945). *How to solve it.* Princeton, NJ: Princeton University Press.

Pope, K. S. (1978). How gender, solitude, and posture influence the stream of consciousness. In K. S. Pope & J. L. Singer (Eds.), *The stream of consciousness.* New York: Plenum.

Pope, K. S., & Singer, J. L. (1978a). Regulation of the stream of consciousness: Toward a theory of ongoing thought. In G. E. Schwartz & D. Shapiro (Eds.), *Consciousness and self regulation: Advances in research* (Vol. 2). New York: Plenum.

Pope, K. S., & Singer, J. L. (Eds.). (1978b). *The stream of consciousness.* New York: Plenum.

Posey, T. B., & and Losch, M. E. (1983). Auditory hallucinations of hearing voices in 375 normal subjects. *Imagination, Cognition and Personality, 3*(2), 99-113.

Pötzl, O. (1960) The relationship between experimentally induced dream images and indirect vision. *Psychological Issues, 3,* Monograph 7, 41-120. New York: International Universities Press. (Original work published 1917)

Pylyshyn, Z. W. (1984). *Computation and cognition: Toward a foundation for cognitive science.* Cambridge, MA: MIT Press.

Quillian, M. R. (1968). Semantic memory. In M. Minsky (Ed.), *Semantic information processing* (pp. 227-270). Cambridge, MA: MIT Press.

Quinn, N. (1981). Marriage is a do-it-yourself project: The organization of marital goals. In *Proceedings of the Third Annual Conference of the Cognitive Science Society* (pp. 31-40), Berkeley, CA.

Randell, B. (1975). System structure for software fault tolerance. *IEEE Transactions on Software Engineering, SE-1*(2), 220-232.

Rapaport, D. (Ed.). (1951). *Organization and pathology of thought.* New York: Columbia University Press.

Rapaport, D. (1960). The structure of psychoanalytic theory: A systematizing attempt. *Psychological Issues, 2*, Monograph 6. New York: International Universities Press.

Rapaport, D. (1974). *The history of the concept of association of ideas.* New York: International Universities Press.

Reed, D. P. (1978). *Naming and synchronization in a decentralized computer system* (Technical Report 205). Cambridge, MA: Massachusetts Institute of Technology, Laboratory for Computer Science.

Rees, J. A., Adams, N. I., & Meehan, J. R. (1984). *The T manual* (4th ed.). New Haven, CT: Yale University, Computer Science Department.

Reik, T. (1948). *Listening with the third ear: The inner experience of a psychoanalyst.* New York: Farrar, Straus, and Giroux.

Reiser, B. J. (1983). *Contexts and indices in autobiographical memory* (Technical Report 24). New Haven, CT: Yale University, Cognitive Science Program.

Ritchie, D. M., & Thompson, K. (1974). The UNIX time-sharing system. *Communications of the ACM, 17*(7), 365-375.

Robinson, J. A. (1965). A machine-oriented logic based on the resolution principle. *Journal of the Association for Computing Machinery, 12*(1), 23-41.

Rosenblatt, F. (1962). *Principles of neurodynamics.* New York: Spartan.

Rosenbloom, P. S. (1983). *The chunking of goal hierarchies: A model of practice and stimulus-response compatibility* (Technical Report 83-148). Doctoral dissertation, Department of Computer Science, Carnegie-Mellon University, Pittsburgh, PA.

Rothenberg, A. (1979). *The emerging goddess: The creative process in art, science, and other fields.* Chicago: University of Chicago Press.

Rulifson, J., Derksen, J., & Waldinger, R. (1972). *QA4: A procedural calculus for intuitive reasoning* (Technical Note 73). Stanford, CA: Stanford Research Institute, Artificial Intelligence Center.

Rumelhart, D. E., Hinton, G. E., & McClelland, J. L. (1986). A general framework for parallel distributed processing. In D. E. Rumelhart, J. L. McClelland, & PDP Research Group (Eds.), *Parallel distributed processing: Explorations in the microstructure of cognition. Volume 1: Foundations* (pp. 45-76). Cambridge, MA: MIT Press.

Rumelhart, D. E., McClelland, J. L., & PDP Research Group (Eds.). (1986). *Parallel distributed processing: Explorations in the microstructure of cognition. Volume 1: Foundations.* Cambridge, MA: MIT Press.

Rumelhart, D. E., & Norman, D. A. (1982). Simulating a skilled typist: A study of skilled cognitive-motor performance. *Cognitive Science, 6*, 1-36.

Rumelhart, D. E., Smolensky, P., McClelland, J. L., & Hinton, G. E. (1986). Schemata and sequential thought processes in PDP models. In J. L. McClelland, D. E. Rumelhart, & PDP Research Group (Eds.), *Parallel distributed processing: Explorations in the microstructure of cognition. Volume 2: Psychological and biological models* (pp. 7-57). Cambridge, MA: MIT Press.

Rumelhart, D. E., & Zipser, D. (1986). Feature discovery by competitive learning. In D. E. Rumelhart, J. L. McClelland, & PDP Research Group (Eds.), *Parallel distributed processing: Explorations in the microstructure of cognition. Volume 1: Foundations* (pp. 151-193). Cambridge, MA: MIT Press.

Russell, B. (1945). *A history of western philosophy*. New York: Simon and Schuster.

Sacerdoti, E. D. (1974). Planning in a hierarchy of abstraction spaces. *Artificial Intelligence, 5*, 115-135.

Sacerdoti, E. D. (1977). *A structure for plans and behavior*. New York: Elsevier.

Salaman, E. (1970). *A collection of moments: A study of involuntary memories*. London: Longman.

Schank, R. C. (1975). *Conceptual information processing*. New York: American Elsevier.

Schank, R. C. (1977). Rules and topics in conversation. *Cognitive Science, 1*(4), 421-441.

Schank, R. C. (1982). *Dynamic memory*. Cambridge: Cambridge University Press.

Schank, R. C. (1986). *Explanation patterns: Understanding mechanically and creatively*. Hillsdale, NJ: Lawrence Erlbaum.

Schank, R. C., & Abelson, R. P. (1977). *Scripts, plans, goals, and understanding*. Hillsdale, NJ: Lawrence Erlbaum.

Schank, R. C., & Riesbeck, C. K. (1981). *Inside computer understanding: Five programs plus miniatures*. Hillsdale, NJ: Lawrence Erlbaum.

Schank, R. C., Wilensky, R., Carbonell, J. G., Kolodner, J. L., & Hendler, J. A. (1978). *Representing attitudes: Some primitive states* (Research Report 128). New Haven, CT: Yale University, Computer Science Department.

Schachter, S., & Singer, J. E. (1962). Cognitive, social and physiological determinants of emotional state. *Psychological Review, 69*, 379-399.

Segal, B., Huba, G., & Singer, J. L. (1980). *Drugs, daydreaming, and personality: A study of college youth*. Hillsdale, NJ: Lawrence Erlbaum.

Selfridge, O. G. (1959). Pandemonium: A paradigm for learning. In *Proceedings of the Symposium on the Mechanization of Thought Processes* (vol. 1). London: H. M. Stationary Office.

Shanon, B. (1981). Thought sequences and the language of consciousness. In *Proceedings of the Third Annual Conference of the Cognitive Science Society* (pp. 234-235), Berkeley, CA.

Shepard, R. N., & Metzler, J. (1971). Mental rotation of three-dimensional objects. *Science, 171,* 701-703.

Shapero, H. (1952). The musical mind. In B. Ghiselin (Ed.), *The creative process.* New York: Mentor. (Original work published 1946)

Shortliffe, E. H. (1976). *Computer-based medical consultations: MYCIN.* New York: American Elsevier.

Shortliffe, E. H., & Buchanan, B. G. (1975). A model of inexact reasoning in medicine. *Mathematical Biosciences, 23,* 351-379.

Silberer, H. (1951). Report on a method of eliciting and observing certain symbolic hallucination-phenomena. In D. Rapaport (Ed.), *Organization and pathology of thought.* New York: Columbia University Press. (Original work published 1909).

Simon, H. (1967). Motivational and emotional controls of cognition. *Psychological Review, 74*(1), 29-39.

Simon, H. (1974). How big is a chunk? *Science, 183,* 482-488.

Singer, J. L. (1966). *Daydreaming.* New York: Random House.

Singer, J. L. (1974). Daydreaming and the stream of thought. *American Scientist, 62,* 244-252.

Singer, J. L. (1975). *The inner world of daydreaming.* New York: Harper & Row.

Singer, J. L. (1978). Experimental studies of daydreaming and the stream of consciousness. In K. S. Pope & J. L. Singer (Eds.), *The stream of consciousness.* New York: Plenum.

Singer, J. L. (1981). Towards the scientific study of imagination. *Imagination, Cognition, and Personality, 1*(1), 5-28.

Singer, J. L., & Antrobus, J. S. (1963). A factor analysis of daydreaming and conceptually related cognitive and personality variables. *Perceptual and Motor Skills,* Monograph supplement, 3-V17.

Singer, J. L., & Antrobus, J. S. (1972). Daydreaming, imaginal processes and personality: A normative study. In P. W. Sheehan (Ed.), *The function and nature of imagery.* New York: Academic Press.

Singer, J. L., & McCraven, V. (1961). Some characteristics of adult daydreaming. *Journal of Psychology, 51,* 151-164.

Singer, J. L., & Pope, K. S. (1978). *The power of human imagination.* New York: Plenum.

Skinner, B. F. (1935). Two types of conditioned reflex and a pseudo type. *Journal of General Psychology, 12,* 66-77.

Sloman, A., & Croucher, M. (1981). Why robots will have emotions. In *Proceedings of the Seventh International Joint Conference on Artificial Intelligence* (pp. 197-202). Los Altos, CA: Morgan Kaufmann.

Small, S., & Rieger, C. (1982). Parsing and comprehending with word experts (a theory and its realization). In W. G. Lehnert & M. H. Ringle (Eds.), *Strategies for natural language processing.* Hillsdale, NJ: Lawrence Erlbaum.

Smirnov, A. A. (1973). *Problems of the psychology of memory.* New York: Plenum.

Smith, B. (1982). *Reflection and semantics in a procedural language* (Technical Report 272). Cambridge, MA: Massachusetts Institute of Technology, Laboratory for Computer Science.

Smith, D. E., Genesereth, M. R., & Ginsberg, M. L. (1986). Controlling recursive inference. *Artificial Intelligence, 30,* 343-389.

Smolensky, P. (1986a). Information processing in dynamical systems: Foundations of harmony theory. In D. E. Rumelhart, J. L. McClelland, & PDP Research Group (Eds.), *Parallel distributed processing: Explorations in the microstructure of cognition. Volume 1: Foundations* (pp. 194-281). Cambridge, MA: MIT Press.

Smolensky, P. (1986b). Neural and conceptual interpretation of PDP models. In J. L. McClelland, D. E. Rumelhart, & PDP Research Group (Eds.), *Parallel distributed processing: Explorations in the microstructure of cognition. Volume 2: Psychological and biological models* (pp. 390-431). Cambridge, MA: MIT Press.

Sofer, K. (1981, October). Art? Or Not Art? *Datamation,* pp. 118-127.

Starker, S. (1982). *Fantastic thought: All about dreams, daydreams, hallucinations, and hypnosis.* Englewood Cliffs, NJ: Prentice-Hall.

Stoller, R. J. (1979). *Sexual excitement: Dynamics of erotic life.* New York: Simon and Schuster.

Stoy, J. E. (1977). *Denotational semantics: The Scott-Strachey approach to programming language theory.* Cambridge, MA: MIT Press.

Suinn, R. M. (1984). Visual motor behavior rehearsal: The basic technique. *Scandinavian Journal of Behaviour Therapy, 13*(3), 131-142.

Suppes, P. & Warren, H. (1975). On the generation and classification of defence mechanisms. *International Journal of Psycho-Analysis, 56,* 405-414.

Sussman, G. J. (1975). *A computer model of skill acquisition.* New York: American Elsevier.

Sussman, G. J., & Steele, G. L., Jr. (1975). *SCHEME: An interpreter for extended lambda calculus* (AI Memo 349). Cambridge, MA: Massachusetts Institute of Technology, Artificial Intelligence Laboratory.

Tarnopolsky, Y. (1986). *Spontaneous thinking as natural selection.* Unpublished manuscript.

Tart, C. T. (Ed.). (1969). *Altered states of consciousness.* New York: John Wiley.

Thompson, K. (1978). UNIX implementation. *The Bell System Technical Journal, 57*(6), Part 2, 1931-1946.

Titchener, E. B. (1912). The schema of introspection. *American Journal of Psychology, 23,* 485-508.

Tomkins, S. S. (1962). *Affect, imagery, consciousness. Vol. I. The positive affects.* New York: Springer.

Tomkins, S. S. (1963). *Affect, imagery, consciousness. Vol. II. The negative affects.* New York: Springer.

Touretzky, D. S. (1986). Boltzcons. In *The Eighth Annual Conference of the Cognitive Science Society.*

Tulving, E. (1972). Episodic and semantic memory. In E. Tulving & W. Donaldson (Eds.), *Organization of memory.* New York: Academic Press.

Tulving, E. (1983). *Elements of episodic memory.* New York: Oxford University Press.

Turing, A. M. (1936). On computable numbers, with an application to the Entscheidungsproblem. In *Proceedings of the London Mathematical Society, 2*(42), 230-265, and *2*(43), 544-546.

Turing, A. M. (1950). Computing machinery and intelligence. *Mind, 54*(236), 433-460.

Varendonck, J. (1921). *The psychology of day-dreams.* London: George Allen & Unwin.

Wallas, G. (1926). *The art of thought.* New York: Harcourt, Brace.

Waltz, D. L., & Boggess, L. (1979). Visual analog representations for natural language understanding. In *Proceedings of the Sixth International Joint Conference on Artificial Intelligence* (pp. 926-934). Los Altos, CA: Morgan Kaufmann.

Waltz, D. L., & Pollack, J. B. (1985). Massively parallel parsing: A strongly interactive model of natural language interpretation. *Cognitive Science, 9*(1), 75-112.

Warren, H. C. (1921). *A history of the association psychology.* New York: Scribner's.

Watkins, M. M. (1976). *Waking dreams.* New York: Gordon and Breach.

Watson, J. B. (1924). *Behaviorism.* New York: W. W. Norton.

Weiner, B. (1980a). A cognitive (attribution)-emotion-action model of motivated behavior: An analysis of judgments of help-giving. *Journal of Personality and Social Psychology, 39*, 186-200.

Weiner, B. (1980b). *Human motivation.* New York: Holt, Rinehart, & Winston.

Weiner, B. (1982). The emotional consequences of causal attributions. In M. S. Clark & S. T. Fiske (Eds.), *Affect and cognition: The 17th Annual Carnegie Symposium on Cognition.* Hillsdale, NJ: Lawrence Erlbaum.

Weizenbaum, J. (1966). ELIZA—A computer program for the study of natural language communication between man and machine. *Communications of the ACM, 9*(1), 36-45.

Weizenbaum, J. (1974). Automating psychotherapy [Letter to the editor]. *Communications of the ACM, 17*(7), 543.

Wertheimer, M. (1945). *Productive thinking.* Chicago: University of Chicago Press.

Whitney, J. (1980). *Digital harmony: On the complementarity of music and visual art.* Peterborough, NH: Byte Books.

Wilensky, R. (1983). *Planning and understanding: A computational approach to human reasoning.* Reading, MA: Addison-Wesley.

Williams, M. D., & Hollan, J. D. (1981). The process of retrieval from very long term memory. *Cognitive Science, 5*, 87-119.

Winston, P. H. (1984). *Artificial intelligence.* Reading, MA: Addison-Wesley.

Winston, P. H., & Horn, B. K. P. (1981). *LISP*. Reading, MA: Addison-Wesley.

Wirth, N. (1971). The programming language PASCAL. *Acta Informatica, 1*, 35-63.

Woodworth, R. S., & Schlosberg, H. (1954). *Experimental psychology* (Revised Edition). New York: Holt, Rinehart, and Winston.

Xenakis, I. (1971). *Formalized music*. Bloomington, Indiana: Indiana University Press.

Zajonc, R. B. (1980). Feeling and thinking: Preferences need no inferences. *American Psychologist, 35*, 151-175.

Appendix A

Annotated Traces from DAYDREAMER

In this appendix, we present annotated traces of some of the daydreams and external experiences produced by DAYDREAMER. Because of space limitations, not all daydreams and experiences produced by the program can be reproduced here. Portions have also been omitted from some of the traces; these are marked by ellipses.

A.1 LOVERS1 Experience

```
Initialize DAYDREAMER
Creating initial reality context Cx.3
State changes from SUSPENDED to DAYDREAMING
```

The program first creates an initial reality context—the context designated as containing the state of the simulated "real world" as seen by the program. Various initial facts are asserted into this context: DAYDREAMER has a job; she is not involved in a romantic relationship; she is romantically interested in Harrison Ford; she is currently at home; she knows where the Nuart Theater is; and so on. DAYDREAMER then starts out in daydreaming mode.

```
******************
Lovers-theme fired as inference in Cx.3
--------------------------------------------------------
IF   self not LOVERS with anyone and
     SEX or LOVE-GIVING or LOVE-RECEIVING or COMPANIONSHIP
     below threshold
THEN activate goal for LOVERS with a person
--------------------------------------------------------
Activate top-level goal #{Ob.1882: (ACTIVE-GOAL obj (LOVERS actor ...)...)}
| I want to be going out with someone.
```

```
| I feel really interested in going out with someone.
Concerns:
#{OB.1882: (ACTIVE-GOAL obj (LOVERS actor ...)...)} (0.9) HALTED
```

DAYDREAMER activates a top-level **LOVERS** goal because she is not involved in such a relationship and one or more of her need states subsumed by the relationship are unsatisfied. A new concern is created and a positive motivating emotion of *interest* is created and associated with the new concern. The intrinsic importance of the **LOVERS** goal is 0.9. This becomes the magnitude of the motivating emotion, as well as the current value for the motivation of the concern.

The top-level goal is stated in terms of a *variable*: The objective of the goal is a relationship whose participants are the daydreamer and *some* appropriate male. A variable is generated in English as "someone" or "something." A concrete value for this variable will have to be found in order to achieve the goal.

Whenever facts are asserted into a context, they are converted into English and produced as output. A simple collection of heuristics associated with classes of representations and particular rules, however, selectively omits generation of certain facts in order to make the English output sound more natural.

The **LOVERS** goal is first invoked as a halted concern; work therefore cannot proceed toward this goal until a serendipity occurs. In the current version of DAYDREAMER, this is done in lieu of having the program go through several plans to achieve this goal, all of which are unsuccessful, and then halting the goal until a new plan is suggested through serendipity or fortuitous success of a subgoal.

```
*****************
Entertainment-theme fired as inference in Cx.3
-----------------------------------------------------------
IF   ENTERTAINMENT need below threshold
THEN activate goal for ENTERTAINMENT
-----------------------------------------------------------
*****************
Activate top-level goal #{OB.1887: (ACTIVE-GOAL obj (ENTERTAINMENT...)...)}
| I want to be entertained.
| I feel interested in being entertained.
Concerns:
#{OB.1887: (ACTIVE-GOAL obj (ENTERTAINMENT...)...)} (0.6)
#{OB.1882: (ACTIVE-GOAL obj (LOVERS actor ...)...)} (0.9) HALTED
```

Next DAYDREAMER activates an **ENTERTAINMENT** goal because its need for entertainment is unsatisfied. Another concern and associated motivating emotion are activated.

```
Running emotion-driven control loop...
Switching to new top-level goal #{OB.1887: (ACTIVE-GOAL obj (ENTERTAINMENT))}
#{CX.3: (CX)} --> #{CX.4: (CX)}
*****************
```

```
Entertainment-plan1 fired as plan
for #{OB.1887: (ACTIVE-GOAL obj (ENTERTAINMENT...)...)}
in Cx.3 sprouting Cx.4
-----------------------------------------------------------
IF   goal for ENTERTAINMENT
THEN subgoal for M-MOVIE
-----------------------------------------------------------
| I have to go see a movie.
```

The top-level control loop is invoked once DAYDREAMER has gotten off the ground by applying inferences and, as a result, creating two concerns. The most highly motivated nonhalted concern is **ENTERTAINMENT** and therefore a unit of planning is performed for this concern. The program has a rule which states that a goal for **ENTERTAINMENT** may be achieved by achieving an **M-MOVIE** subgoal. Thus a new context is sprouted in which the top-level goal for **ENTERTAINMENT** is connected to a subgoal for **M-MOVIE**. (Hereafter, we call this "sprouting a plan.") This context becomes the new reality context.

```
#{CX.4: (CX)} --> #{CX.5: (CX)}
******************
M-movie-alone-plan fired as plan
for #{OB.1897: (ACTIVE-GOAL obj (M-MOVIE actor ...)...)}
in Cx.4 sprouting Cx.5
-----------------------------------------------------------
IF   goal for M-MOVIE alone
THEN subgoals to PTRANS to theater, MTRANS movie, and
     PTRANS back to original location
-----------------------------------------------------------
#{CX.5: (CX)} --> #{CX.6: (CX)}
Fact plan #{OB.286: (AT actor Nuart-theater obj ...)} found
(AT actor Nuart-theater obj ...)
******************
Goal #{OB.1902: (ACTIVE-GOAL obj (AT actor ...)  top-level-goal ...)} succeeds
Instantiating Plan
```

A plan is sprouted for **M-MOVIE**: go to a theater, watch the movie, and go home. The particular theater is not specified by this plan; it is initially a variable. Next, however, a fact is found in the current reality context which unifies with (matches) an active subgoal. This fact gives the location of a specific theater—the Nuart Theater. Therefore, a new context is sprouted in which the active subgoal becomes a succeeded subgoal with the variable replaced by the specific theater. In fact, in this new context, the *entire* plan for **ENTERTAINMENT**—the top-level goal and all descendant subgoals—is instantiated with the variable bindings (values) resulting from the unification.[1] In general, a variable may be propagated from subgoal to further subgoal, so that when a value for that variable is found, it is necessary to replace the variable with its value in any and all subgoals containing that variable.

---

[1]The following optimization is performed in DAYDREAMER: a subgoal is only instantiated if it contains variables.

```
#{CX.6: (CX)} --> #{CX.7: (CX)}
******************
Ptrans-plan fired as plan
for #{OB.1923: (ACTIVE-GOAL obj (PTRANS actor ...)...)}
in Cx.6 sprouting Cx.7
--------------------------------------------------------
IF   goal for person to PTRANS to a location
THEN subgoal for person to KNOW that location
--------------------------------------------------------
******************
Goal #{OB.1933: (ACTIVE-GOAL obj (KNOW actor ...)...)} succeeds
Instantiating plan.
```

A plan is sprouted for the subgoal of going to the Nuart Theater: DAY-DREAMER must know where the theater is located. This subgoal then succeeds because this fact is found in the current reality context.

```
Subgoals of #{OB.1923: (ACTIVE-GOAL obj (PTRANS actor ...)...)} completed
About to perform real action but not in performance mode
Current concern waits #{OB.1887: (ACTIVE-GOAL obj (ENTERTAINMENT...)...)}
No more concerns to process; switching to performance mode
Waking up concern #{OB.1887: (ACTIVE-GOAL obj (ENTERTAINMENT...)...)}
State changes from DAYDREAMING to PERFORMANCE
Subgoals of #{OB.1923: (ACTIVE-GOAL obj (PTRANS actor ...)...)} completed
Perform external action
Perform action goal #{OB.1923: (ACTIVE-GOAL obj (PTRANS actor ...)...)
******************
Goal #{OB.1923: (ACTIVE-GOAL obj (PTRANS actor ...)...)} succeeds
Instantiating plan.
| I go to the Nuart.
******************
At-plan2 fired as inference in Cx.7
--------------------------------------------------------
IF   person PTRANS from one location to another
THEN person AT new location and
     no longer AT old location
--------------------------------------------------------
```

Next, all of the preconditions (subgoals) of the **PTRANS** subgoal have succeeded, and so this action may be performed. However, actions may only be performed in performance mode and DAYDREAMER is currently in daydreaming mode. Therefore, the **ENTERTAINMENT** concern is placed into a waiting condition. Next, the top-level control loop fails to find any concerns to process, since all concerns are either waiting or halted. Therefore, the system enters performance mode. Whenever the system enters performance mode, all waiting concerns are woken up. Thus the **ENTERTAINMENT** concern is woken up.

Now since all of the subgoals of **PTRANS** have succeeded and the system *is* in performance mode, the external action of **PTRANS** is performed. An inference retracts the old location of DAYDREAMER (home) in the current reality context and asserts the new one (the Nuart Theater).

```
Taking optional external input
Enter concepts in #{CX.7: (CX)}
| Input: Harrison Ford is at the Nuart.
******************
Vprox-plan1 fired as inference in Cx.7
----------------------------------------------------------
IF   AT location of person
THEN VPROX that person
----------------------------------------------------------
Serendipity!! (personal goal)
[OB.1958: (AG. (LOVERS actor Me Movie-star1)) EPISODE.1]
   [OB.2180: (AG. (ACQUAINTED actor Me Movie-star1)) EPISODE.2]
      [OB.2199: (AG. (M-CONVERSATION actor Me Movie-star1)) EPISODE.3]
         [OB.2176: (AG. (MTRANS-ACCEPTABLE actor Me Movie-star1)) EPISODE.4]
            [OB.2213: (AG. (MTRANS actor Me from Me to Movie-star1
            obj (ACTIVE-GOAL obj (KNOW actor Me
                                  obj (TIME-OF-DAY))))) EPISODE.5]
               [OB.2221: (AG. (VPROX actor Movie-star1 Me)) EPISODE.6]
                  [OB.2225: (AG. (AT actor Movie-star1 obj Nuart-location))]
                  [OB.2228: (AG. (AT actor Me obj Nuart-location))]
            [OB.2205: (AG. (MTRANS actor Me from Me to Movie-star1
            obj (INTRODUCTION)))]
            [OB.2209: (AG. (MTRANS actor Movie-star1 from Movie-star1 to Me
            obj (INTRODUCTION)))]
      [OB.2184: (AG. (ROMANTIC-INTEREST obj Movie-star1))]
      [OB.2187: (AG. (BELIEVE actor Movie-star1
            obj (ACTIVE-GOAL obj (LOVERS actor Me Movie-star1))))]
   [OB.2192: (AG. (M-DATE actor Me Movie-star1))]
   [OB.2195: (AG. (M-AGREE actor Me Movie-star1
            obj (LOVERS actor Me Movie-star1)))]
#{CX.7: (CX)} --> #{CX.9: (CX)}
Generate surprise emotion
| What do you know!
#{CX.7: (CX)} --> #{CX.10: (CX)}
Concerns:
#{OB.1887: (ACTIVE-GOAL obj (ENTERTAINMENT...)...)} (0.6)
#{OB.1882: (ACTIVE-GOAL obj (LOVERS actor ...)...)} (1.15)
Switching to new top-level goal #{OB.1882: (ACTIVE-GOAL obj (LOVERS actor))}
```

Now DAYDREAMER is provided with an input state: Harrison Ford happens to be at the Nuart Theater. The English sentence is mapped to the corresponding fact by the parser and asserted into the current reality context. The fact that the movie star and DAYDREAMER are **VPROX**—able to communicate—is then inferred.

Next, an input-state-driven serendipity is detected. The input state that Harrison Ford is **AT** the Nuart Theater is applicable to the active **LOVERS** goal of DAYDREAMER, since: (a) in order to be **LOVERS** with someone, one must be **ACQUAINTED** with the person, (b) in order to be **ACQUAINTED** with the person, one may have an **M-CONVERSATION** with the person, (c) in order to have an **M-CONVERSATION** with the person, it must be **MTRANS-ACCEPTABLE** for one to talk to the person, (d) in order for it to be **MTRANS-ACCEPTABLE** to talk to the person, one may **MTRANS** a simple favor to the person, such as knowing the time of day, (e) in order to **MTRANS** to the person, one must be **VPROX** the person, and (f) in order to be **VPROX** to the person, one may be **AT** the same location as the person. Several serendipities are in fact possible here; the one with the longest path—shorter than some maximum length—is chosen.

The serendipity results in an episode suitable for use by the analogical planning mechanism. This episode is associated with the **LOVERS** goal as a suggestion for how to achieve that goal.

As a result of serendipity, the **LOVERS** concern is unhalted and a *surprise* emotion is generated and associated with the concern. The dynamic importance or level of motivation of the **LOVERS** concern thus increases from 0.9 to 1.15. Now the top-level control loop selects the **LOVERS** concern for processing, switching from the previously current **ENTERTAINMENT** concern.

```
Run analogical plan for #{OB.1882: (ACTIVE-GOAL obj (LOVERS actor ...)...)
#{CX.9: (CX)} --> #{CX.11: (CX)}
Apply existing analogical plan
*****************
Lovers-plan fired as analogical plan
for #{OB.1882: (ACTIVE-GOAL obj (LOVERS actor ...)...)}
in Cx.9 sprouting Cx.11
--------------------------------------------------------
IF   goal for LOVERS with a person
THEN subgoals for ACQUAINTED with person and
     ROMANTIC-INTEREST in person and
     person have ACTIVE-GOAL of LOVERS with self and
     self and
     person M-DATE self and
     person M-AGREE to LOVERS
--------------------------------------------------------
#{CX.11: (CX)} --> #{CX.12: (CX)}
*****************
Goal #{OB.2353: (ACTIVE-GOAL obj (ROMANTIC-INTEREST...)...)} succeeds
Instantiating plan.
Apply existing analogical plan
*****************
Acquainted-plan fired as analogical plan
for #{OB.2249: (ACTIVE-GOAL obj (ACQUAINTED...)...)}
in Cx.11 sprouting Cx.12
--------------------------------------------------------
IF   goal to be ACQUAINTED with person
THEN subgoal for M-CONVERSATION with person
--------------------------------------------------------
| I have to have a conversation with him.
#{CX.12: (CX)} --> #{CX.13: (CX)}
Apply existing analogical plan
*****************
M-conversation-plan fired as analogical plan
for #{OB.2269: (ACTIVE-GOAL obj (M-CONVERSATION...)...)}
in Cx.12 sprouting Cx.13
--------------------------------------------------------
IF   goal for M-CONVERSATION between person1 and
```

```
      person2
THEN  subgoals for MTRANS-ACCEPTABLE between person1
      and
      person2 and
      person1 to MTRANS to person2 something and
      person2 to MTRANS to person1 something
-----------------------------------------------------------
#{CX.13: (CX)} --> #{CX.14: (CX)}
Apply existing analogical plan
*****************
Mtrans-acceptable-inf2 fired as analogical plan
for #{OB.2274: (ACTIVE-GOAL obj (MTRANS-ACCEPTABLE...)...)}
in Cx.13 sprouting Cx.14
-----------------------------------------------------------
IF   goal for MTRANS-ACCEPTABLE between self and
     other
THEN subgoal for self to MTRANS to other that self has ACTIVE-GOAL
     to KNOW the time
-----------------------------------------------------------
#{CX.14: (CX)} --> #{CX.15: (CX)}
Apply existing analogical plan
******************
Mtrans-plan2 fired as analogical plan
for #{OB.2290: (ACTIVE-GOAL obj (MTRANS actor ...)...)}
in Cx.14 sprouting Cx.15
-----------------------------------------------------------
IF   goal to MTRANS mental state to person
THEN subgoal to be VPROX that person
-----------------------------------------------------------
#{CX.15: (CX)} --> #{CX.16: (CX)}
Apply existing analogical plan
******************
Vprox-plan1 fired as analogical plan
for #{OB.2295: (ACTIVE-GOAL obj (VPROX actor ...)...)}
in Cx.15 sprouting Cx.16
-----------------------------------------------------------
IF   goal to be VPROX to person
THEN subgoal to be AT location of person
-----------------------------------------------------------
******************
Goal #{OB.2303: (ACTIVE-GOAL obj (AT actor ...)  top-level-goal ...)} succeeds
******************
Goal #{OB.2300: (ACTIVE-GOAL obj (AT actor ...)  top-level-goal ...)} succeeds
******************
Goal #{OB.2300: (ACTIVE-GOAL obj (VPROX actor ...)  top-level-goal ...)} succeeds
Subgoals of #{OB.2369: (SUCCEEDED-GOAL obj (VPROX...)...)} completed
Subgoals of #{OB.2316: (ACTIVE-GOAL obj (MTRANS actor ...)...)} completed
Perform external action
Perform action goal #{OB.2316: (ACTIVE-GOAL obj (MTRANS actor ...)...)
******************
Goal #{OB.2316: (ACTIVE-GOAL obj (MTRANS actor ...)...)} succeeds
| I tell Harrison Ford I would like to know the time.
```

Planning for the **LOVERS** goal then proceeds as suggested by the analogical plan associated with the goal during the above serendipity. DAYDREAMER finally performs the action of asking the movie star for the time. The program will ask anyone for the time, unless (a) a high realism plan to achieve a **LOVERS** goal with the person has been previously daydreamed and incorporated into episodic memory, and (b) this episode has a low desirability (i.e., the outcome is negative). Thus, given a new person, DAYDREAMER will attempt to start a conversation just in case she decides she is interested in the person (for example, if the other person likes the same things that she does). Here, the subgoal for **ROMANTIC-INTEREST** succeeds immediately because DAYDREAMER is already romantically interested in Harrison Ford.

```
******************
Neg-att-inf fired as inference in Cx.17
-----------------------------------------------------------
IF   person is RICH and
     self is AT same location as person and
     self not WELL-DRESSED
THEN person forms NEG-ATTITUDE toward self
-----------------------------------------------------------
| He does not think much of me because I am not well dressed.
******************
Believe-plan1 fired as inference in Cx.17
-----------------------------------------------------------
IF   MTRANS mental state to person
THEN person BELIEVE self mental state
-----------------------------------------------------------

******************
Mtrans-acceptable-inf2 fired as inference in Cx.17
-----------------------------------------------------------
IF   self MTRANS to other that self has ACTIVE-GOAL to
     KNOW the time
THEN MTRANS-ACCEPTABLE between self and
     other
-----------------------------------------------------------

******************
Social-esteem-monitor fired as inference in Cx.17
-----------------------------------------------------------
IF   person has NEG-ATTITUDE toward self and
     self has POS-ATTITUDE toward person
THEN failure of social esteem goal for person to have POS-ATTITUDE
     toward self
-----------------------------------------------------------
| I fail at him thinking highly of me.
Personal goal outcome
| I feel really embarrassed.
```

DAYDREAMER now infers that because she has spoken to the movie star and she is not well dressed, he forms a negative attitude toward her. Thus her **SOCIAL ESTEEM** personal goal fails and an emotional response of *embarrassment* is generated. However, since she has asked him the time, it is now acceptable for the two to communicate.

```
******************
```

```
Reversal-theme fired as inference in Cx.17
--------------------------------------------------------
IF   NEG-EMOTION resulting from a FAILED-GOAL
THEN activate daydreaming goal for REVERSAL of failure
--------------------------------------------------------
******************
Activate top-level goal #{OB.2412: (ACTIVE-GOAL obj (REVERSAL...)...)
Concerns:
#{OB.2412: (ACTIVE-GOAL obj (REVERSAL...)...)} (0.98)
#{OB.1887: (ACTIVE-GOAL obj (ENTERTAINMENT...)...)} (0.6)
#{OB.1882: (ACTIVE-GOAL obj (LOVERS actor ...)...)} (1.15)
#{CX.17: (CX)} --> #{CX.18: (CX)}
******************
Rationalization-theme fired as inference in Cx.17
--------------------------------------------------------
IF   NEG-EMOTION of sufficient strength resulting from
     a FAILED-GOAL
THEN activate daydreaming goal for RATIONALIZATION
     of failure
--------------------------------------------------------
******************
Activate top-level goal #{OB.1420: (ACTIVE-GOAL obj (RATIONALIZATION...)...)
Concerns:
#{OB.1420: (ACTIVE-GOAL obj (RATIONALIZATION...)...)} (0.97)
#{OB.2412: (ACTIVE-GOAL obj (REVERSAL...)...)} (0.98)
#{OB.1887: (ACTIVE-GOAL obj (ENTERTAINMENT...)...)} (0.6)
#{OB.1882: (ACTIVE-GOAL obj (LOVERS actor ...)...)} (1.15)
#{CX.17: (CX)} --> #{CX.19: (CX)}
```

In response to the negative emotion of embarrassment, DAYDREAMER activates a number of daydreaming goals: **RATIONALIZATION** (rationalize the fact that Harrison Ford has a negative attitude toward her), **ROVING** (divert attention from embarrassment to more pleasant thoughts), **REVERSAL** (plan to avoid similar embarrassments in the future), and **RECOVERY** (generate a future plan for Harrison Ford to have a positive, rather than negative, attitude toward her). The level of motivation for these concerns is not as high as the active **LOVERS** concern. However, even if they were higher, they would not be processed until later since daydreaming goals are not processed when the system is in performance mode.

```
******************
Goal #{OB.2331: (ACTIVE-GOAL obj (MTRANS-ACCEPTABLE...)...)} succeeds
#{CX.17: (CX)} --> #{CX.20: (CX)}
******************
Mtrans-plan2 fired as plan
for #{OB.2327: (ACTIVE-GOAL obj (MTRANS actor ...)...)}
in Cx.17 sprouting Cx.20
--------------------------------------------------------
IF   goal to MTRANS mental state to person
THEN subgoal to be VPROX that person
--------------------------------------------------------
******************
Goal #{OB.2444: (ACTIVE-GOAL obj (VPROX actor ...)...)} succeeds
```

```
Subgoals of #{OB.2327: (ACTIVE-GOAL obj (MTRANS actor ...)...)} completed
Perform external action
Perform action goal #{OB.2327: (ACTIVE-GOAL obj (MTRANS actor ...)...)
******************
Goal #{OB.2327: (ACTIVE-GOAL obj (MTRANS actor ...)...)} succeeds
| I introduce myself to him.
******************
Believe-plan1 fired as inference in Cx.20
-------------------------------------------------------
IF   MTRANS mental state to person
THEN person BELIEVE self mental state
-------------------------------------------------------
#{CX.20: (CX)} --> #{CX.21: (CX)}
******************
Mtrans-plan2 fired as plan
for #{OB.2323: (ACTIVE-GOAL obj (MTRANS actor ...)...)}
in Cx.20 sprouting Cx.21
-------------------------------------------------------
IF   goal to MTRANS mental state to person
THEN subgoal to be VPROX that person
-------------------------------------------------------
******************
Goal #{OB.2476: (ACTIVE-GOAL obj (VPROX actor ...)...)} succeeds
Subgoals of #{OB.2323: (ACTIVE-GOAL obj (MTRANS actor ...)...)} completed
Perform other action #{OB.2323: (ACTIVE-GOAL obj (MTRANS actor ...)...)
Enter concepts in #{CX.21: (CX)}
Input: He introduces himself to me.
| He introduces himself to me.
******************
Goal #{OB.2323: (ACTIVE-GOAL obj (MTRANS actor ...)...)} succeeds
```

In order to achieve the **M-CONVERSATION** subgoal, DAYDREAMER introduces herself to the movie star who must then introduce himself to her. When the preconditions (subgoals) succeed for a subgoal whose objective is an action performed by another person, DAYDREAMER waits for an input action. If the input action unifies with (matches) the objective of the subgoal, then the subgoal succeeds. Otherwise the subgoal fails. Here the expected action is performed and the subgoal succeeds.

```
******************
Believe-plan1 fired as inference in Cx.21
-------------------------------------------------------
IF   MTRANS mental state to person
THEN person BELIEVE self mental state
-------------------------------------------------------
******************
M-conversation-plan fired as inference in Cx.21
-------------------------------------------------------
IF   MTRANS-ACCEPTABLE between person1 and
         person2 and
         person1 MTRANS to person1 something and
         person2 MTRANS to person1 something
THEN M-CONVERSATION between person1 and
```

```
     person2
-----------------------------------------------------------
******************
Acquainted-plan fired as inference in Cx.21
-----------------------------------------------------------
IF   M-CONVERSATION with person
THEN ACQUAINTED with person
-----------------------------------------------------------
******************
Goal #{OB.2356: (ACTIVE-GOAL obj (ACQUAINTED...)...)} succeeds
#{CX.21: (CX)} --> #{CX.22: (CX)}
******************
Lovers-theme-plan fired as backward other plan
for #{OB.2348: (ACTIVE-GOAL obj (BELIEVE actor ...)...)}
in Cx.21 sprouting Cx.22
-----------------------------------------------------------
IF   goal for self ACTIVE-GOAL of LOVERS with a person
THEN subgoals for ROMANTIC-INTEREST in person and
     not LOVERS with anyone
-----------------------------------------------------------
| He has to be interested in me.
| He cannot be going out with anyone.
```

DAYDREAMER has the subgoal for Harrison Ford to have the goal to be romantically involved with her. A plan for this subgoal is generated through backward other planning: applying a rule stated in terms of the self to another person. Thus in order for Harrison to have an **ACTIVE-GOAL** of **LOVERS** with DAYDREAMER, he must have **ROMANTIC-INTEREST** toward her and must **BELIEVE** he is not **LOVERS** with anyone.

```
******************
Romantic-interest-plan fired as backward other plan
for #{OB.2527: (ACTIVE-GOAL obj (BELIEVE actor ...)...)}
in Cx.22 sprouting Cx.23
-----------------------------------------------------------
IF   goal to have ROMANTIC-INTEREST in person
THEN subgoal to have POS-ATTITUDE toward person and
     person to be ATTRACTIVE
-----------------------------------------------------------
#{CX.23: (CX)} --> #{CX.24: (CX)}
Subgoal relaxation, #{OB.2539: (ACTIVE-GOAL obj (BELIEVE actor ...)...)} succeeds
| Maybe he thinks I am cute.
#{CX.24: (CX)} --> #{CX.25: (CX)}
#{CX.24: (CX)} --> #{CX.26: (CX)}
```

Again through backward other planning, DAYDREAMER breaks down the subgoal for Harrison to be romantically interested in her into subgoals for him to have a positive attitude toward her and to think she is attractive. There are no plans to make Harrison think she is attractive. Therefore, DAYDREAMER employs the heuristic that a desired mental state of another person may be assumed if there is no information to the contrary. Although this is a heuristic for daydreaming, it is also used in performance mode since the only other alternative would be to declare failure of the subgoal, which would result in a top-level goal failure (since backtracking is not permitted in performance mode).

```
*******************
Positive-attitude-plan2 fired as backward other plan
for #{OB.2543: (ACTIVE-GOAL obj (BELIEVE actor ...)...)}
in Cx.24 sprouting Cx.26
------------------------------------------------------------
IF   goal to have POS-ATTITUDE toward a person
THEN subgoal for person to have POS-ATTITUDE toward self
     MOVIES
------------------------------------------------------------
#{CX.26: (CX)} --> #{CX.27: (CX)}
*******************
Believe-plan1 fired as plan
for #{OB.2565: (ACTIVE-GOAL obj (BELIEVE actor ...)...)}
in Cx.26 sprouting Cx.27
------------------------------------------------------------
IF   goal for person to BELIEVE self mental state
THEN subgoal to MTRANS mental state to person
------------------------------------------------------------
#{CX.27: (CX)} --> #{CX.28: (CX)}
*******************
Mtrans-plan2 fired as plan
for #{OB.2570: (ACTIVE-GOAL obj (MTRANS actor ...)...)}
in Cx.27 sprouting Cx.28
------------------------------------------------------------
IF   goal to MTRANS mental state to person
THEN subgoal to be VPROX that person
------------------------------------------------------------
*******************
Goal #{OB.2575: (ACTIVE-GOAL obj (VPROX actor ...)...)} succeeds
Subgoals of #{OB.2570: (ACTIVE-GOAL obj (MTRANS actor ...)...)} completed
Perform external action
Perform action goal #{OB.2570: (ACTIVE-GOAL obj (MTRANS actor ...)...)
*******************
Goal #{OB.2570: (ACTIVE-GOAL obj (MTRANS actor ...)...)} succeeds
| I tell him I like his movies.
*******************
Believe-plan1 fired as inference in Cx.28
------------------------------------------------------------
IF   MTRANS mental state to person
THEN person BELIEVE self mental state
------------------------------------------------------------
*******************
Goal #{OB.2565: (ACTIVE-GOAL obj (BELIEVE actor ...)...)} succeeds
Subgoals of #{OB.2610: (SUCCEEDED-GOAL obj (BELIEVE...)...)} completed
Subgoals of #{OB.2543: (ACTIVE-GOAL obj (BELIEVE actor ...)...)} completed
*******************
Goal #{OB.2543: (ACTIVE-GOAL obj (BELIEVE actor ...)...)} succeeds
| He thinks highly of me.
Subgoals of #{OB.2527: (ACTIVE-GOAL obj (BELIEVE actor ...)...)} completed
*******************
Goal #{OB.2527: (ACTIVE-GOAL obj (BELIEVE actor ...)...)} succeeds
```

```
| He is interested in me.
#{CX.28: (CX)} --> #{CX.29: (CX)}
Subgoal relaxation, #{OB.2532: (ACTIVE-GOAL obj (BELIEVE actor ...)...)} succeeds
| Maybe he is not going out with anyone.
******************
Goal #{OB.2532: (ACTIVE-GOAL obj (BELIEVE actor ...)...)} succeeds
Subgoals of #{OB.2348: (ACTIVE-GOAL obj (BELIEVE actor ...)...)} completed
******************
Goal #{OB.2348: (ACTIVE-GOAL obj (BELIEVE actor ...)...)} succeeds
| Maybe he wants to be going out with me.
#{CX.29: (CX)} --> #{CX.30: (CX)}
```

DAYDREAMER tells the movie star that she likes his movies in order to get him to have a positive attitude toward her. After subgoal relaxation, DAY-DREAMER believes with low realism (certainty) that Harrison Ford wants to be romantically involved with her.

```
******************
M-date-plan fired as plan
for #{OB.2345: (ACTIVE-GOAL obj (M-DATE actor ...)...)}
in Cx.29 sprouting Cx.30
---------------------------------------------------------
IF   goal for self and
     other to M-DATE
THEN subgoals for other to have POS-ATTITUDE toward activity
     and
     self to have POS-ATTITUDE toward activity and self and
     other to M-AGREE to activity and self and
     other to ENABLE-FUTURE-VPROX and
     to wait until a later time
---------------------------------------------------------
#{CX.30: (CX)} --> #{CX.31: (CX)}
******************
M-agree-plan fired as plan
for #{OB.2671: (ACTIVE-GOAL obj (M-AGREE actor ...)...)}
in Cx.30 sprouting Cx.31
---------------------------------------------------------
IF   goal for self and
     other to M-AGREE to something
THEN subgoals for self to MTRANS to other that self BELIEVE
     ACTIVE-GOAL for that something and
     for other to MTRANS to self that other BELIEVE ACTIVE-GOAL
     for that something
---------------------------------------------------------
#{CX.31: (CX)} --> #{CX.32: (CX)}
******************
Mtrans-plan2 fired as plan
for #{OB.2704: (ACTIVE-GOAL obj (MTRANS actor ...)...)}
in Cx.31 sprouting Cx.32
---------------------------------------------------------
IF   goal to MTRANS mental state to person
THEN subgoal to be VPROX that person
---------------------------------------------------------
******************
```

Goal #{OB.2714: (ACTIVE-GOAL obj (VPROX actor ...)...)} succeeds
Subgoals of #{OB.2704: (ACTIVE-GOAL obj (MTRANS actor ...)...)} completed
Perform external action
Perform action goal #{OB.2704: (ACTIVE-GOAL obj (MTRANS actor ...)...)
****
Goal #{OB.2704: (ACTIVE-GOAL obj (MTRANS actor ...)...)} succeeds
| I tell him I would like to have dinner with him at a restaurant.
****
Believe-plan1 fired as inference in Cx.32

IF   MTRANS mental state to person
THEN person BELIEVE self mental state

#{CX.32: (CX)} --> #{CX.33: (CX)}
****
Mtrans-plan2 fired as plan
for #{OB.2709: (ACTIVE-GOAL obj (MTRANS actor ...)...)}
in Cx.32 sprouting Cx.33

IF   goal to MTRANS mental state to person
THEN subgoal to be VPROX that person

****
Goal #{OB.2758: (ACTIVE-GOAL obj (VPROX actor ...)...)} succeeds
Perform other action #{OB.2709: (ACTIVE-GOAL obj (MTRANS actor ...)...)
Enter concepts in #{CX.33: (CX)}
| Input: He declines.
****
Goal #{OB.2709: (ACTIVE-GOAL obj (MTRANS actor ...)...)} fails in #{CX.33: (CX)}
****
Believe-plan2 fired as inference in Cx.33

IF   MTRANS NOT mental state to person
THEN person BELIEVE self NOT mental state

****
Goal #{OB.2671: (ACTIVE-GOAL obj (M-AGREE actor ...)...)} fails in #{CX.33: (CX)}
Subgoal of #{OB.2345: (ACTIVE-GOAL obj (M-DATE actor ...)...)} failed
****
Goal #{OB.2345: (ACTIVE-GOAL obj (M-DATE actor ...)...)} fails in #{CX.33: (CX)}
Subgoal of #{OB.1882: (ACTIVE-GOAL obj (LOVERS actor ...)...)} failed
****
Goal #{OB.1882: (ACTIVE-GOAL obj (LOVERS actor ...)...)} fails in #{CX.33: (CX)}
Attempting to backtrack
Top-level goal #{OB.1882: (ACTIVE-GOAL obj (LOVERS actor ...)...)} fails
All possibilities failed

****
Goal #{OB.1882: (ACTIVE-GOAL obj (LOVERS actor ...)...)} fails in #{CX.33: (CX)}
| I fail at going out with him.
Terminating planning
Leaf context #{CX.33: (CX)}
[OB.1882: (FG. (LOVERS actor Me Movie-star1))]
   [OB.2518: (SG. (ACQUAINTED actor Me Movie-star1))]
   [OB.2365: (SG. (ROMANTIC-INTEREST obj Movie-star1))]

```
[OB.2648: (SG. (BELIEVE actor Movie-star1
    obj (ACTIVE-GOAL obj (LOVERS actor Me Movie-star1)))))]
  [OB.2622: (SG. (BELIEVE actor Movie-star1
    obj (ROMANTIC-INTEREST obj Me)))]
    [OB.2547: (SG. (BELIEVE actor Movie-star1
      obj (ATTRACTIVE actor Me)))]
    [OB.2617: (SG. (BELIEVE actor Movie-star1
      obj (POS-ATTITUDE obj Me)))]
      [OB.2610: (SG. (BELIEVE actor Movie-star1
      obj (BELIEVE actor Me
                obj (POS-ATTITUDE obj (MOVIES obj Movie-star1))))))]
        [OB.2581: (SG. (MTRANS actor Me from Me to Movie-star1
      obj (POS-ATTITUDE obj (MOVIES obj Movie-star1))))]
          [OB.2577: (SG. (VPROX actor Movie-star1 Me))]
    [OB.2643: (SG. (BELIEVE actor Movie-star1
      obj (NOT obj (LOVERS actor Movie-star1)))))]
  [OB.2807: (FG. (M-DATE actor Me Movie-star1))]
  [OB.2341: (AG. (M-AGREE actor Me Movie-star1
      obj (LOVERS actor Me Movie-star1)))]
Removing motivating emotions
Emotional responses
| I feel really angry at him.
```

In order to go on an **M-DATE** with the movie star, DAYDREAMER has to get the movie star to agree to have dinner at a restaurant. She asks him if he would like to have dinner, but he declines. Thus the subgoal for the movie star to **MTRANS** that he would like to have dinner fails, resulting in failure of the **M-AGREE** subgoal, in turn resulting in failure of the **M-DATE** subgoal, finally resulting in the failure of the top-level **LOVERS** goal. The action of Harrison—his negative **MTRANS**—is noted in the reality context as the cause of the top-level goal failure. The **LOVERS** concern fails, its motivating *interest* emotion is terminated, and an emotional response is generated. The resulting negative emotion is directed toward the movie star, since he is the one who performed the action which caused the personal goal failure. This is generated as an *anger* emotion.

DAYDREAMER bluntly asks Harrison Ford if he would like to have dinner. However, Gibbs and R. A. G. Mueller (1988) have observed that people prefer to preface such a request with an utterance designed to remove the greatest potential obstacle to achieving the request. Thus one might more likely first say "What are you doing Friday night?". In DAYDREAMER, obstacles (or subgoals) involving the other person for which no plans are available are merely hypothesized away (through subgoal relaxation); thus in the trace we have *Maybe he is not going out with anyone* and *Maybe he thinks I am cute.* Although detailed planning of conversational utterances is not performed by DAYDREAMER, avoiding embarrassment resulting from a negative response to a request could be accomplished via the **REVERSAL** daydreaming goal. Rules would have to be added to infer failure of a social regard preservation goal upon a negative response to a request for a date. If no plans were available to ensure a positive response to the request (such as plans for verbally removing obstacles), the system would simply not perform the request at all.[2]

```
#{CX.33: (CX)} --> #{CX.34: (CX)}
Concerns:
#{OB.1420: (ACTIVE-GOAL obj (RATIONALIZATION...)...)} (0.97)
#{OB.2412: (ACTIVE-GOAL obj (REVERSAL...)...)} (0.98)
#{OB.1887: (ACTIVE-GOAL obj (ENTERTAINMENT...)...)} (0.6)
Switching to new top-level goal #{OB.1887: (ACTIVE-GOAL obj (ENTERTAINMENT))}
#{CX.34: (CX)} --> #{CX.35: (CX)}
*****************
Mtrans-movie-plan fired as plan
for #{OB.1919: (ACTIVE-GOAL obj (MTRANS actor ...)...)}
in Cx.34 sprouting Cx.35
------------------------------------------------------------
IF   goal to MTRANS a movie
THEN subgoal to be AT location of movie
------------------------------------------------------------
#{CX.35: (CX)} --> #{CX.36: (CX)}
Fact plan #{OB.286: (AT actor Nuart-theater obj ...)} found
*****************
Goal #{OB.2819: (ACTIVE-GOAL obj (AT actor ...)  top-level-goal ...)} succeeds
Instantiating plan
*****************
Goal #{OB.2826: (ACTIVE-GOAL obj (AT actor ...)  top-level-goal ...)} succeeds
Subgoals of #{OB.1919: (ACTIVE-GOAL obj (MTRANS actor ...)...)} completed
Perform external action
Perform action goal #{OB.1919: (ACTIVE-GOAL obj (MTRANS actor ...)...)
*****************
Goal #{OB.1919: (ACTIVE-GOAL obj (MTRANS actor ...)...)} succeeds
| I watch a movie at the Nuart.
```

After the **LOVERS** concern is terminated, processing shifts back to the **ENTERTAINMENT** concern and DAYDREAMER watches a movie at the Nuart Theater.

```
*****************
Reversal-theme fired as inference in Cx.36
------------------------------------------------------------
IF   NEG-EMOTION resulting from a FAILED-GOAL
THEN activate daydreaming goal for REVERSAL of failure
------------------------------------------------------------
Activate top-level goal #{OB.2851: (ACTIVE-GOAL obj (REVERSAL...)...)
#{CX.36: (CX)} --> #{CX.37: (CX)}
*****************
Rationalization-theme fired as inference in Cx.36
------------------------------------------------------------
IF   NEG-EMOTION of sufficient strength resulting from
     a FAILED-GOAL
THEN activate daydreaming goal for RATIONALIZATION
```

---

[2]DAYDREAMER deactivates a goal if all episodes recalled for achieving that goal have a negative *desirability* rating.

```
of failure
------------------------------------------------------------
******************
Activate top-level goal #{OB.2861: (ACTIVE-GOAL obj (RATIONALIZATION...)...)
******************
Revenge-theme fired as inference in Cx.7
------------------------------------------------------------
IF   NEG-EMOTION toward person resulting from a FAILED-GOAL
THEN activate daydreaming goal to gain REVENGE against
     person
------------------------------------------------------------
Activate top-level goal #{OB.1889: (ACTIVE-GOAL obj (REVENGE obj ...)...)
******************
Roving-theme fired as inference in Cx.7
------------------------------------------------------------
IF   NEG-EMOTION of sufficient strength resulting from
     a FAILED-GOAL
THEN activate daydreaming goal for ROVING
------------------------------------------------------------
******************
Activate top-level goal #{OB.1901: (ACTIVE-GOAL obj (ROVING))
******************
Recovery-theme fired as inference in Cx.3
------------------------------------------------------------
IF   NEG-EMOTION resulting from a FAILED-GOAL
THEN activate daydreaming goal for RECOVERY of failure
------------------------------------------------------------
******************
Activate top-level goal #{OB.1662: (ACTIVE-GOAL obj (RECOVERY...)...)
******************
Lovers-theme fired as inference in Cx.36
------------------------------------------------------------
IF   self not LOVERS with anyone and
     SEX or LOVE-GIVING or LOVE-RECEIVING or COMPANIONSHIP
     below threshold
THEN activate goal for LOVERS with a person
------------------------------------------------------------
******************
Activate top-level goal #{OB.2872: (ACTIVE-GOAL obj (LOVERS actor ...)...)
| I want to be going out with someone.
| I feel really interested in going out with someone.
#{CX.36: (CX)} --> #{CX.39: (CX)}
Concerns:
#{OB.2872: (ACTIVE-GOAL obj (LOVERS actor ...)...)} (0.9)
#{OB.2861: (ACTIVE-GOAL obj (RATIONALIZATION...)...)} (1.0059812499999996)
#{OB.2851: (ACTIVE-GOAL obj (REVERSAL...)...)} (1.0159812499999996)
#{OB.1420: (ACTIVE-GOAL obj (RATIONALIZATION...)...)} (0.97)
#{OB.2412: (ACTIVE-GOAL obj (REVERSAL...)...)} (0.98)
#{OB.1887: (ACTIVE-GOAL obj (ENTERTAINMENT...)...)} (0.6)
#{OB.1901: (ACTIVE-GOAL obj (ROVING...)...)} (0.96)
#{OB.1889: (ACTIVE-GOAL obj (REVENGE...)...)} (0.99)
#{OB.1662: (ACTIVE-GOAL obj (RECOVERY...)...)} (0.99)
```

The negative emotion of *anger* resulting from the personal goal failure activates the following daydreaming goals: **RATIONALIZATION** (rationalize being

turned down), **ROVING** (shift attention from the rejection to more pleasant thoughts), **REVENGE** (imagine getting even with the movie star), **REVERSAL** (plan to prevent similar rejections in the future), and **RECOVERY** (imagine future scenarios in which DAYDREAMER becomes romantically involved with Harrison). Since the **LOVERS** goal is still not achieved and the associated need states are not satisfied, a fresh **LOVERS** goal—not associated with Harrison Ford—is again activated.

```
*******************
Ptrans-plan fired as plan
for #{OB.1916: (ACTIVE-GOAL obj (PTRANS actor ...)...)}
in Cx.36 sprouting Cx.39
-------------------------------------------------------
IF   goal for person to PTRANS to a location
THEN subgoal for person to KNOW that location
-------------------------------------------------------
*******************
Goal #{OB.2901: (ACTIVE-GOAL obj (KNOW actor ...)...)} succeeds
Subgoals of #{OB.1916: (ACTIVE-GOAL obj (PTRANS actor ...)...)} completed
Perform external action
Perform action goal #{OB.1916: (ACTIVE-GOAL obj (PTRANS actor ...)...)
*******************
Goal #{OB.1916: (ACTIVE-GOAL obj (PTRANS actor ...)...)} succeeds
| I go home.
*******************
At-plan2 fired as inference in Cx.39
-------------------------------------------------------
IF   person PTRANS from one location to another
THEN person AT new location and
     no longer AT old location
-------------------------------------------------------
Retracting dependencies
Retract Ob.1943 in Cx.39:
(AT actor Me obj Nuart-location)
Retract Ob.1940 in Cx.39:
(PTRANS actor Me from Home to ...)
Subgoals of #{OB.1897: (ACTIVE-GOAL obj (M-MOVIE actor ...)...)} completed
*******************
Goal #{OB.1897: (ACTIVE-GOAL obj (M-MOVIE actor ...)...)} succeeds
Assert Ob.2943 in Cx.39:
(M-MOVIE actor Me)
*******************
Entertainment-inf. fired as inference in Cx.39
-------------------------------------------------------
IF   M-MOVIE
THEN ENTERTAINMENT need satisfied
-------------------------------------------------------
*******************
Goal #{OB.1887: (ACTIVE-GOAL obj (ENTERTAINMENT...)...)} succeeds
| I succeed at being entertained.
Terminating planning for top-level goal
Leaf context #{CX.39: (CX)}
[OB.1887: (SG. (ENTERTAINMENT strength (UPROC proc 'NEED-SATISFIED?)))]
  [OB.2942: (SG. (M-MOVIE actor Me))]
```

```
[OB.1926: (SG. (AT actor Nuart-theater obj Nuart-location))]
[OB.1939: (SG. (PTRANS actor Me from Home to Nuart-location obj Me))]
  [OB.1935: (SG. (KNOW actor Me obj Nuart-location))]
[OB.2837: (SG. (MTRANS actor Me from Nuart-theater to Me
    obj (MOVIE)))]
  [OB.2829: (SG. (AT actor Nuart-theater obj Nuart-location))]
  [OB.2833: (SG. (AT actor Me obj Nuart-location))]
[OB.2907: (SG. (PTRANS actor Me from Nuart-location to Home obj Me))]
  [OB.2903: (SG. (KNOW actor Me obj Home))]
Removing motivating emotions
Emotional responses
| I feel amused.
Store episode #{OB.1887: (SUCCEEDED-GOAL obj (ENTERTAINMENT...)...)
Assess scenario desirability in #{CX.39: (CX)}
#{OB.1882: (FAILED-GOAL obj (LOVERS actor ...)...)} (1.15)
#{OB.2403: (FAILED-GOAL obj (BELIEVE actor ...)...)} (0.95)
#{OB.1887: (SUCCEEDED-GOAL obj (ENTERTAINMENT...)...)} (0.6)
Scenario desirability = -1.4999999999999996
Activate index #{ENTERTAINMENT-PLAN1: (RULE subgoal (M-MOVIE actor ...)...)}
Activate index #{OB.2563: (MOVIES obj Movie-star1)}
Activate index #{NUART-THEATER: (THEATER name "the Nuart")}
Activate index #{MOVIE-STAR1: (MALE-ACTOR first-name "Harrison"...)}
Activate index #{HOME: (LOCATION name "home")}
Activate index #{NUART-LOCATION: (LOCATION name "the Nuart")}
#{CX.39: (CX)} --> #{CX.40: (CX)}
Concerns:
#{OB.2872: (ACTIVE-GOAL obj (LOVERS actor ...)...)} (0.9)
#{OB.2861: (ACTIVE-GOAL obj (RATIONALIZATION...)...)} (1.0059812499999996)
#{OB.2851: (ACTIVE-GOAL obj (REVERSAL...)...)} (1.0159812499999996)
#{OB.1420: (ACTIVE-GOAL obj (RATIONALIZATION...)...)} (0.97)
#{OB.2412: (ACTIVE-GOAL obj (REVERSAL...)...)} (0.98)
No more goals to run; switching to daydreaming mode
State changes from PERFORMANCE to DAYDREAMING
```

DAYDREAMER successfully completes her plan for **M-MOVIE**. This replenishes the **ENTERTAINMENT** need state and thus the **ENTERTAINMENT** personal goal succeeds. A positive emotion is activated and the concern is terminated. The experience is stored in episodic memory, along with its evaluation of -1.5. All in all, this was not a positive experience for DAYDREAMER.

A.2   REVENGE1 Daydream

```
Switching to new top-level goal #{OB.1889: (ACTIVE-GOAL obj (REVENGE obj ))}
#{CX.8: (CX)} --> #{CX.10: (CX)}
******************
Revenge-plan1 fired as plan
for #{OB.1889: (ACTIVE-GOAL obj (REVENGE obj ...)...)}
in Cx.8 sprouting Cx.10
-----------------------------------------------------
IF    goal to gain REVENGE against person for causing self
      a failed POS-RELATIONSHIP goal
THEN subgoal for person to have failure of same POS-RELATIONSHIP
```

```
-----------------------------------------------------------
#{CX.10: (CX)} --> #{CX.11: (CX)}
******************
Failed-lovers-goal-plan1 fired as plan
for #{OB.1917: (ACTIVE-GOAL obj (BELIEVE actor ...)...)}
in Cx.10 sprouting Cx.11
-----------------------------------------------------------
IF    goal for person to have FAILED-GOAL of LOVERS
THEN  subgoal for person to have ACTIVE-GOAL of LOVERS
      with person and
      then BELIEVE that person does not have ACTIVE-GOAL
      of LOVERS
-----------------------------------------------------------
#{CX.11: (CX)} --> #{CX.12: (CX)}
******************
Lovers-theme-plan fired as plan
for #{OB.1924: (ACTIVE-GOAL obj (BELIEVE actor ...)...)}
in Cx.11 sprouting Cx.12
-----------------------------------------------------------
IF    goal for self ACTIVE-GOAL of LOVERS with a person
THEN  subgoals for ROMANTIC-INTEREST in person and
      not LOVERS with anyone
-----------------------------------------------------------
#{CX.12: (CX)} --> #{CX.13: (CX)}
******************
Romantic-interest-plan4 fired as plan
for #{OB.1938: (ACTIVE-GOAL obj (BELIEVE actor ...)...)}
in Cx.12 sprouting Cx.13
-----------------------------------------------------------
IF    goal to have POS-ATTITUDE toward a person and
      self is STAR
THEN  subgoal for person to be greater STAR
-----------------------------------------------------------
#{CX.13: (CX)} --> #{CX.14: (CX)}
******************
Goal #{OB.1950: (ACTIVE-GOAL obj (BELIEVE actor ...)...)} succeeds
#{CX.14: (CX)} --> #{CX.15: (CX)}
******************
Belief-pers-attr2 fired as plan
for #{OB.1959: (ACTIVE-GOAL obj (BELIEVE actor ...)...)}
in Cx.14 sprouting Cx.15
#{CX.15: (CX)} --> #{CX.16: (CX)}
******************
Star-plan fired as plan
for #{OB.2007: (ACTIVE-GOAL obj (STAR actor ...)...)}
in Cx.15 sprouting Cx.16
-----------------------------------------------------------
IF    goal for self to be a STAR
THEN  subgoal for self to M-STUDY to be an ACTOR
-----------------------------------------------------------
#{CX.16: (CX)} --> #{CX.17: (CX)}
******************
M-study-plan fired as plan
for #{OB.2012: (ACTIVE-GOAL obj (M-STUDY actor ...)...)}
in Cx.16 sprouting Cx.17
```

```
******************
Goal #{OB.2017: (ACTIVE-GOAL obj (RTRUE)  top-level-goal ...)} succeeds
Subgoals of #{OB.2012: (ACTIVE-GOAL obj (M-STUDY actor ...)...)} completed
******************
Goal #{OB.2012: (ACTIVE-GOAL obj (M-STUDY actor ...)...)} succeeds
| I study to be an actor.
Subgoals of #{OB.2007: (ACTIVE-GOAL obj (STAR actor ...)...)} completed
******************
Goal #{OB.2007: (ACTIVE-GOAL obj (STAR actor ...)...)} succeeds
| I am a movie star even more famous than he is.
Subgoals of #{OB.1959: (ACTIVE-GOAL obj (BELIEVE actor ...)...)} completed
******************
Goal #{OB.1959: (ACTIVE-GOAL obj (BELIEVE actor ...)...)} succeeds
******************

...


Subgoals of #{OB.1969: (ACTIVE-GOAL obj (BELIEVE actor ...)...)} completed
******************
Goal #{OB.1969: (ACTIVE-GOAL obj (BELIEVE actor ...)...)} succeeds
Personal goal outcome
Emotional responses
| I feel pleased.
| He is interested in me.
#{CX.17: (CX)} --> #{CX.18: (CX)}
******************
Not-lovers-plan1 fired as plan
for #{OB.1964: (ACTIVE-GOAL obj (BELIEVE actor ...)...)}
in Cx.17 sprouting Cx.18
#{CX.18: (CX)} --> #{CX.19: (CX)}
******************
M-break-up-plan2 fired as plan
for #{OB.2088: (ACTIVE-GOAL obj (M-BREAK-UP...)...)}
in Cx.18 sprouting Cx.19
******************
Goal #{OB.2093: (ACTIVE-GOAL obj (RTRUE)  top-level-goal ...)} succeeds
Subgoals of #{OB.2088: (ACTIVE-GOAL obj (M-BREAK-UP...)...)} completed
******************
Goal #{OB.2088: (ACTIVE-GOAL obj (M-BREAK-UP...)...)} succeeds
| He breaks up with his girlfriend.
******************
Not-lovers-plan1 fired as inference in Cx.19
Subgoals of #{OB.1964: (ACTIVE-GOAL obj (BELIEVE actor ...)...)} completed
******************
Goal #{OB.1964: (ACTIVE-GOAL obj (BELIEVE actor ...)...)} succeeds
Subgoals of #{OB.1982: (ACTIVE-GOAL obj (BELIEVE actor ...)...)} completed
******************
Goal #{OB.1982: (ACTIVE-GOAL obj (BELIEVE actor ...)...)} succeeds
| He wants to be going out with me.
#{CX.19: (CX)} --> #{CX.20: (CX)}
******************
Believe-plan2 fired as plan
for #{OB.1975: (ACTIVE-GOAL obj (BELIEVE actor ...)...)}
in Cx.19 sprouting Cx.20
------------------------------------------------------
IF   goal for person to BELIEVE self NOT mental state
```

```
THEN subgoal to MTRANS NOT mental state to person
-----------------------------------------------------------
#{CX.20: (CX)} --> #{CX.21: (CX)}
*******************
Mtrans-plan2 fired as plan
for #{OB.2142: (ACTIVE-GOAL obj (MTRANS actor ...)...)}
in Cx.20 sprouting Cx.21
-----------------------------------------------------------
IF   goal to MTRANS mental state to person
THEN subgoal to be VPROX that person
-----------------------------------------------------------
#{CX.21: (CX)} --> #{CX.22: (CX)}
#{CX.21: (CX)} --> #{CX.23: (CX)}
#{CX.21: (CX)} --> #{CX.24: (CX)}
Candidates: #{OB.1889: (ACTIVE-GOAL obj (REVENGE obj ...)...)} (1.2)
*******************
Vprox-plan2 fired as plan
for #{OB.2147: (ACTIVE-GOAL obj (VPROX actor ...)...)}
in Cx.21 sprouting Cx.23
-----------------------------------------------------------
IF   goal to be VPROX person
THEN subgoal to M-PHONE person
-----------------------------------------------------------
#{CX.23: (CX)} --> #{CX.25: (CX)}
*******************
M-phone-plan1 fired as plan
for #{OB.2160: (ACTIVE-GOAL obj (M-PHONE actor ...)...)}
in Cx.23 sprouting Cx.25
-----------------------------------------------------------
IF   goal to M-PHONE person
THEN subgoal to KNOW TELNO of person
-----------------------------------------------------------
*******************
Goal #{OB.2171: (ACTIVE-GOAL obj (KNOW actor ...)...)} succeeds
Subgoals of #{OB.2160: (ACTIVE-GOAL obj (M-PHONE actor ...)...)} completed
*******************
Goal #{OB.2160: (ACTIVE-GOAL obj (M-PHONE actor ...)...)} succeeds
| He calls me up.
*******************
Vprox-plan2 fired as inference in Cx.25
-----------------------------------------------------------
IF   M-PHONE person
THEN VPROX person
-----------------------------------------------------------
Subgoals of #{OB.2147: (ACTIVE-GOAL obj (VPROX actor ...)...)} completed
*******************
Goal #{OB.2147: (ACTIVE-GOAL obj (VPROX actor ...)...)} succeeds
Subgoals of #{OB.2142: (ACTIVE-GOAL obj (MTRANS actor ...)...)} completed
*******************
Goal #{OB.2142: (ACTIVE-GOAL obj (MTRANS actor ...)...)} succeeds
| I turn him down.
*******************
Believe-plan2 fired as inference in Cx.25
-----------------------------------------------------------
IF   MTRANS NOT mental state to person
```

```
THEN person BELIEVE self NOT mental state
-----------------------------------------------------------
Subgoals of #{OB.1975: (ACTIVE-GOAL obj (BELIEVE actor ...)...)} completed
******************
Goal #{OB.1975: (ACTIVE-GOAL obj (BELIEVE actor ...)...)} succeeds
Subgoals of #{OB.1989: (ACTIVE-GOAL obj (BELIEVE actor ...)...)} completed
******************
Goal #{OB.1989: (ACTIVE-GOAL obj (BELIEVE actor ...)...)} succeeds
Subgoals of #{OB.1889: (ACTIVE-GOAL obj (REVENGE obj ...)...)} completed
******************
Goal #{OB.1889: (ACTIVE-GOAL obj (REVENGE obj ...)...)} succeeds
| I get even with him.
Emotional response.
| I feel pleased.
Terminating planning for top-level goal
Removing motivating emotions
Store episode #{OB.1889: (SUCCEEDED-GOAL obj (REVENGE...)...)
#{CX.7: (CX)} --> #{CX.26: (CX)}
```

The table-turning strategy for **REVENGE** involves causing a failure for the other person similar to the one which that person caused for DAYDREAMER. When the failed goal is one to form a positive relationship with the other person, **REVENGE** is achieved when the other person experiences the failure of a goal to form a similar relationship with DAYDREAMER.

In order for the movie star to experience a failure of the goal to form a **LOVERS** relationship with DAYDREAMER, he must (a) have such a goal in the first place, and (b) be turned down by her. In order for a person to have a goal to form a **LOVERS** relationship with a second person, the first person must have **ROMANTIC-INTEREST** in the second person and not be in a **LOVERS** relationship with someone else. This rule is applied via backward other planning. **ROMANTIC-INTEREST** in turn requires a **POS-ATTITUDE**. A rule states that a movie star has a **POS-ATTITUDE** toward a person if that person is an even greater star. A low-plausibility plan to become a star is to study to be an actor. Once the movie star is imagined to be interested in DAYDREAMER, she imagines turning him down. This results in a positive emotion. The daydreamed episode is then stored in episodic memory for future use (in REVENGE2 and REVENGE3).

A.3   RATIONALIZATION1 Daydream

```
Switching to new top-level goal #{OB.1420: (ACTIVE-GOAL obj (RATIONALIZATION))}
#{CX.6: (CX)} --> #{CX.7: (CX)}
#{CX.6: (CX)} --> #{CX.8: (CX)}
******************
Rationalization-plan1 fired as plan
for #{OB.1420: (ACTIVE-GOAL obj (RATIONALIZATION...)...)}
in Cx.6 sprouting Cx.7
-----------------------------------------------------------
IF    goal to RATIONALIZE failure
```

```
THEN subgoal for imagined success to LEADTO failure
-----------------------------------------------------------
#{CX.7: (CX)} --> #{CX.9: (CX)}
******************
Leadto-plan1 fired as coded plan
for #{OB.1441: (ACTIVE-GOAL obj (LEADTO ante ...)...)}
in Cx.7 sprouting Cx.9
-----------------------------------------------------------
IF   goal for success to LEADTO failure
THEN hypothesize success and
     explore consequences
-----------------------------------------------------------
| What if I were going out with him?
```

One way to rationalize the failure of a goal is to imagine that had the goal succeeded, it would have resulted in a equal or worse goal failure. One way of generating such a **LEADTO** scenario is to assert the succeeded goal in an alternative past context, and then apply inferences in that context in order to evaluate the consequences of that success. If negative consequences (goal failures) are found, positive emotions will be generated since this is an alternative past. Positive emotions will offset the original negative emotion resulting from failure of the **LOVERS** goal. (Of course, if positive consequences are found, negative emotions will instead result. In the current version of DAYDREAMER, there is no means of directing the outcome one way or another when this particular plan for **LEADTO** is employed.)

```
******************
Other-rule1 fired as inference in Cx.9
-----------------------------------------------------------
IF   LOVERS with a person
THEN initiate forward other planning for person
-----------------------------------------------------------
******************
Acting-job-theme fired as inference in Cx.9
-----------------------------------------------------------
IF   actor does not have an ACTING-EMPLOY
THEN activate goal to have an ACTING-EMPLOY
-----------------------------------------------------------
******************
Activate top-level goal #{OB.1464: (ACTIVE-GOAL obj (ACTING-EMPLOY...)...)}
for #{MOVIE-STAR1: (MALE-ACTOR first-name "Harrison"...)
| He would need work.
```

The first consequence of the hypothesized goal success is that forward other planning is initiated for the movie star. This is performed whenever DAYDREAMER is in a **LOVERS** relationship with another person in order to monitor the mental state of the other person. Once forward other planning begins, the following further consequence is generated: the movie star will have the goal to act in a movie.

```
#{CX.9: (CX)} --> #{CX.10: (CX)}
```

```
Episodic reminding of #{EPISODE.1: (EPISODE rule Episodic-rule.1...)}
| I remember the time he had a job with Paramount Pictures in Cairo.
******************
Episodic-rule.1 fired as analogical plan
for #{OB.1464: (ACTIVE-GOAL obj (ACTING-EMPLOY...)...)}
in Cx.9 sprouting Cx.10
| He would have to be in Cairo.
#{CX.10: (CX)} --> #{CX.11: (CX)}
******************
Rprox-plan fired as plan
for #{OB.1486: (ACTIVE-GOAL obj (RPROX actor ...)...)}
in Cx.10 sprouting Cx.11
---------------------------------------------------------
IF    goal to be RPROX a city
THEN subgoal to PTRANS1 to that city
---------------------------------------------------------
#{CX.11: (CX)} --> #{CX.12: (CX)}
******************
Ptrans1-plan fired as plan
for #{OB.1504: (ACTIVE-GOAL obj (PTRANS1 actor ...)...)}
in Cx.11 sprouting Cx.12
******************
Goal #{OB.1525: (ACTIVE-GOAL obj (RTRUE)  top-level-goal ...)} succeeds
Subgoals of #{OB.1504: (ACTIVE-GOAL obj (PTRANS1 actor ...)...)} completed
******************
Goal #{OB.1504: (ACTIVE-GOAL obj (PTRANS1 actor ...)...)} succeeds
| He would go to Cairo.
******************
Rprox-plan fired as inference in Cx.12
---------------------------------------------------------
IF    PTRANS1 from one city to another
THEN RPROX new city and
     no longer RPROX old city
---------------------------------------------------------
******************
Rprox-plan fired as inference in Cx.12
---------------------------------------------------------
IF    PTRANS1 from one city to another
THEN RPROX new city and
     no longer RPROX old city
---------------------------------------------------------
```

Forward other planning continues for the movie star: DAYDREAMER is reminded of a previous secondhand episode involving Harrison Ford—when he had to go on location in Egypt to shoot a film. Through analogical planning, this episode is applied to the star's current goal: he must be **RPROX** (located in) Cairo to shoot another film. This in turn results in Harrison Ford's **PTRANS1**ing to Cairo.

```
******************
Lovers-p-goal fired as inference in Cx.12
---------------------------------------------------------
IF    LOVERS with person who is RPROX one city and
      self is not RPROX that city
```

```
THEN activate preservation goal on LOVERS
-----------------------------------------------------------
| Our relationship would be in trouble.
*******************
Goal #{OB.1486: (ACTIVE-GOAL obj (RPROX actor ...)...)} succeeds
Subgoals of #{OB.1637: (SUCCEEDED-GOAL obj (RPROX...)...)} completed
Subgoals of #{OB.1464: (ACTIVE-GOAL obj (ACTING-EMPLOY...)...)} completed
*******************
Goal #{OB.1464: (ACTIVE-GOAL obj (ACTING-EMPLOY...)...)} succeeds
#{CX.12: (CX)} --> #{CX.13: (CX)}
#{CX.13: (CX)} --> #{CX.14: (CX)}
*******************
Rprox-plan fired as plan
for #{OB.1688: (ACTIVE-GOAL obj (RPROX actor ...)...)}
in Cx.13 sprouting Cx.14
-----------------------------------------------------------
IF    goal to be RPROX a city
THEN subgoal to PTRANS1 to that city
-----------------------------------------------------------
#{CX.14: (CX)} --> #{CX.15: (CX)}
*******************
Ptrans1-plan fired as plan
for #{OB.1694: (ACTIVE-GOAL obj (PTRANS1 actor ...)...)}
in Cx.14 sprouting Cx.15
*******************
Goal #{OB.1699: (ACTIVE-GOAL obj (RTRUE) top-level-goal ...)} succeeds
Subgoals of #{OB.1694: (ACTIVE-GOAL obj (PTRANS1 actor ...)...)} completed
*******************
Goal #{OB.1694: (ACTIVE-GOAL obj (PTRANS1 actor ...)...)} succeeds
| I would go to Cairo.
*******************
Rprox-plan fired as inference in Cx.15
-----------------------------------------------------------
IF    PTRANS1 from one city to another
THEN RPROX new city and
     no longer RPROX old city
-----------------------------------------------------------
```

The fact that the star is in Cairo and DAYDREAMER is in Los Angeles results in the activation of a preservation goal on the **LOVERS** relationship. Planning is then performed to undo the antecedents for this preservation goal— in particular, the fact that DAYDREAMER is not **RPROX** the same city as the star results in DAYDREAMER **PTRANS1**ing to Cairo. (The other alternative would is for the star to **PTRANS1** back to Los Angeles—or not to go to Cairo in the first place.)

```
*******************
Job-failure fired as inference in Cx.15
-----------------------------------------------------------
IF    have EMPLOYMENT with organization which is RPROX
      one city and
      self is not RPROX that city
THEN goal to have EMPLOYMENT fails
```

```
--------------------------------------------------------
| I would lose my job at May Company.
Personal goal outcome
Emotional responses
| I feel relieved.
*******************
Goal #{OB.1688: (ACTIVE-GOAL obj (RPROX actor ...)...)} succeeds
```

However, since DAYDREAMER is no longer in Los Angeles, she loses her job. Since this moderately important goal failure occurs in an alternative past, it results in the positive emotion of *relief*. This positive emotion is diverted into the original negative emotion, which loses some of its strength as a result. However, the strength of the negative emotion is still not below a certain threshold. Therefore, another rationalization is sought.

A.4 RATIONALIZATION2 Daydream

```
Attempting to backtrack for top-level goal
#{OB.1420: (ACTIVE-GOAL obj (RATIONALIZATION...)...)
Backtracking to next context of #{CX.8: (CX)}
for #{OB.1420: (ACTIVE-GOAL obj (RATIONALIZATION...)...)}
*******************
Rationalization-plan2 fired as plan
for #{OB.1420: (ACTIVE-GOAL obj (RATIONALIZATION...)...)}
in Cx.6 sprouting Cx.8
--------------------------------------------------------
IF   goal to RATIONALIZE failure
THEN subgoal for failure to LEADTO success
--------------------------------------------------------
#{CX.8: (CX)} --> #{CX.16: (CX)}
Episodic reminding of #{EPISODE.5: (EPISODE rule Episodic-rule.4...)}
| I remember the time my being turned down by
| Irving led to a success by being turned down by
| him leading to succeeding at going out with Chris.
```

Another method for rationalization is to imagine future scenarios in which the goal failure leads to an equal or more important goal success. An episode for achieving this goal is recalled: the time DAYDREAMER went to a bar after being turned down by Irving and ended up meeting Chris. Thus through analogical planning, DAYDREAMER imagines going to a bar and meeting someone:

```
*******************
Episodic-rule.4 fired as analogical plan
for #{OB.1448: (ACTIVE-GOAL obj (LEADTO ante ...)...)}
in Cx.8 sprouting Cx.16
#{CX.16: (CX)} --> #{CX.17: (CX)}
Apply existing analogical plan
*******************
Episodic-rule.3 fired as analogical plan
for #{OB.1769: (ACTIVE-GOAL obj (LEADTO ante ...)...)}
in Cx.16 sprouting Cx.17
```

```
#{CX.17: (CX)} --> #{CX.18: (CX)}
Apply existing analogical plan
******************
Episodic-rule.2 fired as analogical plan
for #{OB.1776: (ACTIVE-GOAL obj (LOVERS actor ...)...)}
in Cx.17 sprouting Cx.18
Apply existing analogical plan
******************
At-plan2 fired as analogical plan
for #{OB.1781: (ACTIVE-GOAL obj (AT actor ...)  top-level-goal ...)}
in Cx.18 sprouting Cx.19
-----------------------------------------------------------
IF   goal to be AT a location
THEN subgoal to PTRANS to that location
-----------------------------------------------------------
#{CX.19: (CX)} --> #{CX.20: (CX)}
******************
Ptrans-plan fired as plan
for #{OB.1789: (ACTIVE-GOAL obj (PTRANS actor ...)...)}
in Cx.19 sprouting Cx.20
-----------------------------------------------------------
IF   goal for person to PTRANS to a location
THEN subgoal for person to KNOW that location
-----------------------------------------------------------

******************
Goal #{OB.1794: (ACTIVE-GOAL obj (KNOW actor ...)...)} succeeds
Subgoals of #{OB.1789: (ACTIVE-GOAL obj (PTRANS actor ...)...)} completed
******************
Goal #{OB.1789: (ACTIVE-GOAL obj (PTRANS actor ...)...)} succeeds
| I go to Mom's.
...
Personal goal outcome
Emotional responses
| I feel pleased.
| I am going out with him.
...
```

Generating an imaginary future scenario in which the DAYDREAMER meets someone (Chris or someone similar) at a bar results in the imagined future success of the **LOVERS** goal and corresponding positive emotion. This positive emotion is diverted into the original negative emotion.

A.5   RATIONALIZATION3 Daydream

```
******************
Rationalization-plan3 fired as plan
for #{OB.1420: (ACTIVE-GOAL obj (RATIONALIZATION...)...)}
in Cx.19 sprouting Cx.110
-----------------------------------------------------------
IF   goal to RATIONALIZE failure
THEN subgoal to MINIMIZE failure
-----------------------------------------------------------
******************
```

```
Minimization-plan fired as coded plan
for #{OB.5356: (ACTIVE-GOAL obj (MINIMIZATION...)...)}
in Cx.110 sprouting Cx.111
-------------------------------------------------------
IF   goal for MINIMIZATION of failure
THEN negate and justify antecedents of failure
-------------------------------------------------------
| Anyway, I was well dressed because I was wearing a necklace.
| I feel less embarrassed.
| Anyway, I do not think much of Harrison Ford.
| I feel less embarrassed.
******************
Rationalization-inf1 fired as inference in Cx.111
-------------------------------------------------------
IF   NEG-EMOTION associated with failure less than a
     certain strength or POS-EMOTION associated with
     failure
THEN RATIONALIZATION of failure
-------------------------------------------------------
| I rationalize being turned down by him.
******************
Goal #{OB.1420: (ACTIVE-GOAL obj (RATIONALIZATION...)...)} succeeds
Terminating planning for top-level goal
#{OB.1420: (SUCCEEDED-GOAL obj (RATIONALIZATION...)...)}
Leaf context #{CX.21: (CX)}
[OB.1420:
 (SG. (RATIONALIZATION obj (FAILED-GOAL obj (LOVERS actor Me Movie-star1))))]
   [OB.1855: (SG. (LEADTO ante (FAILED-GOAL obj (LOVERS actor Me Movie-star1))
              conseq (SUCCEEDED-GOAL)))]
      [OB.1845: (SG. (LEADTO ante (FAILED-GOAL obj (LOVERS actor Me Movie-star1))
                 conseq (SUCCEEDED-GOAL obj (LOVERS actor Me Chris)))))]
         [OB.1820: (SG. (LOVERS actor Me Chris))]
            [OB.1810: (SG. (AT actor Me obj Bar1-loc))]
               [OB.1800: (SG. (PTRANS actor Me from Home to Bar1-loc obj Me))]
               [OB.1796: (SG. (KNOW actor Me obj Bar1-loc))]
            [OB.1817: (SG. (AT actor Chris obj Bar1-loc))]
Leaf context #{CX.15: (CX)}
[OB.1420:
 (SG. (RATIONALIZATION obj (FAILED-GOAL obj (LOVERS actor Me Movie-star1))))]
   [OB.1441: (AG. (LEADTO ante (SUCCEEDED-GOAL obj (LOVERS actor Me Movie-star1))
              conseq (FAILED-GOAL)))]
      [OB.1452: (SG. (RTRUE))]
Leaf context #{CX.111: (CX)}
[OB.1420: (SG. (RATIONALIZATION obj (FAILED-GOAL obj (BELIEVE actor Movie-star1
                                    obj (POS-ATTITUDE obj Me))
                         top-level-goal ...))))]
   [OB.5356: (AG. (MINIMIZATION obj (FAILED-GOAL obj (BELIEVE actor Movie-star1
                                    obj (POS-ATTITUDE obj Me))
                         top-level-goal ...))]
      [OB.5360: (SG. (RTRUE))]
Removing motivating emotions of
#{OB.1420: (SUCCEEDED-GOAL obj (RATIONALIZATION...)...)
Store episode #{OB.1420: (SUCCEEDED-GOAL obj (RATIONALIZATION...)...)
```

DAYDREAMER attempts to rationalize her early embarrassment in front of the movie star. The minimization strategy for rationalization involves finding antecedents of the failure; if an antecedent is an attitude, the opposite attitude is asserted, otherwise reasoning chains for the negation of the antecedent are highlighted. In this case, an antecedent for embarrassment in front of someone is that one has a positive attitude toward that person. This attitude is negated, resulting in a reduction of the strength of the resulting embarrassment emotion. Another antecedent is that DAYDREAMER was not well dressed. However, a low realism inference was already made that DAYDREAMER was well dressed because she was wearing a nice necklace. This antecedent is thus negated and the strength of the resulting embarrassment further reduced.

Now the strength of the negative emotion is below the necessary threshold and the **RATIONALIZATION** daydreaming goal succeeds. The various planning episodes are stored in episodic memory for possible future use.

A.6   ROVING1 Daydream

```
Switching to new top-level goal #{OB.1901: (ACTIVE-GOAL obj (ROVING) ...)}
#{CX.9: (CX)} --> #{CX.27: (CX)}
*******************
Roving-plan1 fired as coded plan
for #{OB.1901: (ACTIVE-GOAL obj (ROVING)  top-level-goal ...)}
in Cx.9 sprouting Cx.27
-------------------------------------------------------
IF   daydreaming goal for ROVING
THEN recall pleasant episode
-------------------------------------------------------
Episodic reminding of #{EPISODE.1: (EPISODE rule Induced-rule.1...)}
| I remember the time Steve told me he thought I was wonderful
| at Gulliver's.
Activate index #{INDUCED-RULE.1: (RULE subgoal (RSEQ obj (MTRANS...) ...)...)}
Activate index #{GULLIVERS: (RESTAURANT name "Gulliver's")}
Activate index #{MARINA: (CITY name "the Marina")}
Activate index #{STEVE1: (MALE-PERSON first-name "Steve"...)}
| I feel pleased.
```

The plan for the **ROVING** daydreaming goal is to retrieve a random episode associated with a strong positive emotion. Here a pleasant episode at Gulliver's restaurant, located in Marina del Rey, is recalled, and the associated positive emotion is reactivated.

```
Episodic reminding of #{EPISODE.2: (EPISODE rule Induced-rule.2...)}
| I remember the time I had a job in the Marina.
Activate index #{INDUCED-RULE.2: (RULE subgoal (RPROX actor ...)  goal ...)}
Activate index #{OB.1829: (EMPLOYMENT)}
Index #{MARINA: (CITY name "the Marina")} already active
Activate index #{VENICE: (CITY name "Venice Beach")}
Index #{INDUCED-RULE.1: (RULE subgoal (RSEQ obj (MTRANS...) ...)...)} fades
Episodic reminding of #{EPISODE.3: (EPISODE rule Induced-rule.3...)}
```

```
| I remember the time Steve and I bought sunglasses in
| Venice Beach.
Activate index #{INDUCED-RULE.3: (RULE subgoal (RPROX actor ...)  goal ...)}
Index #{VENICE: (CITY name "Venice Beach")} already active
Subgoals of #{OB.1901: (ACTIVE-GOAL obj (ROVING)  top-level-goal ...)} completed
******************
Goal #{OB.1901: (ACTIVE-GOAL obj (ROVING)  top-level-goal ...)} succeeds
Terminating planning for top-level goal
#{OB.1901: (SUCCEEDED-GOAL obj (ROVING...)...)}
Removing motivating emotions
#{CX.26: (CX)} --> #{CX.28: (CX)}
```

Whenever an episode is retrieved via some of its indices, its remaining indices
are activated. (The number of indices required for the retrieval of an episode
depends on the episode. Some may be retrieved using only one index; others
require several indices to be active before they can be retrieved.) The Gulliver's
episode is indexed under Marina del Rey and so this index is activated. This
index is sufficient to result in the recall of a job DAYDREAMER once had in
Marina del Rey. Venice, which is near Marina del Rey, is another index for
this second episode and so it is activated. The active indices of Venice and
Peter together result in the retrieval of an episode involving both. (Retrieval
of episodes involving more than one index is accomplished through a one-level
intersection search.)

A.7   RECOVERY2 Daydream

```
Switching to new top-level goal #{OB.1662: (ACTIVE-GOAL obj (RECOVERY...)...)}
******************
Recovery-plan fired as plan
for #{OB.1662: (ACTIVE-GOAL obj (RECOVERY...)...)}
in Cx.4 sprouting Cx.5
----------------------------------------------------------------
IF   goal for RECOVERY from a LOVERS failure
THEN subgoal to ask out person again
----------------------------------------------------------------
| I have to ask him out.
```

In general, to achieve **RECOVERY** of a failed **LOVERS** goal, one must imagine ways of achieving that same **LOVERS** goal in the future. After LOVERS1,
DAYDREAMER considers possible ways to go out with the movie star in the
future. For simplicity, instead of generating an entire future plan for achieving
the **LOVERS** goal (which is demonstrated by other traces), only the problem of
contacting the movie star to ask him out is considered here.

```
******************
Mtrans-plan2 fired as plan
for #{OB.1711: (ACTIVE-GOAL obj (MTRANS actor ...)...)}
in Cx.5 sprouting Cx.6
----------------------------------------------------------------
```

```
IF    goal to MTRANS mental state to person
THEN subgoal to be VPROX that person
----------------------------------------------------------
*******************
Vprox-plan2 fired as plan
for #{OB.1716: (ACTIVE-GOAL obj (VPROX actor ...)...)}
in Cx.6 sprouting Cx.8
----------------------------------------------------------
IF    goal to be VPROX person
THEN subgoal to M-PHONE person
----------------------------------------------------------
| I have to call him.
*******************
M-phone-plan1 fired as plan
for #{OB.1729: (ACTIVE-GOAL obj (M-PHONE actor ...)...)}
in Cx.8 sprouting Cx.9
----------------------------------------------------------
IF    goal to M-PHONE person
THEN subgoal to KNOW TELNO of person
----------------------------------------------------------
| I have to know his telephone number.
*******************
Know-plan2 fired as plan
for #{OB.1735: (ACTIVE-GOAL obj (KNOW actor ...)...)}
in Cx.9 sprouting Cx.10
----------------------------------------------------------
IF    goal for person to KNOW TELNO of another
THEN subgoal for someone to MTRANS TELNO from to that person
----------------------------------------------------------
| He has to tell me his telephone number.
*******************
Mtrans-plan1 fired as plan
for #{OB.1741: (ACTIVE-GOAL obj (MTRANS actor ...)...)}
in Cx.10 sprouting Cx.11
----------------------------------------------------------
IF    goal for person2 to MTRANS info from object to person1
THEN subgoals for object to KNOW info and
     for person1 to be VPROX person2 and
     for person2 to be VPROX object and
     for person2 to BELIEVE that person1 has ACTIVE-GOAL
        to KNOW info and
     for person2 to have ACTIVE-GOAL for person1 to KNOW
        info
----------------------------------------------------------
...
Attempting to backtrack for top-level goal
#{OB.1662: (ACTIVE-GOAL obj (RECOVERY...)...)
Top-level goal #{OB.1662: (ACTIVE-GOAL obj (RECOVERY...)...)} fails
All possibilities failed for #{OB.1662: (ACTIVE-GOAL obj (RECOVERY...)...)
```

DAYDREAMER fails to generate a scenario of high realism in which she is able to get in touch with the movie star. Therefore, DAYDREAMER attempts to generate mutations of failed action goals which might enable her to get in touch with the star:

```
Action mutations for #{OB.1662: (ACTIVE-GOAL obj (RECOVERY...)...)}
Mutating action goal #{OB.1741: (ACTIVE-GOAL obj (MTRANS actor ...)...)}
| Suppose he tells someone else his telephone number.
Serendipity!! (daydreaming goal)
[OB.1989: (AG. (KNOW actor Me
      obj (TELNO actor Movie-star1))) EPISODE.7]
  [OB.2041: (AG. (MTRANS actor ?Person6 from ?Person6 to Me
        obj (TELNO actor Movie-star1))) EPISODE.8]
   [OB.2052: (AG. (KNOW actor ?Person6
      obj (TELNO actor Movie-star1))) EPISODE.9]
     [OB.2058: (AG. (MTRANS actor Movie-star1 from Movie-star1 to ?Person6
      obj (TELNO actor Movie-star1)))]
   [OB.2035: (AG. (BELIEVE actor ?Person6
       obj (ACTIVE-GOAL obj (KNOW actor Me
             obj (TELNO actor Movie-star1))))))]
   [OB.2028: (AG. (BELIEVE actor ?Person6
       obj (BELIEVE actor Me
                   obj (ACTIVE-GOAL obj (KNOW actor Me
                             obj (TELNO actor Movie-star1))))))))]
   [OB.2025: (AG. (VPROX actor Me ?Person6))]
   [OB.2022: (AG. (VPROX actor ?Person6 ?Person6))]
| What do you know!
Apply existing analogical plan
*****************
Know-plan3 fired as analogical plan
for #{OB.2063: (ACTIVE-GOAL obj (KNOW actor ...)...)}
in Cx.26 sprouting Cx.27
--------------------------------------------------------
IF   goal for person to KNOW TELNO of another
THEN subgoal for someone to MTRANS TELNO from to that person
--------------------------------------------------------
| This person has to tell me his telephone number.
```

The action "he tells me his telephone number" is mutated into "he tells
someone else his telephone number." The serendipity mechanism detects that
this mutated action is applicable to the goal of finding out the star's telephone
number: If the star tells someone else his number, this other person will know his
number, who can then tell it to DAYDREAMER. Next, the planning mechanism
fleshes out this possibility. In particular, a value for the variable which represents
the other person must be found. Initially, DAYDREAMER proceeds by analogy
to the episode produced as a suggestion by the serendipity mechanism:

```
Apply existing analogical plan
*****************
Mtrans-plan1 fired as analogical plan
for #{OB.2090: (ACTIVE-GOAL obj (MTRANS actor ...)...)}
in Cx.27 sprouting Cx.28
--------------------------------------------------------
IF   goal for person2 to MTRANS info from object to person1
THEN subgoals for object to KNOW info and
      for person1 to be VPROX person2 and
      for person2 to be VPROX object and
      for person2 to BELIEVE that person1 has ACTIVE-GOAL
```

```
            to KNOW info and
            for person2 to have ACTIVE-GOAL for person1 to KNOW
            info
------------------------------------------------------------
Apply existing analogical plan
*****************
Know-plan2 fired as analogical plan
for #{OB.2095: (ACTIVE-GOAL obj (KNOW actor ...)...)}
in Cx.28 sprouting Cx.29
------------------------------------------------------------
IF   goal for person to KNOW TELNO of another
THEN subgoal for someone to MTRANS TELNO from to that person
------------------------------------------------------------
| He has to tell this person his telephone number.
Non-empty bd (#{OB.1403: (KNOW actor Movie-star1 obj ...)} (PERSON6...))
*****************
Mtrans-plan1 fired as plan
for #{OB.2118: (ACTIVE-GOAL obj (MTRANS actor ...)...)}
in Cx.29 sprouting Cx.30
------------------------------------------------------------
IF   goal for person2 to MTRANS info from object to person1
THEN subgoals for object to KNOW info and
     for person1 to be VPROX person2 and
     for person2 to be VPROX object and
     for person2 to BELIEVE that person1 has ACTIVE-GOAL
     to KNOW info and
     for person2 to have ACTIVE-GOAL for person1 to KNOW
     info
------------------------------------------------------------
*****************
Jiffy-rule1 fired as plan
for #{OB.2187: (ACTIVE-GOAL obj (BELIEVE actor ...)...)}
in Cx.30 sprouting Cx.31
------------------------------------------------------------
IF   goal for person to have ACTIVE-GOAL to KNOW TELNO
     of another person
THEN subgoal for person to have ACTIVE-GOAL of LOVERS
     with person
------------------------------------------------------------
| He has to want to be going out with this person.
*****************
Lovers-theme-plan fired as plan
for #{OB.2298: (ACTIVE-GOAL obj (BELIEVE actor ...)...)}
in Cx.31 sprouting Cx.32
------------------------------------------------------------
IF   goal for self ACTIVE-GOAL of LOVERS with a person
THEN subgoals for ROMANTIC-INTEREST in person and
     not LOVERS with anyone
------------------------------------------------------------
| He has to be interested in this person.
| He has to believe that he is not going out with anyone.
*****************
Romantic-interest-plan fired as plan
for #{OB.2397: (ACTIVE-GOAL obj (BELIEVE actor ...)...)}
in Cx.32 sprouting Cx.33
```

```
-----------------------------------------------------------
IF   goal to have ROMANTIC-INTEREST in person
THEN subgoal to have POS-ATTITUDE toward person and
     person to be ATTRACTIVE
-----------------------------------------------------------
| He has to believe that this person is attractive.
```

In order for the star to tell someone else his number, he must have the goal for the other person to know his number. The following rule is employed here: One has the goal to tell someone else one's number if one has the goal to go out with that person. Other rules (such as having the goal to give someone one's number when one wants a job from that person) would lead to other daydreams. In order for the star to have the goal to go out with this unknown person, he must be interested in her. Thus he must have a positive attitude toward this person and she must be attractive. A suitable value for the variable representing this unknown person is therefore DAYDREAMER's rich and attractive friend Karen:

```
*****************
Belief-pers-attr3 fired as plan
for #{OB.2513: (ACTIVE-GOAL obj (BELIEVE actor ...)...)}
in Cx.33 sprouting Cx.34
Fact plan #{OB.1606: (ATTRACTIVE obj Karen)} found
*****************
Subgoals of #{OB.2640: (ACTIVE-GOAL obj (BELIEVE actor ...)...)} completed
*****************
| He believes that Karen is attractive.
...
Subgoals of #{OB.2742: (ACTIVE-GOAL obj (BELIEVE actor ...)...)} completed
*****************
| He is interested in her.
*****************
for #{OB.2645: (ACTIVE-GOAL obj (BELIEVE actor ...)...)}
in Cx.37 sprouting Cx.38
*****************
M-break-up-plan2 fired as plan
for #{OB.2842: (ACTIVE-GOAL obj (M-BREAK-UP...)...)}
in Cx.38 sprouting Cx.39
*****************
Subgoals of #{OB.2842: (ACTIVE-GOAL obj (M-BREAK-UP...)...)} completed
*****************
| He breaks up with his girlfriend.
*****************
Subgoals of #{OB.2869: (SUCCEEDED-GOAL obj (BELIEVE...)...)} completed
Subgoals of #{OB.2650: (ACTIVE-GOAL obj (BELIEVE actor ...)...)} completed
*****************
| He wants to be going out with her.
*****************
Jiffy-rule1 fired as inference in Cx.39
-----------------------------------------------------------
IF   person to have ACTIVE-GOAL of LOVERS with person
THEN person have ACTIVE-GOAL to KNOW TELNO of person
-----------------------------------------------------------
*****************
```

```
Activate top-level goal #{OB.2884: (ACTIVE-GOAL obj (KNOW actor ...)...)}
for #{MOVIE-STAR1: (MALE-ACTOR first-name "Harrison"...)
| He wants her to know his telephone number.
...
Subgoals of #{OB.2674: (ACTIVE-GOAL obj (MTRANS actor ...)...)} completed
*****************
| He tells her his telephone number.
*****************
Know-plan3 fired as inference in Cx.43
*****************
Believe-plan1 fired as inference in Cx.43
-----------------------------------------------------------
IF   MTRANS mental state to person
THEN person BELIEVE self mental state
-----------------------------------------------------------
*****************
Enable-future-vprox-plan1 fired as inference in Cx.43
-----------------------------------------------------------
IF   person1 KNOW TELNO of person2
THEN person1 and
     person2 ENABLE-FUTURE-VPROX
-----------------------------------------------------------
```

Karen asks the movie star for his telephone number, and he gives it to her.
Now DAYDREAMER must get the telephone number from Karen:

```
*****************
Subgoals of #{OB.3005: (SUCCEEDED-GOAL obj (KNOW actor ...)...)} completed
Subgoal relaxation, #{OB.2716: (ACTIVE-GOAL obj (BELIEVE actor ...)...)} succeeds
*****************
Believe-plan1 fired as plan
for #{OB.2722: (ACTIVE-GOAL obj (BELIEVE actor ...)...)}
in Cx.44 sprouting Cx.45
-----------------------------------------------------------
IF   goal for person to BELIEVE self mental state
THEN subgoal to MTRANS mental state to person
-----------------------------------------------------------
| I have to tell her that I want to know his telephone number.
*****************
Mtrans-plan2 fired as plan
for #{OB.3022: (ACTIVE-GOAL obj (MTRANS actor ...)...)}
in Cx.45 sprouting Cx.46
-----------------------------------------------------------
IF   goal to MTRANS mental state to person
THEN subgoal to be VPROX that person
-----------------------------------------------------------
*****************
Vprox-plan2 fired as plan
for #{OB.3027: (ACTIVE-GOAL obj (VPROX actor ...)...)}
in Cx.46 sprouting Cx.48
-----------------------------------------------------------
IF   goal to be VPROX person
THEN subgoal to M-PHONE person
-----------------------------------------------------------
*****************
```

```
M-phone-plan1 fired as plan
for #{OB.3040: (ACTIVE-GOAL obj (M-PHONE actor ...)...)}
in Cx.48 sprouting Cx.49
-------------------------------------------------------
IF   goal to M-PHONE person
THEN subgoal to KNOW TELNO of person
-------------------------------------------------------
*****************
Subgoals of #{OB.3040: (ACTIVE-GOAL obj (M-PHONE actor ...)...)} completed
*****************
| I call her.
*****************
Vprox-plan2 fired as inference in Cx.49
-------------------------------------------------------
IF   M-PHONE person
THEN VPROX person
-------------------------------------------------------
*****************
Subgoals of #{OB.3072: (SUCCEEDED-GOAL obj (VPROX...)...)} completed
*****************
Subgoals of #{OB.3022: (ACTIVE-GOAL obj (MTRANS actor ...)...)} completed
*****************
| I tell her that I want to know his telephone number.
*****************
Believe-plan1 fired as inference in Cx.49
-------------------------------------------------------
IF   MTRANS mental state to person
THEN person BELIEVE self mental state
-------------------------------------------------------
*****************
Subgoals of #{OB.3104: (SUCCEEDED-GOAL obj (BELIEVE...)...)} completed
*****************
Vprox-reflexivity fired as plan
for #{OB.2732: (ACTIVE-GOAL obj (VPROX actor ...)...)}
in Cx.49 sprouting Cx.50
*****************
Subgoals of #{OB.2732: (ACTIVE-GOAL obj (VPROX actor ...)...)} completed
*****************
Subgoals of #{OB.2702: (ACTIVE-GOAL obj (MTRANS actor ...)...)} completed
*****************
| She tells me his telephone number.
*****************
Know-plan3 fired as inference in Cx.50
*****************
Believe-plan1 fired as inference in Cx.50
-------------------------------------------------------
IF   MTRANS mental state to person
THEN person BELIEVE self mental state
-------------------------------------------------------
*****************
Enable-future-vprox-plan1 fired as inference in Cx.50
-------------------------------------------------------
IF   person1 KNOW TELNO of person2
THEN person1 and
     person2 ENABLE-FUTURE-VPROX
```

```
-----------------------------------------------------------
*****************
Mtrans-acceptable-inf3 fired as inference in Cx.50
-----------------------------------------------------------
IF    other MTRANS anything to self
THEN MTRANS-ACCEPTABLE between self and
     other
-----------------------------------------------------------
```

DAYDREAMER calls Karen and gets the star's number. Now she can call him and ask him out:

```
*****************
Subgoals of #{OB.3166: (SUCCEEDED-GOAL obj (KNOW actor ...)...)} completed
Subgoals of #{OB.2693: (ACTIVE-GOAL obj (M-PHONE actor ...)...)} completed
*****************
| I call him.
*****************
Vprox-plan2 fired as inference in Cx.50
-----------------------------------------------------------
IF    M-PHONE person
THEN VPROX person
-----------------------------------------------------------
*****************
Subgoals of #{OB.3191: (SUCCEEDED-GOAL obj (VPROX...)...)} completed
Subgoals of #{OB.2683: (ACTIVE-GOAL obj (MTRANS actor ...)...)} completed
*****************
| I ask him out.
*****************
Believe-plan1 fired as inference in Cx.50
-----------------------------------------------------------
IF    MTRANS mental state to person
THEN person BELIEVE self mental state
-----------------------------------------------------------
Subgoals of #{OB.1662: (ACTIVE-GOAL obj (RECOVERY...)...)} completed
*****************
Terminating planning for top-level goal
Leaf context #{CX.50: (CX)}
[OB.1662:
 (SG. (RECOVERY obj (FAILED-GOAL obj (LOVERS actor Me Movie-star1))))]
  [OB.3196: (SG. (MTRANS actor Me from Me to Movie-star1
           obj (ACTIVE-GOAL obj (LOVERS actor Me Movie-star1))))]
    [OB.3191: (SG. (VPROX actor Movie-star1 Me))]
      [OB.3172: (SG. (M-PHONE actor Me to Movie-star1))]
        [OB.3166: (SG. (KNOW actor Me
         obj (TELNO actor Movie-star1)))]
          [OB.3125: (SG. (MTRANS actor Karen from Karen to Me
           obj (TELNO actor Movie-star1)))]
            [OB.3005: (SG. (KNOW actor Karen
           obj (TELNO actor Movie-star1)))]
              [OB.2969:
           (SG. (MTRANS actor Movie-star1 from Movie-star1 to Karen
           obj (TELNO actor Movie-star1)))]
                [OB.2657: (SG. (KNOW actor Movie-star1
           obj (TELNO actor Movie-star1)))]
```

```
        [OB.2900: (SG. (BELIEVE actor Movie-star1
obj (ACTIVE-GOAL obj (KNOW actor Karen
                        obj (TELNO actor Movie-star1)))))]
        [OB.2876: (SG. (BELIEVE actor Movie-star1
 obj (ACTIVE-GOAL obj (LOVERS actor Movie-star1 Karen))))]
        [OB.2831: (SG. (BELIEVE actor Movie-star1
obj (ROMANTIC-INTEREST obj Karen)))]
            [OB.2759: (SG. (BELIEVE actor Movie-star1
obj (ATTRACTIVE obj Karen)))]
              [OB.2755: (SG. (ATTRACTIVE obj Karen))]
            [OB.2825: (SG. (BELIEVE actor Movie-star1
obj (POS-ATTITUDE obj Karen)))]
              [OB.2806: (SG. (BELIEVE actor Movie-star1
obj (RICH actor Movie-star1)))]
              [OB.2820: (SG. (BELIEVE actor Movie-star1
obj (RICH actor Karen)))]
                [OB.2816: (SG. (RICH actor Karen))]
          [OB.2869: (SG. (BELIEVE actor Movie-star1
obj (NOT obj (LOVERS actor Movie-star1))))]
            [OB.2852: (SG. (M-BREAK-UP actor Movie-star1))]
            [OB.2849: (SG. (RTRUE))]
      [OB.2948: (SG. (BELIEVE actor Movie-star1
obj (BELIEVE actor Karen
        obj (ACTIVE-GOAL obj (KNOW actor Karen
                        obj (TELNO actor Movie-star1))))))]
        [OB.2927: (SG. (MTRANS actor Karen from Karen to Movie-star1
obj (ACTIVE-GOAL obj (KNOW actor Karen
                        obj (TELNO actor Movie-star1)))))]
          [OB.2920: (SG. (VPROX actor Movie-star1 Karen))]
          [OB.2923: (SG. (VPROX actor Karen Movie-star1))]
          [OB.2965: (SG. (VPROX actor Movie-star1 Movie-star1))]
            [OB.2962: (SG. (RTRUE))]
      [OB.3013: (SG. (BELIEVE actor Karen
obj (ACTIVE-GOAL obj (KNOW actor Me
                        obj (TELNO actor Movie-star1)))))]
      [OB.3104: (SG. (BELIEVE actor Karen
obj (BELIEVE actor Me
        obj (ACTIVE-GOAL obj (KNOW actor Me
                        obj (TELNO actor Movie-star1))))))]
        [OB.3081: (SG. (MTRANS actor Me from Me to Karen
obj (ACTIVE-GOAL obj (KNOW actor Me
                        obj (TELNO actor Movie-star1)))))]
          [OB.3072: (SG. (VPROX actor Karen Me))]
          [OB.3053: (SG. (M-PHONE actor Me to Karen))]
          [OB.3048: (SG. (KNOW actor Me
obj (TELNO actor Karen)))]
          [OB.3077: (SG. (VPROX actor Me Karen))]
          [OB.3121: (SG. (VPROX actor Karen Karen))]
            [OB.3118: (SG. (RTRUE))]


...
Removing motivating emotions
Store episode #{OB.1662: (SUCCEEDED-GOAL obj (RECOVERY...)...)
```

A.8   RECOVERY3 Daydream

```
******************
Entertainment-theme fired as inference in Cx.4
-----------------------------------------------------------
IF    ENTERTAINMENT need below threshold
THEN activate goal for ENTERTAINMENT
-----------------------------------------------------------

******************
Activate top-level goal #{OB.1835: (ACTIVE-GOAL obj (ENTERTAINMENT...)...)
| I want to be entertained.
| I feel interested in being entertained.
Switching to new top-level goal #{OB.1835: (ACTIVE-GOAL obj (ENTERTAINMENT))}
******************
Entertainment-plan2 fired as plan
for #{OB.1835: (ACTIVE-GOAL obj (ENTERTAINMENT...)...)}
in Cx.4 sprouting Cx.7
-----------------------------------------------------------
IF    goal for ENTERTAINMENT
THEN subgoal to MTRANS mail
-----------------------------------------------------------
```

The level of satisfaction of DAYDREAMER's need for entertainment has
again fallen below the threshold, and so an **ENTERTAINMENT** goal is activated. This time, DAYDREAMER selects the plan of getting and reading the
mail.

```
******************
Poss-plan2 fired as plan
for #{OB.1857: (ACTIVE-GOAL obj (POSS actor ...)...)}
in Cx.7 sprouting Cx.8
-----------------------------------------------------------
IF    goal for person to POSS object
THEN subgoal to GRAB object
-----------------------------------------------------------

******************
Grab-plan fired as plan
for #{OB.1880: (ACTIVE-GOAL obj (GRAB actor ...)...)}
in Cx.8 sprouting Cx.9
-----------------------------------------------------------
IF    goal to GRAB object
THEN subgoal to be AT location of object
-----------------------------------------------------------
Fact plan #{OB.528: (AT actor Mail1 obj Outside)} found
******************
Goal #{OB.1902: (ACTIVE-GOAL obj (AT actor ...)  top-level-goal ...)} succeeds
******************
At-plan2 fired as plan
for #{OB.1909: (ACTIVE-GOAL obj (AT actor ...)  top-level-goal ...)}
in Cx.10 sprouting Cx.11
-----------------------------------------------------------
IF    goal to be AT a location
THEN subgoal to PTRANS to that location
-----------------------------------------------------------
```

```
******************
Ptrans-plan fired as plan
for #{OB.1938: (ACTIVE-GOAL obj (PTRANS actor ...)...)}
in Cx.11 sprouting Cx.12
----------------------------------------------------------
IF   goal for person to PTRANS to a location
THEN subgoal for person to KNOW that location
----------------------------------------------------------
******************
Goal #{OB.1943: (ACTIVE-GOAL obj (KNOW actor ...)...)} succeeds
Subgoals of #{OB.1938: (ACTIVE-GOAL obj (PTRANS actor ...)...)} completed
Current top-level goal waits #{OB.1835: (ACTIVE-GOAL obj (ENTERTAINMENT...)...)}
No more goals to run; switching to performance mode
Waking up top-level goal #{OB.1835: (ACTIVE-GOAL obj (ENTERTAINMENT...)...)}
State changes from DAYDREAMING to PERFORMANCE
Subgoals of #{OB.1938: (ACTIVE-GOAL obj (PTRANS actor ...)...)} completed
Perform external action
******************
Goal #{OB.1938: (ACTIVE-GOAL obj (PTRANS actor ...)...)} succeeds
| I go outside.
******************
Believe-believe-inf fired as inference in Cx.12
******************
At-plan2 fired as inference in Cx.12
----------------------------------------------------------
IF   person PTRANS from one location to another
THEN person AT new location and
     no longer AT old location
----------------------------------------------------------
Subgoals of #{OB.1909: (ACTIVE-GOAL obj (AT actor ...)  top-level-goal ...)}
completed
******************
Goal #{OB.1909: (ACTIVE-GOAL obj (AT actor ...)  top-level-goal ...)} succeeds
Subgoals of #{OB.1912: (ACTIVE-GOAL obj (GRAB actor ...)...)} completed
Perform external action
******************
Goal #{OB.1912: (ACTIVE-GOAL obj (GRAB actor ...)...)} succeeds
| I grab the mail.
******************
Poss-plan2 fired as inference in Cx.12
******************
Vprox-plan3 fired as inference in Cx.12
----------------------------------------------------------
IF   POSS and object
THEN VPROX that object
----------------------------------------------------------
******************
Poss-mail-inf fired as inference in Cx.12
| I have the UCLA Alumni directory.
******************
Vprox-plan3 fired as inference in Cx.12
----------------------------------------------------------
IF   POSS and
     object
THEN VPROX that object
```

```
----------------------------------------------------------
| Input: Carol Burnett went to UCLA.
| Input: Carol's telephone number is in the UCLA Alumni directory.
Adding rule #{INDUCED-RULE.1: (RULE subgoal (COLLEGE actor ...)...)}
Serendipity!! (daydreaming goal)
[OB.2035: (AG. (MTRANS actor Me from Me to Movie-star1
          obj (ACTIVE-GOAL obj (LOVERS actor Me Movie-star1)))) EPISODE.4]
  [OB.2124: (AG. (VPROX actor Movie-star1 Me)) EPISODE.5]
    [OB.2128: (AG. (M-PHONE actor Me to Movie-star1)) EPISODE.6]
      [OB.2132: (AG. (KNOW actor Me
      obj (TELNO actor Movie-star1))) EPISODE.7]
        [OB.2114: (AG. (MTRANS actor Me
        from (ALUMNI-DIR obj ?Var.1731:ORGANIZATION)
        to Me
        obj (TELNO actor Movie-star1))) EPISODE.8]
          [OB.2140: (AG. (KNOW actor (ALUMNI-DIR obj ?Var.1731:ORGANIZATION)
          obj (TELNO actor Movie-star1))) EPISODE.9]
        [OB.2146: (AG. (COLLEGE actor Movie-star1 obj ?Var.1731:ORGANIZATION))]
          [OB.2111: (AG. (VPROX actor Me Me))]
          [OB.2107: (AG. (VPROX actor Me
            (ALUMNI-DIR obj ?Var.1731:ORGANIZATION)))]
          [OB.2100: (AG. (BELIEVE actor Me
          obj (BELIEVE actor Me
                      obj (ACTIVE-GOAL obj (KNOW actor Me
                                   obj (TELNO actor Movie-star1))))))]
          [OB.2094: (AG. (BELIEVE actor Me
          obj (ACTIVE-GOAL obj (KNOW actor Me
                           obj (TELNO actor Movie-star1)))))]
| What do you know!
```

DAYDREAMER receives an Alumni directory from her college, UCLA, in the mail. The program is then provided with world states as input: The directory contains the telephone number of Carol Burnett, who happens also to have gone to UCLA. A new rule is induced from this input: An Alumni directory for a given college contains the telephone number of a person if that person went to that college. The serendipity mechanism recognizes that this new rule is applicable to the active goal of getting in touch with the movie star. In particular, DAYDREAMER needs to obtain the Alumni directory of the college the movie star attended. However, since DAYDREAMER is still in performance mode, it must first complete the processing of its **ENTERTAINMENT** goal:

```
Subgoals of #{OB.1924: (ACTIVE-GOAL obj (POSS actor ...)...)} completed
*****************
Goal #{OB.1924: (ACTIVE-GOAL obj (POSS actor ...)...)} succeeds
Fact plan #{OB.2021: (KNOW actor Alumni-dir1 obj ...)} found
*****************
Goal #{OB.1921: (ACTIVE-GOAL obj (KNOW actor ...)...)} succeeds
*****************
Mtrans-plan1 fired as plan
for #{OB.2471: (ACTIVE-GOAL obj (MTRANS actor ...)...)}
in Cx.17 sprouting Cx.18
----------------------------------------------------------
IF    goal for person2 to MTRANS info from object to person1
```

```
THEN subgoals for object to KNOW info and
     for person1 to be VPROX person2 and
     for person2 to be VPROX object and
     for person2 to BELIEVE that person1 has ACTIVE-GOAL
     to KNOW info and
     for person2 to have ACTIVE-GOAL for person1 to KNOW
     info
-----------------------------------------------------------
*****************
Goal #{OB.2492: (ACTIVE-GOAL obj (VPROX actor ...)...)} succeeds
...
Perform external action
*****************
Goal #{OB.2650: (ACTIVE-GOAL obj (MTRANS actor ...)...)} succeeds
| I read her telephone number in the UCLA Alumni directory.
*****************
Know-plan fired as inference in Cx.23
-----------------------------------------------------------
IF   someone MTRANS info from some object to person
THEN person KNOW info
-----------------------------------------------------------
*****************
Enable-future-vprox-plan1 fired as inference in Cx.23
-----------------------------------------------------------
IF   person1 KNOW TELNO of person2
THEN person1 and
     person2 ENABLE-FUTURE-VPROX
-----------------------------------------------------------
*****************
Entertainment-inf2 fired as inference in Cx.23
Subgoals of #{OB.1835: (ACTIVE-GOAL obj (ENTERTAINMENT...)...)} completed
*****************
Goal #{OB.1835: (ACTIVE-GOAL obj (ENTERTAINMENT...)...)} succeeds
| I succeed at being entertained.
Terminating planning for top-level goal
#{OB.1835: (SUCCEEDED-GOAL obj (ENTERTAINMENT...)...)}
Removing motivating emotions.
Emotional responses.
| I feel amused.
Store episode #{OB.1835: (SUCCEEDED-GOAL obj (ENTERTAINMENT...)...)
No more goals to run; switching to daydreaming mode
State changes from PERFORMANCE to DAYDREAMING
```

Now that DAYDREAMER has completed its **ENTERTAINMENT** goal, focus shifts to **RECOVERY** and getting in touch with the star:

```
Switching to new top-level goal #{OB.1810: (ACTIVE-GOAL obj (RECOVERY...)...)}
Apply existing analogical plan
*****************
Mtrans-plan2 fired as analogical plan
for #{OB.2150: (ACTIVE-GOAL obj (MTRANS actor ...)...)}
in Cx.14 sprouting Cx.25
-----------------------------------------------------------
IF   goal to MTRANS mental state to person
THEN subgoal to be VPROX that person
```

```
----------------------------------------------------------
Apply existing analogical plan
******************
Vprox-plan2 fired as analogical plan
for #{OB.2742: (ACTIVE-GOAL obj (VPROX actor ...)...)}
in Cx.25 sprouting Cx.26
----------------------------------------------------------
IF    goal to be VPROX person
THEN subgoal to M-PHONE person
----------------------------------------------------------
Apply existing analogical plan
******************
M-phone-plan1 fired as analogical plan
for #{OB.2747: (ACTIVE-GOAL obj (M-PHONE actor ...)...)}
in Cx.26 sprouting Cx.27
----------------------------------------------------------
IF    goal to M-PHONE person
THEN subgoal to KNOW TELNO of person
----------------------------------------------------------
Apply existing analogical plan
******************
Know-plan fired as analogical plan
for #{OB.2753: (ACTIVE-GOAL obj (KNOW actor ...)...)}
in Cx.27 sprouting Cx.28
----------------------------------------------------------
IF    goal for person to KNOW info
THEN subgoal for someone to MTRANS info from some object
     to that person
----------------------------------------------------------
| I have to read Harrison Ford's telephone number in the Alumni directory.
Apply existing analogical plan
******************
Mtrans-plan1 fired as analogical plan
for #{OB.2759: (ACTIVE-GOAL obj (MTRANS actor ...)...)}
in Cx.28 sprouting Cx.29
----------------------------------------------------------
IF    goal for person2 to MTRANS info from object to person1
THEN subgoals for object to KNOW info and
     for person1 to be VPROX person2 and
     for person2 to be VPROX object and
     for person2 to BELIEVE that person1 has ACTIVE-GOAL
     to KNOW info and
     for person2 to have ACTIVE-GOAL for person1 to KNOW
     info
----------------------------------------------------------
Apply existing analogical plan
******************
Induced-rule.1 fired as analogical plan
for #{OB.2765: (ACTIVE-GOAL obj (KNOW actor ...)...)}
in Cx.29 sprouting Cx.30
******************
Inverse-college-plan fired as plan
for #{OB.2788: (ACTIVE-GOAL obj (COLLEGE actor ...)...)}
in Cx.30 sprouting Cx.31
| I have to know where he went to college.
```

In order for DAYDREAMER to find out the star's telephone number, she must read his telephone number in the Alumni directory of the college he attended. Therefore, she must first know what college he attended (if any).

```
Episodic reminding of #{EPISODE.3: (EPISODE rule Know-plan goal ...)}
| I remember the time I knew where Brooke Shields went to
| college by reading that she went to Princeton University in
| People magazine.
```

DAYDREAMER recalls an episode in which she was able to find out what college Brooke Shields was attending. This plan is then applied by analogy:

```
******************
Know-plan fired as analogical plan
for #{OB.2855: (ACTIVE-GOAL obj (KNOW actor ...)...)}
in Cx.31 sprouting Cx.32
----------------------------------------------------------
IF   goal for person to KNOW info
THEN subgoal for someone to MTRANS info from some object
     to that person
----------------------------------------------------------
| I have to read where he went to college in People magazine.
...
Store episode #{OB.1810: (SUCCEEDED-GOAL obj (RECOVERY...)...)...)
```

This episode is then stored for possible future use in achieving a **LOVERS** goal with the movie star (or other movie stars).

A.9   REVENGE3 Daydream

DAYDREAMER goes to Westward Ho to buy a newspaper, learns of an opening at the Broadway, interviews for the job and is hired. After all personal goal concerns have completed, DAYDREAMER switches to daydreaming mode and generates a **REVENGE** daydream which was pending:

```
Switching to new top-level goal
#{OB.1749: (ACTIVE-GOAL obj (REVENGE actor ...)...)}
Emotion #{OB.1768: (POS-EMOTION strength 0.14273477802205323...)} below threshold.
Episodic reminding.
| I remember the time I got even with Harrison Ford for turning me down
| by studying to be an actor, being a star even more famous than he is,
| him calling me up, him asking me out, and me turning him down.
******************
Revenge-plan1 fired as analogical plan
for #{OB.1749: (ACTIVE-GOAL obj (REVENGE actor ...)...)}
in Cx.10 sprouting Cx.41
----------------------------------------------------------
IF   goal to gain REVENGE against person for causing self
     a failed POS-RELATIONSHIP goal
THEN subgoal for person to have failure of same POS-RELATIONSHIP
----------------------------------------------------------
```

DAYDREAMER recalls its previous revenge daydream in response to
LOVERS1. A single rule from this daydream (*Revenge-plan1*) is applied to
the current situation. Other rules of the old daydream are not applicable to the
EMPLOYMENT situation; thus the remainder of the daydream is generated
through regular planning:

```
******************
Employment-theme-plan fired as plan
for #{OB.2925: (ACTIVE-GOAL obj (BELIEVE actor ...)...)}
in Cx.41 sprouting Cx.43
******************
Pos-att-employ-plan1 fired as plan
for #{OB.2952: (ACTIVE-GOAL obj (BELIEVE actor ...)...)}
in Cx.43 sprouting Cx.44
Subgoal relaxation, #{OB.2958: (ACTIVE-GOAL obj (BELIEVE actor))} succeeds
| Say I am a powerful executive.
******************
Goal #{OB.2958: (ACTIVE-GOAL obj (BELIEVE actor ...)...)} succeeds
Subgoals of #{OB.2952: (ACTIVE-GOAL obj (BELIEVE actor ...)...)} completed
******************
Goal #{OB.2952: (ACTIVE-GOAL obj (BELIEVE actor ...)...)} succeeds
Subgoals of #{OB.2925: (ACTIVE-GOAL obj (BELIEVE actor ...)...)} completed
******************
Goal #{OB.2925: (ACTIVE-GOAL obj (BELIEVE actor ...)...)} succeeds
******************
Mtrans-plan2 fired as analogical plan
for #{OB.2930: (ACTIVE-GOAL obj (MTRANS actor ...)...)}
in Cx.45 sprouting Cx.46
---------------------------------------------------------
IF   goal to MTRANS mental state to person
THEN subgoal to be VPROX that person
---------------------------------------------------------
******************
Goal #{OB.2987: (ACTIVE-GOAL obj (VPROX actor ...)...)} succeeds
Subgoals of #{OB.2930: (ACTIVE-GOAL obj (MTRANS actor ...)...)} completed
******************
Goal #{OB.2930: (ACTIVE-GOAL obj (MTRANS actor ...)...)} succeeds
| Agatha offers me a job.
...
```

A.10   COMPUTER-SERENDIPITY

```
| Input: Computer.
Serendipity!! (personal goal)
| What do you know!
Episodic reminding of #{EPISODE.8}
| I remember the time Harold and I broke the ice by
| me being a member of the computer dating service, and by him
| being a member of the computer dating service. I knew his
| telephone number by the dating service employee
| telling me Harold's telephone number.
```

DAYDREAMER has an active goal to initiate a **LOVERS** relationship with someone when she happens to stare at an object in the room: a computer. The serendipity mechanism, using this physical object as one index and the active **LOVERS** goal as another, retrieves an episode which contains a plan for achieving the **LOVERS** goal. In general, the episode which is retrieved need not be directly related to, or indexed under, the **LOVERS** goal; rather, it is sufficient for the episode to contain a rule which is applicable to *any potential subgoal* of the **LOVERS** goal. An intersection search in the space of rule connections is employed in order to recognize such a relationship, and unification is employed to verify applicability of a path to the current concrete goal. The episode produced by the serendipity mechanism is then applied through analogical planning:

```
Apply existing analogical plan
*****************
Lovers-plan fired as analogical plan
for #{OB.2038}
in Cx.25 sprouting Cx.26
-----------------------------------------------------------
IF   goal for LOVERS with a person
THEN subgoals for ACQUAINTED with person and
     ROMANTIC-INTEREST in person and
     person have ACTIVE-GOAL of LOVERS with self and
     self and
     person M-DATE self and
     person M-AGREE to LOVERS
-----------------------------------------------------------
Apply existing analogical plan
*****************
Acquainted-plan fired as analogical plan
for #{OB.2339}
in Cx.26 sprouting Cx.27
-----------------------------------------------------------
IF   goal to be ACQUAINTED with person
THEN subgoal for M-CONVERSATION with person
-----------------------------------------------------------
Apply existing analogical plan
*****************
M-conversation-plan fired as analogical plan
for #{OB.2367}
in Cx.27 sprouting Cx.28
-----------------------------------------------------------
IF   goal for M-CONVERSATION between person1 and
     person2
THEN subgoals for MTRANS-ACCEPTABLE between person1
     and
     person2 and
     person1 to MTRANS to person2 something and
     person2 to MTRANS to person1 something
-----------------------------------------------------------
Apply existing analogical plan
*****************
Induced-rule.2 fired as analogical plan
for #{OB.2371}
```

```
in Cx.28 sprouting Cx.29
| I have to be a member of the dating service.
| He has to be a member of the dating service.
******************
Induced-rule.1 fired as analogical plan
for #{OB.2382}
in Cx.29 sprouting Cx.30
| I have to pay the dating service employee.
...
```

Appendix B

English Generation for Daydreaming

This appendix describes the English generator employed in DAYDREAMER in detail.

B.1 The Generator

Whenever a concept is asserted into a context, and in certain other situations, it is eligible for conversion into English by the *generator*. The generator takes the following arguments: (a) con: the concept to be generated, (b) bp: the current belief path, (c) switches: various specifications such as the tense, case, and so on, of the phrase to be generated, and (d) context: the context in which to generate the concept.

B.1.1 The Form of Generational Knowledge

Associated with each type of concept are *generational templates* which specify how to produce an English phrase for that concept in different situations. Generational templates consist of: (a) an optional subject concept, (b) an optional verb, (c) literal text, and (d) other concepts. Subjects are generated according to a special procedure for subject generation which in turn recursively invokes the generator. Verbs are generated by a procedure which contains fairly extensive morphological knowledge. Literal text is produced as English output exactly as it appears in the template. Other concepts are generated through recursive invocation of the generator. Modifications to the switches may be specified following a concept: this enables specification of a different tense or case.

Here is a sample generational template:

actor(con) **want** obj(con)+INF

This template specifies to: (a) invoke the special subject generation procedure on the actor slot of the concept, (b) generate an appropriate form of the verb **want**, (c) recursively invoke the generator on the value of the obj slot of the concept, and employ an INF (infinitive) in generating that concept. Given the concept

(**WANT** actor *John*
        obj (**PTRANS** actor *John*
                to *Movies*))

the template might produce:

John would have wanted to go to the movies.

The template for obj(con) might have been:

actor(con) **go** to to(con)

Notice, however, that the subject in this template was not generated. That is, the generator did not produce:

John would have wanted John to go to the movies.

This is because the special procedure for subject generation does not generate a subject if it is the same as the subject of the phrase in which it is embedded.

If there is no generational template associated with the type of a concept to be generated, the parent of this type is consulted, the parent of the parent of this type is consulted, and so on, until a template is found.

B.1.2   Subject Generation

The method for generating a subject s' of a phrase S' embedded in a phrase S is as follows: If s' is the same as the subject s of S and +INF (infinitive) or +GER (gerund) have been specified for s', do not generate s' at all. Otherwise, if +INF or +GER have been specified, s' is generated in the possessive or objective case, depending on whether +POSS-SUBJ (possessive subject) has been specified. Otherwise, if +REFL (reflexive) has been specified, s' is generated in the reflexive case. Otherwise, s' is generated in the nominative case.

B.2   Generational Templates

This section presents the generational templates for conversion of concepts into English in DAYDREAMER.

B.2.1 Mental Transfer

Actions of type **MTRANS** are generated different ways, depending on the values of the various slots. A variety of phrases are employed when one person **MTRANS**es to another that the former person has or does not have a goal to be in a positive relationship (such as **LOVERS** and **EMPLOYMENT**) with the latter person.

Such an **MTRANS** is generated according to the following template:

?Person1 verb ?Person2 [text]

and according to the following table:

relationship	verb	text
pos **LOVERS**	**ask**	out
neg **LOVERS**	**turn**	down
neg existing **LOVERS**	**dump**	
pos **EMPLOYMENT**	**offer**	a job
neg **EMPLOYMENT**	**pass**	up
neg existing **EMPLOYMENT**	**fire**	

Other **MTRANS**es are generated as follows: If obj(con) is an **INTRODUCTION**, **MTRANS** is generated as:

actor(con) **introduce** actor(con)+REFL to to(con)

Otherwise, if obj(con) is a **MOVIE**, **MTRANS** is generated as:

actor(con) **watch** obj(con) [at from(con)]

Otherwise, if the actor and from of *con* are the same, **MTRANS** is generated as:

actor(con) **tell** to(con) obj(con)

Otherwise, if actor(con) and to(con) are the same and from(con) is not a **PERSON**, **MTRANS** is generated as:

actor(con) **read** obj(con) in from(con)

(If obj(con) is a **TELNO**, **look up** is used instead of **read**.)

B.2.2 Other Actions and States

A number of actions and states concepts are generated according to the following template:

actors(con) *verb text [slot(con)]*

These concepts are specified by the following tables:

type	verb	text	slot
ACQUAINTED	be	acquainted	
AT	be	at	obj
ATTRACTIVE	be	cute	
COLLEGE	go	to school at	obj
EXECUTIVE	be	a powerful executive	
HURT	be	injured	
INSURED	be	insured	

M-AGREE	agree		obj+INF
M-BEAT-UP	beat	up	obj
M-BREAK-UP	break	up with his boyfriend	
M-DATE	go	out on a date	
M-KISS	kiss		
M-LOGIN	log	into	obj
M-MOVIE	go	see a movie	
M-PHONE	call		to
M-PUTON	put	on	obj
M-STUDY	study	to be	obj
M-WORK	work		obj

POSS	have		obj
PTRANS	go	to	to
PTRANS1	go	to	to
RICH	be	rich	
RPROX	be	in	location
SELLS	sell		obj
STAR	be	a movie star	
UNDER-DOORWAY	be	under a doorway	
WELL-DRESSED	be	dressed to kill	

An **M-PURCHASE** is generated as:

actors(con) **buy** obj(con) from from(con)

Several concepts are generated according to the following template:

head(bp) *verb text1 others* [*text2*] [*slot*(con)]

where *others* is actors(con) without head(bp) (or anyone [negative verb] or someone [positive verb] if *others* would be empty). These concepts are specified by the following table:[1]

---

[1] +PROGR refers to progressive.

type	verb	text1	text2	slot
EMPLOYMENT	have	a job with	at	org.
LOVERS	go+PROGR	out with		
M-CONVERSATION	have	a conv. with		
M-RESTAURANT	have	dinner with	at a rest.	
MTRANS-ACCEPTABLE	break	the ice with		

An exception is that if the concept is a **VPROX** and any element of actors(con) is not a **PERSON** and head(bp) is an element of actors(con), then the **VPROX** is generated as:

head(bp) **be near** *others*

where *others* is actors(con) without head(bp).

An **ATRANS** is generated as:

actors(con) **pay to**(con)

if obj(con) is **CASH**, otherwise it is generated as:

actors(con) **give** obj(con) **to** to(con)

An **EARTHQUAKE** is generated as:

obj(con) **has an earthquake**

An **EARTHQUAKE-ONSET** is generated as:

an earthquake **start** in obj(con)

A **KNOW** is generated as follows: If actor(con) is not a **PERSON**, then it is generated as:

obj(con) **be in** actor(con)

otherwise, it is generated as:

actor(con) **know** [that] obj(con)

where that is not generated if obj(con) is a **OBJECT, LOCATION,TIME-OF-DAY, VPROX,** or **TELNO**.

Several states are generated using the template:[2]

slot(con)+POSS *text*

according to the following table:

concept	text	slot
TELNO	telephone number	actor
MOVIES	movies	obj
CLOTHES	clothes	obj

---

[2] +POSS refers to possessive.

A daydreaming goal objective is generated according to the following template:

head(bp) *verb* [*text*] [*slot*(con)]

and the following table:

dd goal	verb	text	slot
RATIONALIZATION	**rationalize**		obj+GER
ROVING	**think**	about something else	
REVENGE	**get**	even with	to
REVERSAL	**reverse**		obj+GER
RECOVERY	**recover**	from	obj+GER
REHEARSAL	**rehearse**		obj+GER
REPERCUSSIONS	**consider**		obj+GER

Some personal goal objectives are generated as:

head(bp) *verb* *text*

according to the following table:

type	verb	text
ENTERTAINMENT	**be**	entertained
MONEY	**have**	enough money

A **POSSESSIONS** is generated as:

head(bp) **keep** head(bp)+POSS belongings

B.2.3   Goals

A **SUCCEEDED-GOAL** is generated as:

head(bp) **succeed** at obj(con)+GER

A **FAILED-GOAL** is generated differently, depending on obj(con), the objective of the goal. If the objective is an existing **LOVERS** relationship, the failed goal is generated as:

head(bp) **lose** (actors(obj(con)) - head(bp))

Otherwise, if the objective is an existing **EMPLOYMENT**, the failed goal is generated as:

head(bp) **lose** head(bp)+POSS job

Otherwise, it is generated as:

head(bp) **fail** at obj(con)+GER

An **ACTIVE-GOAL** is generated as follows: If obj(con) is an **EMPLOYMENT** and head(bp) is an **ACTOR**, then it is generated as:

head(bp) **need work**

Otherwise, if obj(con) is **EMPLOYMENT**, it is generated as:

head(bp) *verb* **a job** [**with** *other*]

where (a) *verb* is **want** if it is a top-level goal and **need** if it is a subgoal, and (b) *other* is actors(obj(con)) without head(bp), if any. Otherwise, if the concept is a top-level goal, it is generated as:

head(bp) **want** obj(con)+INF

Otherwise, if the concept is a subgoal, it is generated as:

obj(con)+HAVETO

In each case above, **would** like is substituted for **want** if the concept is embedded within an **MTRANS**. Thus, depending on whether a goal is a subgoal, a top-level goal, or embedded within an **MTRANS**, respectively, it is generated as:

I have to go to the store.
I want to go to the store.
I would like to go to the store.

B.2.4 Emotions

In DAYDREAMER, the generation of an emotion depends on: (a) whether it is a positive, negative, or surprise emotion, (b) whether it is directed toward someone, (c) whether it resulted from an alternative past scenario, and (d) the goals to which it is connected.

With the exception of **SURPRISE** emotions, which are simply generated as What do you know!, emotions are generated according to the following template:

head(bp) **feel** [strength(con)] *emotion* [to(con)] [*reason*]

For example, in

I feel interested in being entertained.

the head of the belief path (*Me* in this case) is generated as I, the verb is generated as feel, no strength is generated, the *emotion* is generated as interested, no to is generated, and the *reason* is generated as in being entertained.

A strength less than .3 is generated as a bit and a strength greater than .7 is generated as really; otherwise, no strength is generated. The *emotion* and *reason*, and whether to generate the to, depend on the particular emotion, as specified in the following table. The conditions are given on the left side of the table and the corresponding generational components on the right side. The table is scanned starting from the top. Once a line is encountered whose conditions are satisfied, the emotion is generated according to the specified components.

to	alt?	to gl	from gl		*pos*	*neg*	to?	*reason*
y		y			relieved	regretful		about fm-gl
		y	n		interested			in to-gl
		y	ddg concern		hopeful	worried		about fm-gl
y			social esteem		proud	humiliated		
y					grateful to	angry at	y	
			social esteem		poised	embarrassed		
			self-esteem		proud	ashamed		
			existing lovers			broken hearted		
			friends or lovers			rejected		
			entertainment		amused			
			food		satiated	starved		
					pleased	displeased		about fm-gl

The conditions are: (a) the to of the emotion specifying a person, if any, toward whom the emotion is directed, (b) the altern? of the emotion which specifies whether the emotion resulted from an alternative past scenario, (c) the active top-level goal (or concern), if any, associated with the emotion, and (d) the failed or succeeded goal, if any, associated with the emotion. An additional condition for the application of a given line in the table is the presence of an *emotion* entry corresponding to the type of the emotion to be generated.

The notation "ddg concern" means that the concern of the goal is a daydreaming goal. Thus, I feel worried about losing my possessions is generated when an emotion results from an imagined future failure of the POSSESSIONS goal generated within a REPERCUSSIONS daydreaming goal concern (which in turn motivates another REVERSAL goal to prevent that failure).

The notation "existing LOVERS" refers to a relationship that was already true when the goal was first activated. Thus when Harrison Ford turns down DAYDREAMER, I feel rejected is generated, while when DAYDREAMER is dumped by Guy, I feel broken hearted is generated.

If the to? is marked "yes," the to slot of the emotion is generated. If a reason specifies a word and a goal, then a reason is generated as follows: If the goal is a SUCCEEDED-GOAL or ACTIVE-GOAL, then the following is generated as the reason:

> word obj(goal)+GER

Otherwise, the following is generated as a reason.

> word goal+GER

For example:

> I feel interested in being entertained.
> I feel displeased about failing at being entertained.

(However, a positive emotion corresponding to success of an ENTERTAIN-MENT goal is generated as I feel amused.)

B.2.5 Attitudes

A **POS-ATTITUDE** is generated as:

head(bp) **like** obj(con)

If obj(con) is a **PERSON**, a **NEG-ATTITUDE** is generated as:

head(bp) **think**+NEG?=negate(NEG?) much of obj(con)

(If the concept is negative, a positive verb is generated, otherwise a negative verb is generated.) Otherwise, a **NEG-ATTITUDE** is generated as:

head(bp) **dislike** obj(con)

A **ROMANTIC-INTEREST** is generated as:

head(bp) **be interested in** obj(con)

B.2.6 Beliefs

A **BELIEVE** is generated as follows: If obj(con) is a **MENTAL-STATE** and not a **KNOW** or **BELIEVE**, it is generated as:

obj(con)+BP=cons(actor(con),BP)

(That is, the new belief path consists of actor(con) added to the head of the old belief path.) Otherwise, it is generated as:

actor(con) *verb* obj(con)+SIMPLIFY-TENSE+BP=tail(BP)

The *verb* is think when obj(con) is **ATTRACTIVE**, otherwise it is believe.

B.2.7 Objects and Locations

Concepts of the following types are generated by through use of a literal string specified in the name slot of the concept: **OBJECT**, **LOCATION**, and **CITY**. If there is no name slot, the concept is generated as some company, some city, and so on.

MOVIE is generated as a movie; **FASHIONABLE-CLOTHES** is generated as my cute outfit; **TIME-OF-DAY** is generated as the time.

ALUMNI-DIR is generated as:

the obj(obj(con)) Alumni Directory

Concepts which are variables are generated by retrieving an *exemplar* object associated with the type of the variable, and then generating that object.

B.2.8 Persons

A global variable called *REFERENCES* enables generation of pronouns. Any concept, once generated, of type **OBJECT**, but not an exemplar, and not *Me*, and not already contained in *REFERENCES* is added to *REFERENCES* after first deleting any references of the same gender or same type as the concept. Whenever a new paragraph is generated as a result of backtracking or switching concerns, *REFERENCES* is cleared.

Persons contained in *REFERENCES* are generated as pronouns. (The concept *Me*, which corresponds to the daydreamer, is always generated as a pronoun.) Pronouns are generated as follows, depending on the concept and the specified case (which defaults to the objective case if not otherwise specified by, for example, the subject generation algorithm):

concept	nominative	possessive	reflexive	objective
Me	I	my	myself	me
FEMALE-PERSON	she	her	herself	her
MALE-PERSON	he	his	himself	him

Otherwise, a **PERSON** is generated as follows:

first-name(con) [last-name(con)] ['s]

The 's is employed if +POSS is specified. If there is no first-name(con) or last-name(con), someone or someone's (if +POSS) is generated. An exemplar, corresponding to a variable, is generated as this person

B.2.9 Other Concepts

A **LIST** whose rest(con) is a **LIST** is generated as:

first(con) and rest(con)

Otherwise, a **LIST** is generated as:

first(con) rest(con)

Lisp lists containing *Me* are generated as:

element1 element2 ... elementn and *Me*

Otherwise, lists are generated as:

element1 element2 ... elementn-1 and *elementn*

A **NOT** is generated as:

obj(con)+NEG?=negate(NEG?)

A **HYPOTHESIZE** is generated as:

head(bp) **imagine** state(con)+GER

A **LEADTO** is generated as:

ante(con)+POSS-SUBJ+GER **lead** to conseq(con)+GER

An **ACTIVE-P-GOAL** is generated as:

Our relationship **be** in trouble

A **PRESERVATION** is generated as:

head(bp) **continue** obj(con)+INF

B.3 Global Generation Alterations

Various global modifications to a sentence are performed by a collection of strategies:

The generator produces a particle (such as up and on) after a pronoun and before a non-pronoun. For example:

Karen took him on.

but

Karen took on Thomas.

The generator capitalizes the first word of the sentence. Normally, sentences are ended with a period. If a concept is a **SURPRISE** emotion, the sentence is ended with an exclamation point.

A new paragraph is begun (in English-only output traces) whenever back-tracking or a shift to a new concern occurs.

Concepts asserted into a context marked as an alternative past context are generated by default in the CONDITIONAL-PRESENT-PERFECT tense. For example:

I would have told him I would like to know the time.

If the strength of a concept is less than .3 (and it is not a **SURPRISE, NEG-SURPRISE**, or **OVERALL-EMOTION**), it is generated as:

possibly con

If a concept resulted from subgoal relaxation, it is generated as:

maybe con

in performance mode, and

> say con

in daydreaming mode. Concepts resulting from subgoal relaxation are generated in the PAST-SUBJ tense if asserted into a context marked as an alternative past context. For example:

> Say he thought I was cute.

When initial hypothetical concepts are asserted, they are generated as:

> what if con+PAST-SUBJ+QUEST

For example:

> What if I were going out with him?
> What if there were an earthquake in Los Angeles?

B.4 Generation Pruning

Much of the work in producing natural-sounding English narration of daydreaming is deciding when concepts should *not* be generated. Rules have optional inf-no-gen and plan-no-gen slots. These slots enable the programmer to add pruning heuristics directly to inference and planning rules. (Hand-coded episode definitions may also specify a plan-no-gen list to be associated with induced rules.) Facts inferred by an inference rule are only generated if the inf-no-gen of the rule is NIL. The value of a plan-no-gen slot is a list of flags, one for each subgoal. The meaning of the flags is as follows:

flag	meaning
NIL	may generate subgoal
T	never generate subgoal
'ACTIVATE	generate subgoal only when activated
'OUTCOME	generate subgoal only on outcome

Whenever a goal is activated, its gen-advice slot is set to the value of the associated flag for future use. The generation of activated goals and goal outcomes is then determined according to the following strategies:

- An activated goal is only generated if its objective is not already satisfied in the context and its associated gen-advice flag is either NIL or 'ACTIVATE.

- A goal outcome is only generated if it is not a daydreaming goal and its gen-advice is either NIL or 'OUTCOME.

- A goal success is not generated if the objective of the goal had already been satisfied in the context when the goal was activated.

- An activated top-level goal is only generated if it is not a daydreaming goal.

However, input states (such as *Harrison Ford is at the Nuart*) are always generated.

B.5 Additional Generation

Concepts are generated in various situations other than when a concept is asserted into a context. These section describes these situations.

When a negated fact is bolstered in the minimization strategy for rationalization, the negated fact and its justification are generated as:

anyway con+PAST because *causes*

where *causes* is determined by finding the source leaves of the inference chain. For example:

Anyway, I was well dressed because I was wearing a necklace.

When an attitude is inverted in the minimization strategy for rationalization, the new attitude is generated as:

anyway con

Anyway, I don't think much of him.

After minimization has altered the strengths of emotions, these emotions are generated.

Whenever retrieval of an episode (reminding) occurs, it is generated as:

[I remember the time] goal+PAST by subgoal1+GER subgoal2+GER
. . . and by subgoaln+GER

where *goal* is the goal of the episode, and *subgoal1* through *subgoaln* are the subgoals of that goal. Each subgoal is then recursively generated according to the same formula, until leaves are reached. The phrase I remember the time is only generated for the top goal of the episode. Subgoals are not generated if indicated by the associated rule.

Author/Subject Index

a-lists, 169
Abelson, Robert P., 32, 33, 34, 35, 50, 67,
    77, 80, 82, 95, 151, 152, 177,
    206, 245, 276, 283, 297
Abrahams, Paul W., 169, 291
abreaction, 217
absolute meaning, 240
abstract sameness of algorithms, 260
abstract wish, 227
ABSTRIPS 139, 164
    DAYDREAMER vs., 139
accessibility of knowledge, 100
accident, 13, 122, 132, 138
accidental by-product, 81
achievement goals, 32
ACQUAINTED, 157
action mutation 48, 132, 194, 223, 263
    function of, 132
    strategies for, 134
    traces of, 334, 335
action mutation procedure, 48, 194
action preconditions, 143
action subgoals, 176
actions 2, 3, 4, 31, 34, 87, 115, 146, 179,
    183, 306
    English generation of, 353
    failure of, 183
    input, 56
    of others, 183
    performing, 176, 179
    representation of, 152
    rules for, 175
activation 198, 243, 278
    decay of, 199
    stable pattern of, 203, 208
active concern, 130
active goal, 179
active indices, 209
ACTIVE-P-GOAL
    English generation of, 361
activities, 34
actor slot, 152, 153
actors, 280
Adams, James L., 15, 123, 133, 266, 283

Adams, Norman I., 169, 296
adaptiveness, 62, 69, 79, 82, 217, 271
add-name (operation), 170
add-slot-values (operation), 170
adding rule conditions, 116
Adler, Alfred, 227
age differences, 20
age of imagination, 219
aggression reduction, 68
agreements, 160
AI-complete problem, 282
aims, 31
algorithms 235, 258
    sameness of, 260
algorithms for daydreaming, 178
alien biologies, 234
alpha EEG, 230
altered state of consciousness, 19
alternative past contexts
    English generation and, 361
alternative past daydreams, 8
alternative plans 98, 179
    selecting from among, 98
alternative worlds, 179
AM 14, 43, 146
    agenda mechanism of, 43, 146
    DAYDREAMER vs., 43, 147
    success of, 147
amusement, 55
anagrams, 135
analogical daydream generation, 60
analogical planning 30, 46, 86, 89, 90, 97,
    98, 117, 118, 133, 143, 189, 224,
    264, 308, 327, 335, 347, 349
    examples of, 91
    multiple roles of, 282
    previous work in, 117
    procedures for, 46, 186
    recursive, 91, 118
analogical rule application procedure, 47,
    189
analogy
    finding an, 118
    learning by, 117

365